Mathematics Experiences for the Early Childhood Years

Linda Barron

George Peabody College for Teachers

CHARLES E. MERRILL PUBLISHING COMPANY
A Bell & Howell Company
Columbus Toronto London Sydney

Published by
Charles E. Merrill Publishing Company
A Bell and Howell Company
Columbus, Ohio 43216

This book is dedicated to my mother
Virginia S. Smith.

This book was set in Caledonia and Futura.
The production editors were Elizabeth A. Martin and Laura Wallencheck Gustafson.
The cover was prepared by Will Chenoweth.

Cover photo by Mariam G. Lasher. Photos on pp. 19 and 20 courtesy of
American Printing House for the Blind.

Library of Congress Catalog Card Number: 78–71996

International Standard Book Number: 0–675–08284–6

Printed in the United States of America

1 2 3 4 5 6 7 8 9 10 / 85 84 83 82 81 80 79

Contents

Preface

Mathematics Experiences for the Early Childhood Years will assist the reader to:

- master basic mathematics content for kindergarten through third grade to facilitate success in teaching children at these levels,
- design mathematics activities appropriate for students at different levels of intellectual development,
- sequence learning tasks in the early childhood curriculum—knowledge essential to the diagnosis of a child's progress and prescription of subsequent instructional tasks,
- construct strategies for teaching various mathematical topics,
- use manipulative and other instructional materials and apply them to the development of mathematical concepts in young children,
- unify learning experiences so that students learn mathematics in ways which underscore structure, and
- provide appropriate mathematics instruction for exceptional children.

The text can be used:

- in a college course for early childhood, elementary education, or special education majors,
- by inservice teachers in planning their class work and giving input into the design of the mathematics curriculum in their schools,
- by interested parents or private tutors of young children.

Mathematics content is examined so the reader can better understand the mathematics, but suggestions for methods and materials for teaching this content to young children are given utmost consideration. Learning theory is used as a vehicle for developing teaching strategies and sequencing mathematics instruction according to children's developmental needs. Attention is given to the various learning styles of exceptional children in order to help mainstream and resource teachers instruct such children.

Many of the suggested activities are presented in detail to help the inexperienced teacher gain confidence in teaching mathematics. Other activities are only sketched to encourage the teacher to creatively fit them to various students and classroom settings. A nucleus of ideas for teaching mathematics to young children is presented, but many additional ones can be devised by the thoughtful teacher. Also, the reader may find the assignments that require reading of current reference material in mathematics education both enjoyable and beneficial. One special feature of the book are the laboratory exercises which actively involve the reader in studying alternative techniques and materials designed to help students learn mathematics. In each laboratory activity, preparation, objectives, materials needed, and the procedure are given; the evaluation is left to the discretion of the college instructor.

A balance of the modern (emphasis on "why") and traditional (emphasis on "how") approaches to teaching mathematics is the basic philosophy throughout the book. Children should first develop an understanding of concepts and principles in mathematics and then, when they understand the basis of what they are doing, develop the skills needed to use these ideas effectively in classroom situations and day-to-day life.

There is no substitute for an informed, caring, and enthusiastic teacher. If she or he has a good attitude toward mathematics, students will more likely adopt a similar attitude. Mathematics is a part of everyday life, and a healthy attitude toward it should be developed early. Mathematics is not difficult to learn and retain if the learner understands the organization and utility of the subject matter. Children's positive reactions in the affective domain may very well mean that the teacher is doing the job well.

I extend special appreciation to my husband, Allen, for his continued encouragement and assistance in the preparation of this book.

1

Foundation

Mathematics experiences at the early childhood level provide a base for children's future work in mathematics. In order to design developmentally and psychologically sound experiences which promote the learning of mathematics, the early childhood educator needs to understand how children learn mathematics. To help toward this end, this chapter presents conceptualizations of Jean Piaget, Jerome Bruner, and Robert Gagné, contemporary investigators whose ideas have greatly influenced mathematics education. Also, these ideas provide a developmental and psychological foundation for the teaching strategies discussed throughout this book.

Piaget's Developmental Theory

Jean Piaget, a Swiss biologist, psychologist, philosopher, and logician, and his co-workers in Geneva have conducted many clinical interviews with children in which each child was given cognitive tasks to perform and questioned on what he was thinking as he attempted to carry out the tasks.* From the analysis of these responses, Piaget identified four stages of intellectual development. His theory holds

* Throughout this book, the child is referred to as "he" and the teacher as "she," merely as a matter of convenience. Also, "teacher" includes a classroom teacher, parent, tutor, teacher's aide, and any other individual involved in helping young children experience mathematics.

that all children pass through the same stages at approximately the same age. Piaget's research on mental development beginning at age four is especially enlightening.[1] A description of the development stages given by Piaget follows.†

Sensorimotor Stage

This stage, which generally extends from birth to eighteen months or two years, begins with reflex actions such as crying, sucking, grasping, and smiling. The infant refines spontaneous movements as he reacts to his environment, repeats actions that are pleasing (the beginning of memory and intelligence), and forms habits such as thumb sucking. Toward the end of the stage, language begins. The entire period is characterized by the child's exploring his environment and responding and adapting to it.

Preoperational Stage

Generally, children who are between eighteen months or two years old and six or seven years old function intellectually at the preoperational level. The preoperational child cannot reason from another's point of view, a trait Piaget called *egocentrism,* and is limited to perceptual judgments. The fact that a child at this level is perceptually oriented is seen when he attempts to perform conservation tasks of number, quantity, length, area, time, and volume. For example, if marbles are dropped simultaneously one at a time into two identical jars and then the marbles in one of the jars are poured into a container of different size or shape,

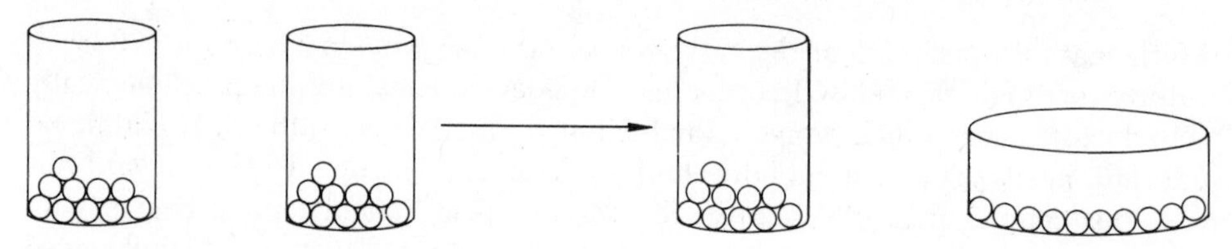

the change in appearance of the container may lead the child to conclude the number of marbles has changed. The dominance of the child's thought processes by perception and the inability to mentally reverse the transformation to the beginning situation when the same number of marbles was in each of the two like containers prevent the child from conserving number at this point.

During this stage, language develops and language and thought become coordinated. Representational thought is evidenced in play activities such as "keeping house" with dolls and "let's pretend." Toward the end of this stage, a transitional period emerges in which a child's thoughts become more logical. He may be able to perform a conservation task but exhibits considerable trial-and-error thinking. The characteristics of a child in the preoperational stage of mental development are especially significant to teachers of nursery school, kindergarten, and first grade.

† To examine in detail the Piagetian stages of mental development, read *The Psychology of the Child* by Jean Piaget and Barbel Inhelder, published by Basic Books, Inc., New York, 1969.

Concrete-Operational Stage

This stage, which generally applies to children between the ages of six or seven years and eleven or twelve years, is characterized by the emergence of logical thought; i.e, a child's thinking becomes operational. A student at this level forms mathematical concepts from the manipulation of concrete objects and thus contrasts with the preoperational child who plays with objects and experiences physical learning about the objects themselves, but derives no mathematical ideas from the manipulation. The concrete-operational student, not limited by perceptual judgments, can reason that the number of marbles in the experiment described previously is the same regardless of the shape or size of the container. Reversibility of thought, the mental capability to observe a transformation of an object or objects and revert the object or objects to the original state, occurs during this period. This capability allows the student to conserve number, quantity, length, area, time, weight, and volume (and in that order, according to Piagetian research). A child at this level of thought can also classify and seriate logically. He can reason that an object can have more than one property; e.g., a woman can be a wife, mother, and teacher at the same time.

Understanding this level of thought is very important to early childhood educators because some kindergarten and first-grade pupils and many second- and third-grade students function intellectually at this level. The Piagetian tasks given in chapter 4 are aimed at determining if a child has reached the concrete-operational thinking stage.

Formal-Operational Stage

At about the age of eleven or twelve years, children's thinking becomes formal-operational. Students' reasoning during this stage is no longer tied to concrete objects. Children can make predictions and check hypotheses without having to resort to objects in the physical world. The learner can reason deductively and understand hypothetical situations. Formal logic, combinations and permutations, probability theory, and deductive Euclidean geometry are appropriate curricular experiences for a formal-operational thinker. Only gifted students in the early childhood years will reach this level of thought.

Some Implications of Piaget's Research for Teachers

Piaget's work, which covers several decades, provides valuable information on how children acquire knowledge. Even though the work of Piaget and his colleagues concentrates on how children learn rather than on how they should be taught, it can be used as a tool for diagnosis (see chapter 4), for understanding how students think, and for dealing with curricular and instructional problems in mathematics. For example, Piaget's research indicates that primary-level students may not be mentally equipped to fully understand measurement of time on a clockface and the concept of zero.[2] These topics, however, are usually introduced in first grade or even kindergarten.

Lavatelli has developed an early childhood curriculum which applies Piagetian theory. The following are some considerations she suggests for implementing such a curriculum:

1. *A student must be active in the process of learning.* Young children need concrete materials in their environment, and by manipulation of these materials and proper teacher questioning they learn mathematical concepts.
2. *If a student is reasoning incorrectly, the teacher should provide feedback and guidance.* A spatial-perspective task in the Lavatelli program involves a student's studying a place setting on one side of a table and setting a place across the table for a friend. If the student makes errors in completing this task, Lavatelli suggests that the teacher simply ask the student to go to the other side of the table and see if the place setting for his friend looks the same as his own. It may also help to have another child sit across the table from the student and to ask the children to place their left hands to the left of their plates. Have each child tell what eating utensil is next to his left hand, what is next to that utensil, and so on. In this way, each child may be able to understand that "left to right" on his side becomes "right to left" as he views the same arrangement on the opposite side of the table.
3. *The teacher should guide the children in their learning but should not force them to parrot the correct answers.* Lavatelli points out that ". . . giving him the right answer will not convince the child. *He* must be convinced by *his own* actions."[3] She also advocates that the early childhood teacher spend part of the day leading small groups of children in structured activities and during the remainder of the day allow students to engage in self-activity in which she reinforces the concepts and skills taught in the directed activities.[4]

Piaget indicates that language plays a secondary role in logical development, and Copeland points out that the language of children often reflects their level of thinking.[5] Thus, as students explore mathematical ideas, you, as teacher, should encourage them to discuss their thoughts and findings with their peers and with you. You may further enhance language development in mathematical activity by creating interest centers and by engaging students in problem-solving activities in a laboratory setting, both of which should promote exchange of ideas among students. These situations also lend themselves to having individuals or groups report on their work to the class. The ideal, early childhood classroom is bustling with activity and conversation as young children participate in hands-on experiences and communicate their discoveries to others.

Bruner's Levels of Learning

Just as Piaget has described four levels of intellectual development, Jerome Bruner, who has interpreted much of Piaget's work, has identified in a corollary way three levels of learning: enactive, iconic, and symbolic.[6] At the *enactive (concrete) level,* the student manipulates concrete objects to investigate mathematical concepts and principles. At the *iconic (pictorial) level,* pictures of objects, diagrams, or images in the mind's eye are used instead of the objects themselves. A child functioning at the *symbolic level* uses symbolism (words or mathematical symbols) to represent action that occurred at levels 1 and 2.

Two examples of Bruner's three modes of instruction follow:

Example 1: The three levels applied to the basic multiplication fact $4 \times 2 = 8$ can be illustrated this way:

a. *Concrete.* Make four circles with rope on the floor. Ask two students to stand in one circle. Repeat for the other three sets. Emphasize that there are four sets with two children in each, and let the children tell the number in the combined sets. They count or add two four times to determine there are eight in four sets of two.

b. *Pictorial.* Instruct the children to draw a picture of four sets with two objects in each and then find the number of objects in the sets when they are joined together. Again the child finds eight members altogether.

c. *Symbolic.* Record and have the children record and read the mathematical sentence $4 \times 2 = 8$ represented in the two previous levels.

Example 2: When a more advanced task in multiplication arises, these same modes are repeated. For example, the sequence for teaching $2 \times 14 = \square$, an example of problems typically taught in third grade, could go this way:

a. *Concrete.* Select a student to enter 14 on an abacus twice. Then ask the children to read the product 28 from the abacus.

b. *Pictorial.* Have the children outline a 2×14 array on grid paper. Let them suggest how to get the product from the array. Many may suggest to count the square units in the array. Guide them to separate the array into two smaller arrays so that two easy multiplications can replace the counting.

14

10 4

2 2×10 2×4

The student reasons that $20 + 8 = 28$, the product sought.

c. *Symbolic.* Construct a series of open sentences to guide the child's thinking. (He fills in the blanks.)

$$\begin{aligned} 2 \times 14 &= 2 \times (\_\_ + 4) \\ &= (2 \times 10) + (2 \times \_\_) \\ &= 20 + \_\_ \\ &= \_\_. \end{aligned}$$

Notice how the pictorial model in step (b) represents this symbolic solution.

Any one activity may include all three of these levels, two of them, or only one. The objective of the activity and needs of the learner determine which level or levels is (are) best.

Bruner's levels of learning are significant to the teacher in planning instruction and adjusting her expectations of different children. For example, mathematically gifted children readily advance to the symbolic level. The amount of time they need to spend at each level on mathematical topics may be approximated by this time line:

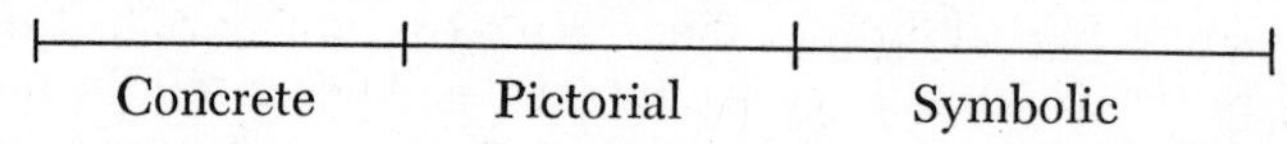

Slow learners need extensive experience at the concrete and pictorial levels. Their curriculum may be represented this way:

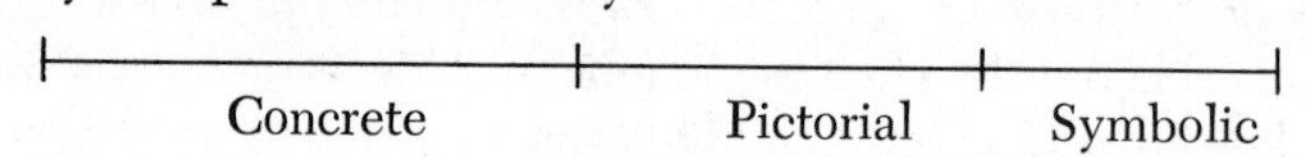

"Average" learners at the early childhood level need about the same amount of time at each level, with the abstract manipulation of symbols being related to concrete and pictorial models. Their curriculum is suggesed by this time line:‡

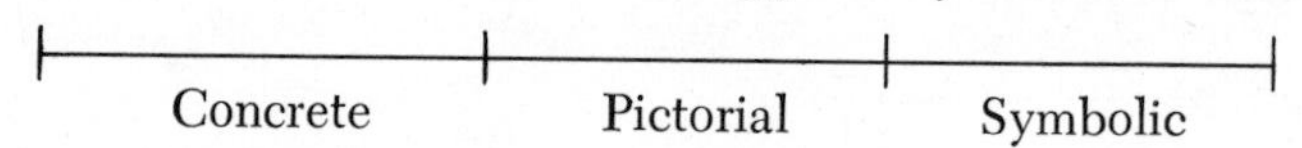

As a student grows older and matures mathematically, the degree to which he needs concrete experiences is lessened.

Gagné's Learning Hierarchy

In his book, *The Conditions of Learning,* Gagné places cognitive learning into eight categories: signal learning, stimulus-response learning, chaining, verbal association, discrimination learning, concept learning, rule learning, and problem solving.§ The capabilities of the learner and the instruction or instructional sequence that causes the learning to be acquired influence the learning. Gagné emphasizes that prerequisite learning is important in acquiring a given behavior. It is in this respect that his work becomes especially useful in planning, providing, and evaluating instruction.

Type 1—Signal Learning

Signal learning, the most basic form of learning, is a generalized emotional response essentially involuntary in nature. A classic example of this type of learning is the

‡ This representation is a modification of an idea obtained in Robert G. Underhill, *Teaching Elementary School Mathematics,* 2nd ed. (Columbus, Ohio: Charles E. Merrill Publishing Company, 1977), p. 18.

§ This discussion of the eight categories of cognitive learning that make up the learning hierarchy was adapted and reprinted from Robert M. Gagné, *The Conditions of Learning,* 2nd ed. (New York: Holt, Rinehart and Winston, Inc., 1970) by permission of the publisher and the author.

Pavlov experiment in which a dog salivates to the stimulus of the sounding of a bell after meat powder and the bell are presented in close time succession a number of times. Responses of fear are also examples of signal learning. Gagné indicates that this type of learning can hardly be regarded as cognitive learning. It does have some significance, however, to you as a mathematics teacher in that you should be alert to the fact that some children have a general fear of mathematics, possibly because of their poor past performance or of a parent's or sibling's dislike of the subject, while other students are naturally motivated and enthusiastic. Success-oriented experiences are in order for those children who need to overcome a general dread of mathematics.

Type 2—Stimulus-Response Learning

In stimulus-response learning, some discrimination is necessary in that the stimulus causing a certain response is singled out. Reward for the correct response and repetition are important conditions for learning at this level to occur. When a young child grasps a bottle, toy, or spoon or smiles at people he knows, he exhibits this type of learning. In mathematics, stimulus-response learning is evidenced when the child can pronounce common mathematical terms such as number names and the names of Euclidean shapes. The child, however, may not know the meaning of the term he is using.

Type 3—Chaining

When chaining, the learner sequences two or more stimulus-response situations. This learning is characterized by motor behavior. A child demonstrates chaining when he puts on his jacket or sharpens his pencil. In mathematics, chaining includes making the numerals 0 through 9 and tracing a Euclidean shape such as a square or triangle. Reward correct responses as you guide the student to sequence the chain of motor behavior in the proper order, and have him practice this action.

Type 4—Verbal Association

Verbal association is characterized by a verbal rather than a motor sequence. This level includes tasks such as memorizing poems and the basic facts of addition, subtraction, multiplication, and division with little regard to meaning. Some adult self-improvement courses emphasize this type of learning when suggesting arbitrary verbal associations to help one remember lists of items such as groceries to be bought. Important mathematical experiences at this level during early childhood are counting aloud, saying the word associated with numerical symbols (e.g., saying "five" when "5" is seen), saying the name of a given shape, and skip counting (counting by twos, threes, and so on).

Type 5—Discrimination

Discrimination learning involves one's being able to respond correctly to each of several stimuli. Some mathematical experiences at this level during the early childhood years include:

a. naming numbers correctly when numerals are given in random order,

Example: Name these numerals in the order shown.

5 2 7 9 8

b. discriminating among symbols such as "3" and "E," "6" and "9," "7" and "T," "5" and "S," and "2" and "Z,"

c. distinguishing among open and closed curves, simple and nonsimple curves, and various Euclidean shapes,

Example: Touch all the triangles in this set of shapes.

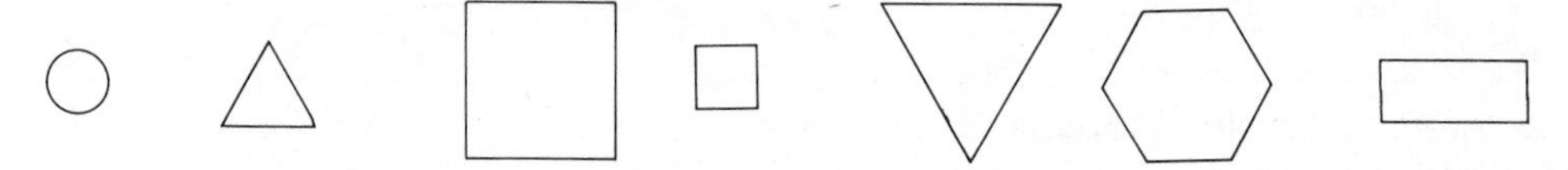

Notice that in this example the student must associate the correct shape with the word "triangle." An easier directive would be "Show me all the figures that have this shape: △." Children who have certain learning disabilities often have difficulty in making such discriminations.

d. making distinctions among the operation symbols $(+, -, \times, \div)$ and among relation symbols $(<, =, >)$,

e. discriminating among the coins in our money system, and

f. associating the correct name with the lengths centimeter, decimeter, and meter.

If a student has difficulty in making discriminations, return to learning of earlier types and isolate and reteach the chain or stimulus-response link causing the trouble. If you minimize this type of learning, your students will likely have trouble with higher-level tasks and with reading the language of mathematics.

Type 6—Concept Learning

When learning concepts, the learner categorizes objects in his environment into classes such as plants and animals, natural and synthetic products, and plants which are edible and those which are not. Generally, one learns a concept by observing many examples and recognizing a common property such as shape, color, or number among the examples. Once the concept is learned, the learner can recognize other examples of it in new learning situations. Learning formal definitions is also a kind of concept learning.

A typical mathematics activity at this level is presenting students with a set of blocks and asking them to place the blocks in piles (sets) so that all of the objects in each pile are alike in some way. This time, unlike the discrimination learning just described in which the child selects the triangles from a set of shapes, he must scan the objects and make a logical decision as to what criterion (shape, color, or size) to use in the classification. A child who places the objects

   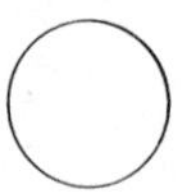 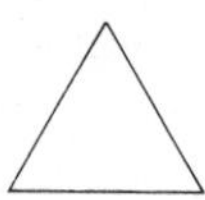 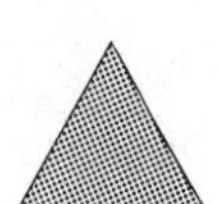

into the two sets

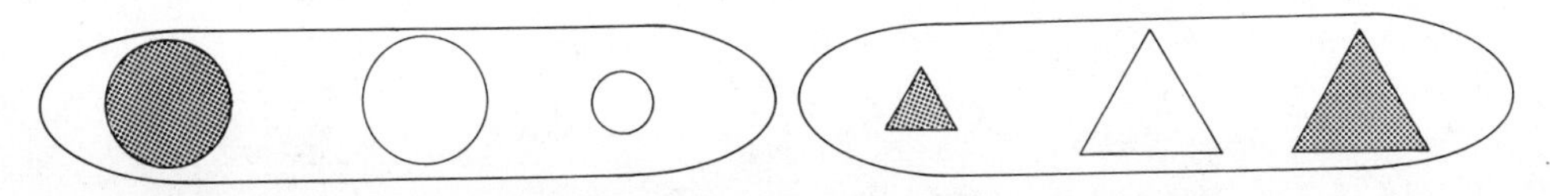

has classified by shape; into the sets

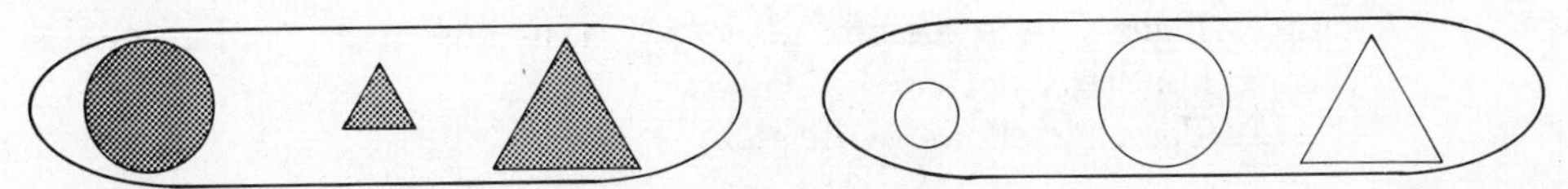

has used the attribute color; and into the sets

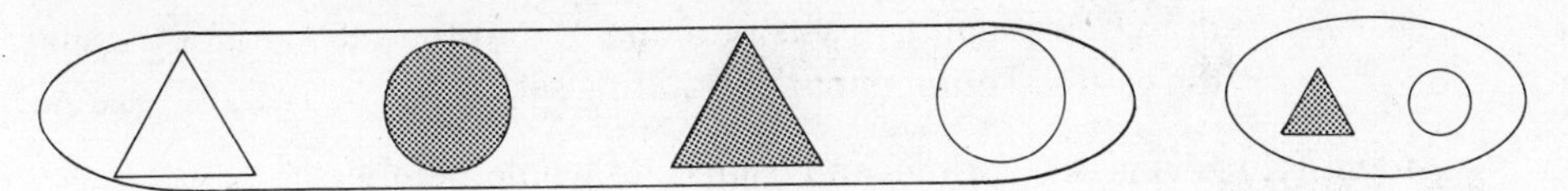

has sorted by size. The logical thought used to solve an example such as this one is the kind of thinking demanded in the Piagetian logical-classification task given in chapter 4. Other mathematical examples of concept learning are:

a. identifying the cardinal number of a set,

Example: Determine the cardinal number of the set.

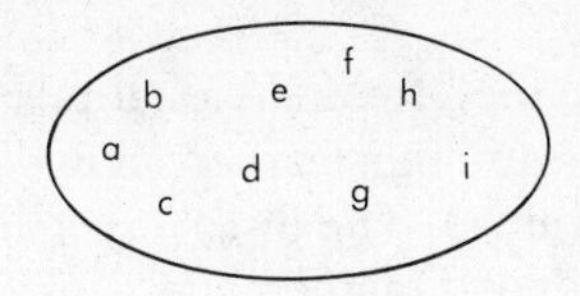

The child must order the elements and count nine objects.

b. identifying the ordinal number associated with an element of a set,

Example: What is the fourth letter in this list?

s k a t e

c. identifying the shortest object, longest object, middle element, object nearest a given one, etc., in a set of objects,

Example: Point to the object nearest the apple ( ) in this set.

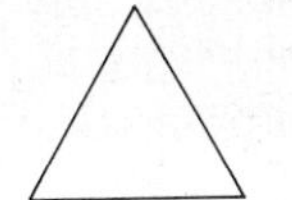 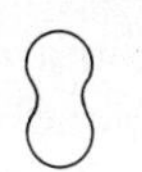  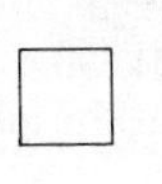 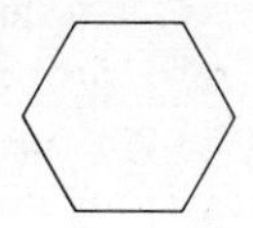 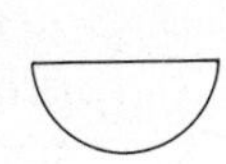

d. recognizing congruent figures,

Example: Place an X over the two congruent figures among these:

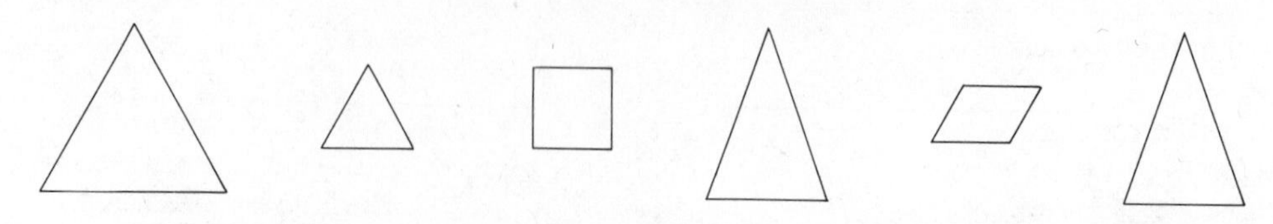

The student should select the two large isosceles triangles.

e. recognizing place value,

Example: Place an 8 in the tens place, a 3 in the ones place, and a 4 in the hundreds place in the blanks: ___ ___ ___
The child should write 483.

f. performing whole number operations, and

Example: Solve $3 + 4 = \square$ by drawing a picture of sets. The student must draw representative sets of 3 and 4, join the sets, and determine the cardinal number of the resulting set.

g. identifying even, odd, prime, and composite whole numbers.

Example: Write the prime numbers between 10 and 20. The student must know the definition of a prime number (remember, definitions are a part of concept learning) to list the required primes, 11, 13, 17, and 19.

Type 7—Rule Learning

When learning rules, or principles, the individual relates two or more concepts. In order for this learning to occur, the learner should have the prerequisite conceptual structure (type 6 learning). The discovery and use of principles are an important part of the early childhood mathematics curriculum. Do not, however, expect young children to verbalize principles in mathematics too early. You should supply this step until your students can state the generalization verbally.

Gagné points out that rules can be taught by the discovery method whereby the student does his own chaining of concepts but the teacher can use language effectively to give cues for guidance, summarize results, and state the rule discovered. He cautions the teacher, however, to teach the concepts before verbal generalizations are given or else the student may be able to parrot a rule but not understand it (i.e., his learning would be type 4, verbal association).

Some early childhood mathematics tasks at this level include:

a. using the commutative, associative, and identity properties of addition and multiplication, the zero property of multiplication, the distributive property of multiplication over addition and subtraction, and certain properties of subtraction and division,
b. generating a set of fractional numbers equivalent to a given one by multiplying numerator and denominator by the same nonzero number,
c. determining which sums and products of even and odd whole-number combinations are even and which are odd.

Type 8—Problem Solving

When an individual exhibits problem-solving learning, he relates rules to form higher-level principles and solves problems in his environment. Early childhood activities in mathematics at this level include using algorithms in which facts and properties of the operation and place-value principles are applied, solving word problems, and solving discovery situations. Algorithms are discussed in chapters 9 and 10 and verbal problems and discovery situations, in chapter 15. When a student exercises creativity and inventiveness, he is operating at this level of learning.

Learning types 3 through 8 are all a part of the early childhood mathematics curriculum with the most time being spent on concept, rule, and problem-solving learning. Bright children can successfully skip or combine steps whereas retarded learners make small-step progressions at a slow rate.

To help you see Gagné's learning types applied to the early childhood mathematics curriculum, the following categorization of a small segment of related mathematics content is given.

Type of Cognitive Learning	**Task**	**Approximate Grade Level**
chaining	Student copies or traces this figure: □.	Preschool, K
verbal association	Student names the shape □ and draws the correct shape when asked to draw a square.	Preschool, K
discrimination	Student selects all the squares from among several different shapes.	K
concept	Student learns the concept of "squareness" by examining many examples of squares. He learns a definition which gives exactness of description.	K, 1st
rule	Child learns that area in square units of a rectangular region can be found by covering the region with squares and multiplying the number of squares down with the number of squares across.	2nd, 3rd

$$A = 2 \times 4$$
$$= 8 \text{ square units}$$

problem solving	Child is given an irregularly shaped region such as 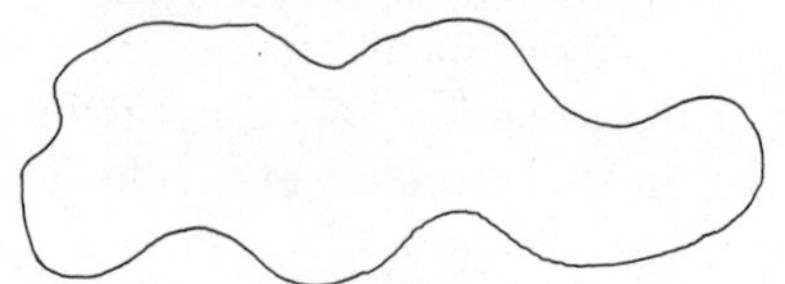 square-unit grid paper, and scissors and asked to approximate the area of the region. He cuts out the squares, covers the region with them, puts together parts of squares on the periphery, counts the squares, and thus arrives at the approximate area.	3rd

Gagné's learning types provide general guidelines for curriculum planners and teachers to sequence mathematics instruction. There are times, however, when the sequence of instruction does not progress in the order of Gagné's types of learning. In the preceding example, many three- and four-year-old children can associate the word "square" with the figure □ but do not have the motor coordination to copy a square. Also, a concept is often introduced before the symbolism associated with it is presented. For example, children can work with fractional parts of an object (concept learning) before they read and write fractional numerals (verbal association and discrimination learning). Another example in which mathematics instruction proceeds from concept to verbal association and discrimination learning is the practicing of basic facts of addition, subtraction, multiplication, and division after the concepts of these operations have been formed. Children must learn to respond quickly to examples such as $5 + 4 = \square$, $12 - 5 = \square$, $6 \times 8 = \square$, and $30 \div 6 = \square$ with the correct answer. If concept learning has already occurred, practice for mastery becomes meaningful and is not rote learning.

Bruner's and Gagné's Levels of Learning Related to Selected Mathematics Content

The following list shows Bruner's levels of representation and Gagné's learning types related to some mathematical topics appropriate for young children.

Mathematics content	Gagné's type of learning	Bruner's mode of instruction
1. classifying objects to generate sets	discrimination, concept	concrete, pictorial
2. seriating objects of a set	discrimination, concept	concrete, pictorial

3. saying the number name when given the numeral	verbal association	symbolic
4. counting	verbal association	symbolic
5. identifying the cardinal number of a set	concept	concrete, pictorial
6. identifying the ordinal number of an element in a series	concept	concrete, pictorial
7. determining if one number is $>, =, <$ another number	concept	symbolic
8. adding cardinal numbers using union of sets	concept	concrete, pictorial
9. using the commutative, associative, and identity properties of addition	rule	symbolic
10. memorizing the basic facts of addition	verbal association	symbolic
11. subtracting cardinal numbers using set separation	concept	concrete, pictorial
12. representing place value with base blocks, an abacus, money, etc.	concept	concrete, pictorial
13. writing numerals in expanded notation	concept, rule	symbolic
14. multiplying cardinal numbers using Cartesian product	concept	concrete, pictorial
15. illustrating the distributive property of multiplication with respect to addition on an array	rule	concrete, pictorial
16. multiplying factors which are multiples of 10	rule	symbolic
17. using the vertical addition algorithm, no regrouping required, two-digit plus two-digit case	problem solving	symbolic
18. using the subtraction algorithm, regrouping required, two-digit minus two-digit case	problem solving	symbolic
19. dividing whole numbers using the number line	concept	pictorial

20. representing fractional numbers with paper folding	concept	concrete
21. representing Euclidean shapes	verbal association, discrimination	pictorial
22. learning that 10 centimeters = 1 decimeter	concept, rule	symbolic
23. demonstrating equivalences in values of coins	concept, rule	concrete, pictorial
24. measuring the width of a student desk to the nearest centimeter	rule, problem solving	concrete
25. writing the mathematical sentence associated with a story problem	problem solving	symbolic
26. generalizing that the sum of two odd whole numbers is even	rule	symbolic

Closing

The teacher is challenged to use the research of Piaget, Bruner, and Gagné to structure an environment in the classroom that reflects an understanding of the students' levels of intellectual growth and to design sequenced, unified learning experiences. The developmental and learning theory discussed in this chapter must be coordinated with the mathematics content to produce a working model in the classroom. Sequencing of mathematical content with respect to the psychological views presented in this chapter and the nature of the particular learning task will be treated in succeeding chapters.

Notes

1. Celia Stendler Lavatelli, *Piaget's Theory Applied to an Early Childhood Curriculum*, A Center for Media Development, Inc. Book (Boston: American Science and Engineering, Inc., 1970), pp. 28–29.
2. Richard W. Copeland, *How Children Learn Mathematics*, 2nd ed. (New York: Macmillan Co., 1974), p. 62.

3. Lavatelli, *Piaget's Theory,* p. 48.
4. Lavatelli, *Piaget's Theory,* p. 24.
5. Copeland, *How Children,* pp. 44–45.
6. Jerome S. Bruner, *Toward a Theory of Instruction* (Cambridge: Harvard University Press, Belknap Press, 1963), p. 28.

Assignment

1. Examine the Lavatelli program, the Piagetian early childhood curriculum discussed in this chapter. Make a list of ideas offered in the program you think will be useful in your teaching.
2. Select an area of mathematics content and use it to illustrate Bruner's three levels of learning.

Laboratory Session

Preparation:

1. Study chapter 1 of this textbook.
2. Consult elementary mathematics textbooks, and identify at least ten learning tasks for K-3 different from those given at the end of this chapter. Identify the type of learning (Gagné's) and the mode of instruction (Bruner's) with which each task is associated.

Objective: Identify mathematics content appropriate for early childhood education and relate it to the learning levels theorized by Bruner and Gagné.

Materials: Several K-3 elementary mathematics textbook series.

Procedure: In groups of four or five, compare your lists of mathematics content and tell the learning type and mode of instruction with which each task is associated.

Bibliography

Beilin, Harry. "The Training and Acquisition of Logical Operations." In *Piagetian Cognitive-Development Research and Mathematical Education,* edited by Myron F. Rosskopf, Leslie P. Steffe, and Stanley Taback. Washington, D.C.; National Council of Teachers of Mathematics, 1971.

Bingham-Newman, Ann M., and Saunders, Ruth A. "Take a New Look at Your Classroom with Piaget as a Guide." *Young Children* 32 (May 1977): 62–72.

Bruner, Jerome S. *The Process of Education.* New York: Vintage Books, 1960.

———. *Toward a Theory of Instruction.* Cambridge: Harvard University Press, Belknap Press, 1963.

Copeland, Richard W. *How Children Learn Mathematics,* 2nd ed. New York: Macmillan Company, 1974.

Gagné, Robert M. *The Conditions of Learning,* 2nd ed. New York: Holt, Rinehart and Winston, Inc., 1970.

———. "Contributions of Learning to Human Development." *Psychological Review* 75 (1968): 177–91.

Lavatelli, Celia Stendler. *Piaget's Theory Applied to an Early Childhood Curriculum,* A Center for Media Development, Inc. Book. Boston: American Science and Engineering, Inc., 1970.

Reisman, Fredricka K. *A Guide to the Diagnostic Teaching of Arithmetic,* 2nd ed. Columbus, Ohio: Charles E. Merrill Publishing Company, 1972.

Underhill, Robert G. *Teaching Elementary School Mathematics,* 2nd ed. Columbus, Ohio: Charles E. Merrill Publishing Company, 1977.

2

Mainstreaming and Exceptional Children

Mainstreaming, a major educational trend, is the inclusion of exceptional children in public school "regular" classrooms. Some categories of exceptionalities, which are by no means mutually exclusive, are the visually handicapped, hearing impaired, mentally retarded, learning disabled, emotionally disturbed, physically impaired, and gifted. Some research supports the idea of special separate classes for handicapped children while other studies indicate that special education does little to improve the students' achievement in academic areas.[1] Such special classes place the handicapped student in an isolated and a restrictive environment. In addition to the "inclusion" feature, mainstreamed classrooms must have available appropriate services for the exceptional children.

Providing a "free appropriate education" for handicapped children aged 3 to 21 by September 1, 1980, is now mandated by federal Public Law 94–142, 1975. Accountability rests with each state, which must establish a policy for educating handicapped youngsters in order to receive federal funds. The federal law recognizes that mainstreaming offers an environment which is least restrictive for handicapped children. Moreover, it specifies that states provide a system of inservice training of general and special education instructional and support personnel to ensure that appropriately trained educators are serving handicapped children. Programs, consisting of college and school educators, are being developed to provide inservice teachers with additional skills in individualizing experiences for a diverse student population. Colleges having teacher education programs are requiring students to take special education courses as a part of their preparation to teach in the elementary school. More often than not, the considerations given to an exceptional child in a mainstreamed classroom equally apply to "normal" students because, after all, all children need a teacher who respects them, understands basic ideas of child development, provides materials and activities that foster their growth in the cognitive and affective domains, and displays a positive attitude toward children and teaching.

In the following descriptions of various handicapped conditions, attention is given to the characteristics of the exceptionality, some studies which indicate how the exceptionality affects the learning of mathematics, and what the regular classroom teacher can do to meet the needs and enhance the mathematical abilities of each exceptional student.

The Visually Impaired Student

Students who are visually impaired can be classified as partially sighted or as legally blind. A partially sighted child can read large print but possibly needs the assistance of a magnifying device. You should encourage these students to use their residual vision to the fullest extent possible. A legally blind student needs to use braille or auditory materials.[2] Studies show that the performance of blind students on standardized achievement tests indicates some retardation in mathematics.

Visually deficient students can be mainstreamed into regular classrooms with only a few modifications of instructional materials. For example, you could use large letters and numerals when preparing written materials and use black ink instead of colored inks when duplicating handouts or darken dim copies with a black felt-tip pen. Other commonsense considerations you should make include seating a visually handicapped child near you as you introduce a concept with materials or write on the chalkboard, verbalizing what you write on the chalkboard or overhead projector, choosing concrete teaching aids that help visually impaired and normal students understand the concept, and consistently storing materials in a designated place in the classroom.

Many visually deficient children enter school with far fewer quantitative experiences than their sighted peers. Sighted children make visual judgments that daddy's shoes are larger than mother's, the tree in the front yard is taller than the tree in the back yard, an ant is smaller than a bee, a Cadillac is larger than a Volkswagen, and so on. The curriculum of low-vision and blind children should be rich in experiences such as comparing the height, length, and area of objects; classifying objects by shape and size; ordering objects according to size and sets according to number of elements; and identifying cardinal and ordinal numbers. Ideas for teaching many of these topics are given in chapter 5.

To promote concept development, the mode for teaching mathematics to all children should begin with the concrete level. The importance of manipulative experience for blind students is implied from the fact that concept formation was the problem most cited by teachers of blind students when responding to a survey on problems that visually handicapped students face in learning mathematics.[3] Furthermore, Nolan cited evidence that computational achievement of visually impaired students can increase when concrete experience accompanies the computational process.[4] Teachers should be especially sensitive to the need of visually impaired students to spend adequate time at the exploratory level engaged in hands-on activities.

The American Printing House for the Blind* publishes materials that are particularly relevant to visually impaired young children, but many are similar to materials commonly used by all students. Some of these materials are:

1. the Cranmer Abacus, a pocket-sized calculating device (In a 1970 survey, this apparatus was ranked by teachers of blind students as being the most valuable computational device to blind students.[5]);
2. the Large Abacus, an abacus larger than the Cranmer Abacus and better suited for young visually handicapped children;
3. Geometric Forms, a primary-level device for introducing topological figures, Euclidean shapes, and solids;
4. Graphic Aid for Mathematics, a device similar to a geoboard and useful for making geometric configurations;
5. Tactile Ruler Familiarization Unit, an aid for teaching measurement of length in the metric and English systems;
6. Clockface with Raised Print and Braille Numbers, a device for teaching blind children to tell time;
7. APH Number Line Device, a concrete aid for modeling operations of arithmetic; and
8. Fractional Parts to Wholes, a tactile aid which parallels materials commonly used for teaching sighted students fractional numbers.

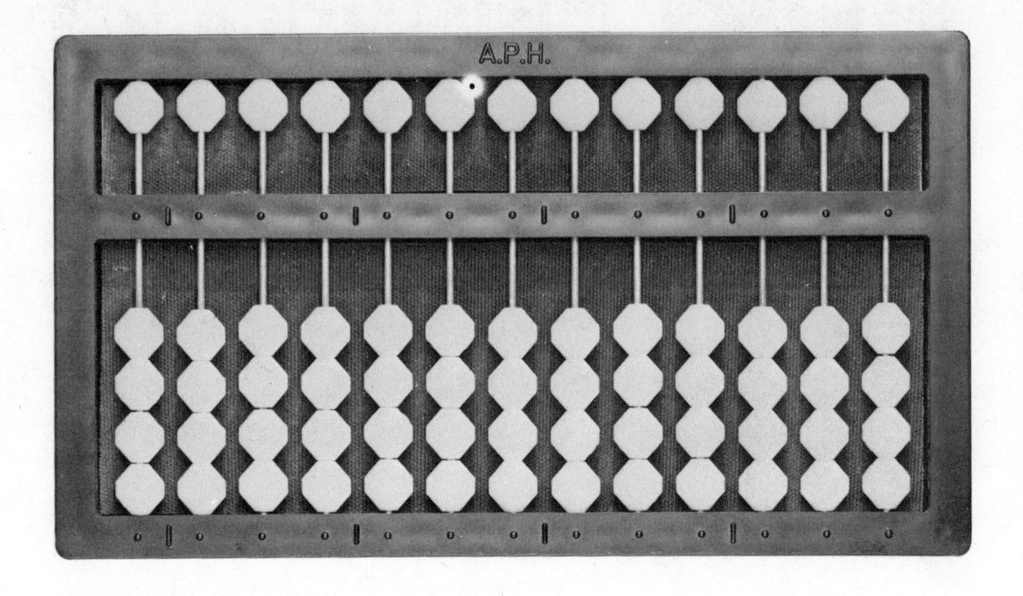

The Large Abacus

* The address is 1839 Frankfort Avenue, P.O. Box 6085, Louisville, Kentucky 40206.

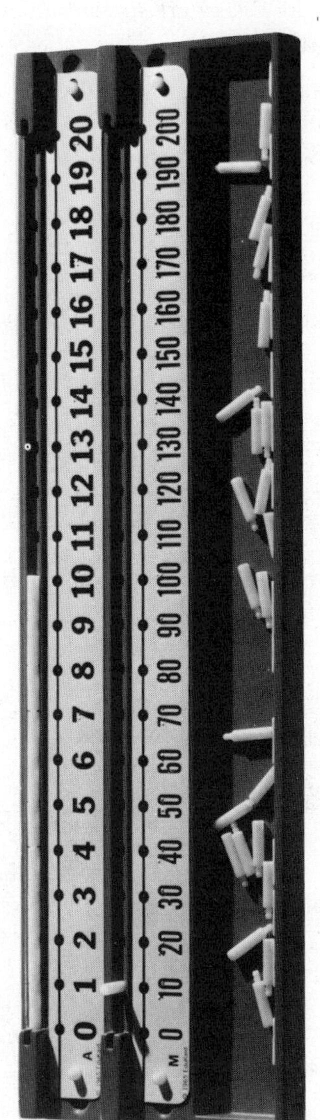

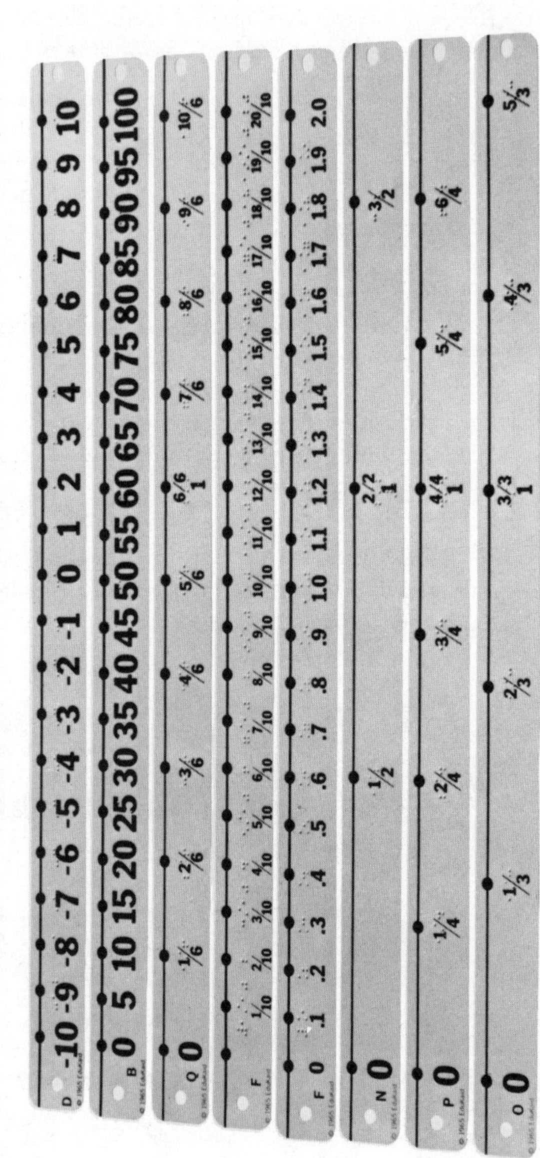

Number Line Device

A special educator should be available in schools with mainstreamed classrooms to give students with visual impairments instruction in topics such as braille and calculation on an abacus.

The Hearing Impaired Child

There are some general guidelines for teaching a deaf or hard-of-hearing child in a regular classroom. You as teacher should ascertain all you can about the hearing loss of the child, discuss in a sensitive way the nature of the hearing problem with the other class members, and become knowledgeable of the functions and care of a hearing aid. It may help to employ the "buddy" system in which a classmate is assigned to the hearing handicapped child for a few weeks to ensure that he understands assignments and is following class discussions. Also, to assist in speech reading, you should face the class when speaking, speak distinctly, and encourage the other class members to do likewise. It often helps to meet briefly with the hearing impaired student before an activity is done in class, explain new materials that will be used, and give him an outline of the upcoming activity.[6] The fact that some research indicates that the hearing impaired student tires easily implies that activities should be kept short and that these students' attention may drift.[7]

Very little research is available in regard to achievement of deaf students in mathematics. One study showed that deaf children may be weak in ability to reason but function close to normal on computational techniques. Suppes, who has engaged in extensive research on deaf children, indicates that these students perform as well on cognitive tasks as normal hearing students when verbal skills are not central to the task. Furthermore, Suppes has compiled evidence that the deaf and hearing students perform at about the same level on mathematical tasks when computer-assisted instruction is the mode of instruction.[8]

The Educable Mentally Retarded Student

Generally educable mentally retarded children have intelligence quotients which range from 50 to 75 or 80 and perform at least two years below grade level. These children often have the same general appearance as "normal" students of the same age, come from deprived home environments, possess poor language skills, have poor memories, and are low in perceptual and conceptual abilities.[9] These children show weakness in short-term retention, but once they acquire material it is retained at an almost normal level.[10]

Much of the current research indicates that educable mentally retarded students perform up to mental-age expectancy on computational skills and functional areas such as time and money but do poorly on verbal problems and situations dealing with concept development and reasoning. This situation may be due to the fact that computational and functional topics have been emphasized in the special-class curriculum of these children.[11] When educable mentally retarded students perform computations, however, they generally make more careless errors and resort to more elementary means of computation such as using markers or counting on their fingers than younger normal students of the same mental age.[12]

Since educable mentally retarded students usually do only about half as much academically in a school year as average children, an elementary teacher who has one of these children in her classroom should make a checklist of what material in mathematics during the year is essential to the subsequent year of study and what mathematics is important in real-life situations. In deciding which topics are essential, it may be helpful to examine the next year's mathematics textbooks or ascertain the opinion of a teacher of the next grade. Once the decision is made on what concepts and skills are critical, you as the teacher can select appropriate material from the mathematics textbook and from programmed-instruction workbooks or design your own worksheets. All of the instruction should be in small step sequences designed to produce success. For example, when an educable mentally retarded child is ready for symbolic work in addition, you might have him select the correct sum in

$$2+5=\begin{matrix}9\\7\end{matrix} \quad \text{and then} \quad 2+5=\begin{matrix}9\\8\\7\\6\end{matrix}$$

before you present him with the more abstract notation $2 + 5 = \square$.

Individualized instruction should be coupled with ample opportunity for the student to work with concrete materials and to join in small group and class activities dealing with concept formation. Physical activity should be a part of these students' instructional experiences. The use of concrete, action-oriented teaching aids by slow students is supported by Piaget, Brueckner, Grossnickle, Reckzeh, and Dienes.[13] Furthermore, reading should be kept minimal in arithmetic activities, standards of evaluation should be reasonable, and diagnostic and evaluative techniques should be applied frequently.[14]

The Learning Disabled Child

A student with a learning disability, unlike a slow learner, often has the mental ability to perform grade-level or above-grade-level tasks. His growth pattern is irregular, however.[15] A learning disabled child often has a poor memory, low attention span, poor motor coordination, and difficulty with visual-perceptual tasks. You may suspect that a learning disability is involved if you, when administering the Piagetian-task interviews in chapter 4, notice that a child's motor and perceptual development is considerably behind grade level. You should seek help from special education facilities to diagnose the nature of the learning disability if one exists. A student with visual-perception problems will have difficulty in classifying objects by shape, size, and color; in seriating objects according to relative differences; in writing and reading numerals; and in aligning numerals vertically when performing algorithmic computations. One who has visual-spatial problems may have trouble with geometric relationships such as congruence and similarity and with spatial aspects of quantity, length, capacity, and area.

You can assist a learning disabled child to perform up to his potential by:

1. giving him duplicated sheets instead of requiring him to copy problems from the chalkboard, overhead projector, or book;
2. showing him a flowchart of the procedure for solving computational problems or giving him a sample problem already worked out;
3. letting a child with a memory problem use a composite table of basic facts as he performs advanced algorithms;
4. underlining pertinent information and the question to be answered in verbal problems; and
5. reading directions and story problems orally.[16]

Research indicates that a teacher of a child with a learning disability should stress body movement and the use of manipulative aids when teaching mathematical processes, allow counting on the fingers, and emphasize mathematical language.[17] Kurtz and Spiker describe the task of teaching the student with a learning disability this way:

> The teacher who is creative in the use of concrete materials, skilled at breaking complex skills down into smaller sequential units, and who views clear auditory verbalizations as a necessity in working with all children will be able to work successfully with the learning disabled child.[18]

The Orthopedically Handicapped Student

Adjustments needed for a student who is crippled are primarily physical and emotional in nature rather than educational. The physical plant may require some modification, and the teacher may need to rearrange desks and tables, adjust the equipment needed by the handicapped student, and assign a student helper to the handicapped child for a few weeks. Sometimes these students have difficulty in performing tasks requiring coordination, perception, and cognition, but many have the same range of abilities as their nonhandicapped classmates.

Some suggestions for instruction of these students are to:

1. demonstrate the procedure instead of giving excessive verbal directions for activities,
2. plan activities that exploit as many senses as possible,
3. provide slow, deliberate, and concrete instruction,
4. involve the student in individual, structured activities until his self-concept reaches the point where group work will be beneficial,
5. keep periods for practice brief,
6. be supportive, but firm, and
7. praise the student for doing genuinely good work.[19]

The Emotionally Disturbed Child

Studies generally show that students who have emotional disorders are underachievers in mathematics. In a longitudinal analysis of the relationship of classroom behavior to academic achievement, mathematics achievement of students with behavior disorders was found to be significantly lower than that of their peers who exhibited socially approved behavior.[20] Whether problems in mathematics contribute to the emotional disorder or vice versa has not been determined, however. In their review of the current research on the relationship between emotional disturbances in students and arithmetic disability, Callahan and Glennon concluded that the academic performance of emotionally disturbed children may be benefitted by a classroom environment which is humanely structured and which rewards academic performance.[21]

The Mathematically Gifted Child

The following are some characteristics which distinguish mathematically talented children from those who are less able mathematically:

1. Sensitivity to, awareness of, and curiosity regarding quantity and the quantitative aspects of things within the environment.
2. Quickness in perceiving, comprehending, understanding, and dealing effectively with quantity and the quantitative aspects of things within the environment.
3. Ability to think and work abstractly and symbolically when dealing with quantity and quantitative ideas.
4. Ability to communicate quantitative ideas effectively to others, both orally and in writing, and readily to receive and assimilate quantitative ideas in the same ways.
5. Ability to perceive mathematical patterns, structures, relationships, and inter-relationships.
6. Ability to think and perform in quantitative situations in a flexible rather than a stereotyped manner; with insight, imagination, creativity, originality, self-direction, independence, eagerness, concentration, and persistence.
7. Ability to think and reason analytically and deductively; ability to think and reason inductively and to generalize.
8. Ability to transfer learning to new or novel "untaught" quantitative situations.
9. Ability to apply mathematical learning to social situations, to other curricular areas, and the like.
10. Ability to remember and retain that which has been learned.[22]

One study showed that high achievers in arithmetic demonstrate highly desirable traits such as a healthy ego, ability to get along well with others, creativity, and ability to deal with abstractions.[23] Not all gifted students, however, will exhibit all of these attributes.

The teacher generally has two avenues for individualizing instruction for a gifted student: vertical extension (acceleration) and horizontal extension (enrichment). In the former approach, a student advances to mathematics content usually taught at higher grade levels after he completes the work for his grade level. This approach can be easily implemented by having the student work from textbooks for succeeding school years. According to Glennon, however, vertical extension ". . . . views the talented child merely as a fast learner of average content. Such a perception of the talented child and his mathematics program is wholly insufficient and inadequate."[24] If the teacher and student select only the higher level topics that interest the child, the most stimulating aspects of the subject may be "drained off," thus causing the student to become bored in subsequent years when the less interesting content is left to be studied.

In horizontal extension, or enrichment, the gifted student examines in greater depth the topics the average child at the same grade level studies. He should delve into the structure of mathematics and do independent research on topics treated in class. Most of the current elementary mathematics textbooks contain suggestions for enrichment; other sources for ideas include books of activities, laboratory manuals, methods books, and articles in periodicals such as *The Arithmetic Teacher, Instructor,* and *Early Years.* You should develop a card file of ideas which extend the topics taught in the regular program. There are varying opinions on which type of extension is most advantageous; it may be best, therefore, to use a combination of vertical and horizontal extension to meet the needs of gifted learners.

To help a mathematically precocious student to develop creativity and inventiveness, you should keep these points in mind:

1. Provide challenging problems but choose problems within the student's level of comprehension.
2. Challenge him to find alternative and original solutions to problems.
3. Lead him to understand the importance of dependable data.
4. Encourage the use of disciplined thought which is important in the study of mathematics.
5. Steadily prod him to expand his creative thought.[25]

Closing

In reviewing the groups of exceptionalities described in this chapter, it is apparent that many handicapped students are slow learners in mathematics, and instruction should, accordingly, be adapted to accommodate their rate of learning. Students with certain learning disabilities, however, can be expected to perform at grade level if instructional procedures are adjusted to meet their needs. Although all children at the early childhood level need the opportunity to manipulate concrete materials in order to discover and conceptualize mathematical ideas, the amount of time slow learners need to spend at this level will be considerably more than that required by superior learners. Exceptional children should be treated in essentially the same way as their classmates, and evaluation of them should be in line with standards set for the other students.

In order to be prepared for situations that may arise in a mainstreamed classroom, you as teacher should study the types of exceptionalities and the implications for classroom instruction. As you learn about exceptional children, you can more readily identify students who have special needs, be more adept in meeting the needs of these children, and generally be less overwhelmed with the idea of mainstreaming. Actually good instruction for special students may result in improved instruction for all students, each of whom is special in his own unique way.

Notes

1. Kathleen H. Dunlop, "Mainstreaming: Valuing Diversity in Children," *Young Children* 32 (May 1977): 28.
2. Bill R. Gearheart and Mel W. Weishahn, "Strategies and Alternatives for Educating the Visually Impaired," *The Handicapped Child in the Regular Classroom* (St. Louis: The C. V. Mosby Company, 1976), p. 52.
3. Leroy G. Callahan and Vincent J. Glennon, *Elementary School Mathematics: A Guide to Current Research* (Washington, D. C.: Association for Supervision and Curriculum Development, 1975), p. 67.
4. Carson Y. Nolan, "Implications from Education of the Visually Handicapped for Early Childhood Education," in *Children with Special Needs: Early Development and Education,* Howard H. Spicker, Nicholas J. Anastasiow, and Walter L. Hodges, eds. (Minneapolis, Minn.: Leadership Training Institute/Special Education, 1976), p. 117.
5. Marian Lewis, "Teaching Arithmetic Computation Skills," *Education of the Visually Handicapped* 2 (1970): 72.
6. Gearheart and Weishahn, "Strategies and Alternatives," p. 42.
7. Nancy Allen et al., "Mainstreaming the Hearing Impaired Child" (unpublished booklet, TEMPO Project, George Peabody College for Teachers, 1977), p. 13.
8. Callahan and Glennon, *Elementary School Mathematics,* pp. 68–69.
9. Samuel A. Kirk, *Educating Exceptional Children,* 2nd ed. (Boston: Houghton Mifflin Company, 1972), pp. 195–96.
10. Austin J. Connolly, "Research in Mathematics Education and the Mentally Retarded," *The Arithmetic Teacher* 20 (October 1973): 493.
11. John F. Cawley and John O. Goodman, "Interrelationships among Mental Abilities, Reading, Language Arts, and Arithmetic with the Mentally Handicapped," *The Arithmetic Teacher* 15 (November 1968): 631.
12. Charlotte W. Junge, "Adjustment of Instruction (Elementary School)," in William C. Lowry, ed. *The Slow Learner in Mathematics,* Thirty-fifth Yearbook (Reston, Virginia: National Council of Teachers of Mathematics, 1972), p. 131.
13. Connolly, "Research in Mathematics Education," p. 495.
14. Gregory R. Baur and Linda Olsen George, *Helping Children Learn Mathematics* (Menlo Park, Calif.: Cummings Publishing Company, Inc., 1976), p. 42.
15. Ray Kurtz and Joan Spiker, "Slow or Learning Disabled—Is There a Difference?" *The Arithmetic Teacher* 23 (December 1976): 618–20.
16. Kurtz and Spiker, "Slow or Learning Disabled," p. 621.
17. Callahan and Glennon, *Elementary School Mathematics,* pp. 65–66.
18. Kurtz and Spiker, "Slow or Learning Disabled," p. 621.

19. Nelle Liggett et al., "Mainstreaming the Physically Handicapped" (unpublished booklet, TEMPO Project, George Peabody College for Teachers, 1977), p. 4.
20. John F. Feldhusen, John R. Thurston, and James J. Benning, "Aggressive Classroom Behavior and School Achievement," *Journal of Special Education* 4 (1970): 435–37.
21. Callahan and Glennon, *Elementary School Mathematics*, p. 67.
22. J. Fred Weaver and Cleo Fisher Brawley, "Enriching the Elementary School Mathematics Program for More Capable Children," *Journal of Education* 142 (October 1959): 6–7.
23. Ernest A. Haggard, "Socialization, Personality, and Academic Achievement in Gifted Children," *School Review* 65 (1957): 408.
24. Vincent J. Glennon, "Some Perspectives in Education," in Julius H. Hlavaty, ed. *Enrichment Mathematics for the Grades*, Twenty-seventh Yearbook (Washington, D.C.: National Council of Teachers of Mathematics, 1963), p. 26.
25. Charlotte W. Junge, "Depth Learning in Arithmetic—What Is It?" *The Arithmetic Teacher* 7 (November 1960): 342.

Assignment

1. Study chapters 4, 5, 7, and 8 of *The Slow Learner in Mathematics*, Thirty-fifth Yearbook, published by the National Council of Teachers of Mathematics. Prepare a card file of at least ten teaching ideas you find there which are appropriate for slow learners (most are suitable for all students) at the primary level.
2. Try to observe a mainstreamed classroom and interview a regular classroom teacher who has mainstreamed a handicapped child into her classroom. What was the nature of the exceptionality? What did the teacher do to modify materials and activities in mathematics? Does she think the attempt at mainstreaming has been successful?
3. Read and make a note of main points in these articles in *The Arithmetic Teacher*:
   a) "Mathematics and the Low Achiever" by Arnold M. Chandler, March 1970, pp. 196–98.
   b) "Mathematics, Questions, and 'Schools without Failure'" by Teri Perl, December 1974, pp. 669–73.
   c) "Problems Associated with the Reading of Arithmetic" by Leroy Barney, February 1972, pp. 131–33.

Laboratory Session

Preparation:

1. Study Chapter 2 of this textbook.
2. Read at least one article on teaching mathematics to each of the exceptional children discussed in this chapter. Prepare a card file of teaching ideas you find.

3. Select a grade level among K-3. Consult elementary mathematics textbooks for that grade level and mathematics activity and methods books and make a card file of at least ten suggestions for horizontal enrichment in mathematics at your chosen grade level.

Objectives: Exchange ideas on enrichment in mathematics and teaching exceptional children mathematics.

Materials: Ideas the students have collected.

Procedure: Form small groups and exchange the ideas you have collected on teaching exceptional children mathematics and enrichment in mathematics.

Bibliography

Allen, Nancy et al. "Mainstreaming the Hearing Impaired Child." Unpublished booklet, TEMPO Project, George Peabody College for Teachers, 1977.

Baur, Gregory R., and George, Linda Olsen. *Helping Children Learn Mathematics.* Menlo Park, California: Cummings Publishing Company, 1976.

Callahan, Leroy G., and Glennon, Vincent J. *Elementary School Mathematics: A Guide to Current Research.* Washington, D.C.: Association for Supervision and Curriculum Development, 1975.

Cawley, John F., and Goodman, John O. "Interrelationships among Mental Abilities, Reading, Language Arts, and Arithmetic with the Mentally Handicapped." *The Arithmetic Teacher* 15 (November 1968): 631–36.

Connolly, Austin J. "Research in Mathematics Education and the Mentally Retarded." *The Arithmetic Teacher* 20 (October 1973): 491–97.

Dunlop, Kathleen H. "Mainstreaming: Valuing Diversity in Children." *Young Children* 32 (May 1977): 26–32.

Feldhusen, John F.; Thurston, John R.; and Benning, James J. "Aggressive Classroom Behavior and School Achievement." *Journal of Special Education* 4 (1970): 431–39.

Gearheart, Bill R., and Weishahn, Mel W. *The Handicapped Child in the Regular Classroom.* St. Louis: The C. V. Mosby Company, 1976.

Haggard, Ernest A. "Socialization, Personality, and Academic Achievement in Gifted Children." *School Review* 65 (1957): 388–414.

Hlavaty, Julius H., ed. *Enrichment Mathematics for the Grades.* Twenty-seventh Yearbook. Washington, D.C.: National Council of Teachers of Mathematics, 1963.

Jackson, Roberta; MacKay, Louise; and Cunningham, Joe. "Mainstreaming the Mildly Retarded." Unpublished booklet, TEMPO Project, George Peabody College for Teachers, 1977.

Junge, Charlotte W. "Depth Learning in Arithmetic—What Is It?" *The Arithmetic Teacher* 7 (November 1960): 341–46.

Kirk, Samuel A. *Educating Exceptional Children,* 2nd ed. Boston: Houghton Mifflin Company, 1972.

Kurtz, Ray, and Spiker, Joan. "Slow or Learning Disabled—Is There a Difference?" *The Arithmetic Teacher* 23 (December 1976): 617–22.

Lewis, Marian. "Teaching Arithmetic Computation Skills." *Education of the Visually Handicapped* 2 (1970): 66–72.

Liggett, Nelle et al. "Mainstreaming the Physically Handicapped." Unpublished booklet, TEMPO Project, George Peabody College for Teachers, 1977.

Lowry, William C. ed. *The Slow Learner in Mathematics,* Thirty-fifth Yearbook. Reston, Virginia: National Council of Teachers of Mathematics, 1972.

Nolan, Carson Y. "Implications from Education of the Visually Handicapped for Early Childhood Education," in Howard H. Spicker, Nicholas J. Anastasiow, and Walter L. Hodges, eds. *Children with Special Needs: Early Development and Education.* Minneapolis, Minnesota: Leadership Training Institute/Special Education, 1976, pp. 113–19.

Patterson, Beverly; MacKay, Louise; and Dunlop, Kathleen. "Mainstreaming the Visually Handicapped Child." Unpublished booklet, TEMPO Project, George Peabody College for Teachers, 1977.

Spitzer, Herbert F. *What Research Says to the Teacher, Teaching Elementary School Mathematics.* Washington, D.C.: Association of Classroom Teachers, National Education Association, 1970.

Weaver, J. Fred, and Brawley, Cleo Fisher. "Enriching the Elementary School Mathematics Program for More Capable Children." *Journal of Education* 142 (October 1959): 1–40.

3

Instructional Procedures and Materials

This chapter deals with a sequence for teaching mathematical concepts and skills, methods of instruction, and instructional materials. Also, since the laboratory approach is useful in teaching mathematics, some suggestions for conducting an activity in a laboratory setting and a sample laboratory activity are included.

A Teaching Sequence

As you plan your yearly, unit, and daily work in mathematics, consider this instructional sequence: determining and providing readiness, teaching the concept or skill, furnishing practice activities, and providing opportunities for application of the concept or skill.

Readiness

Views of investigators such as Piaget, Bruner, and Gagné differ on readiness. Piaget favors providing an environment that enhances the development of intellectual capacities in a child but not the speeding up of this process through educational processes.[1] Piaget contends that if a child is taught, for example, that objects retain the same quantity, however they are shaped, before he is developmentally ready, he has learned a particular task but may not be able to generalize and relate the learning to similar but untaught situations.

Unlike Piaget and other investigators who view interaction with the environment as the factor that determines when certain intellectual capabilities appear in a child, Gagné holds that readiness depends on previously learned skills which relate to the new learning situation. To help a child acquire a new capability, one must determine what the child already knows and does not know and then help him learn the prerequisite concepts and skills. In this process, according to Gagné, learning accumulates by transfer and new potentialities for learning result.[2]

Bruner also has disagreed with Piaget's stand on developmental readiness as evidenced in this statement:

> . . . instruction in scientific ideas, even at the elementary level, need not follow slavishly the natural course of cognitive development in the child. It can also lead intellectual development by providing challenging but usable opportunities for the child to forge ahead in his development. Experience has shown that it is worth the effort to provide the growing child with problems that tempt him into next stages of development.[3]

In view of the contrasting ideas on readiness of leading investigators, you may at this point be asking, "Well, what should I do about readiness in teaching mathematics?" It seems reasonable to conclude that the developmental level of the student should be considered in readiness, but an analysis of the given mathematical content to determine prerequisite learning is important also. Let us determine the prerequisite concepts and skills for the task of assigning a cardinal number to a set (how many elements the set has), a typical objective for kindergarten. Consider the collection

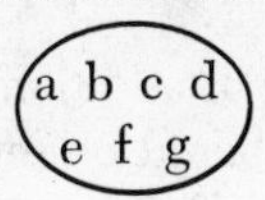

for which the student is to tell the cardinal number. Readiness includes:

1. ordering the elements of a set so that each member is counted exactly one time. For example, the student should not count "a" at one time and again later or point to "a" and say "1, 2, 3," then move to "b" and say "4, 5," and so on,
2. counting to at least ten, and
3. putting sets into one-to-one correspondence.

These steps applied to the example take this form: The student orders the elements of

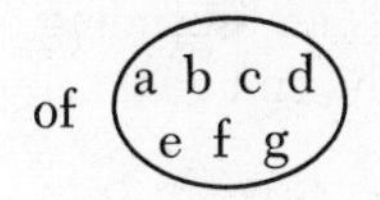

This step is suggested by the notation {a, b, c, d, e, f, g}.

Then he counts the ordered elements by placing {a, b, c, d, e, f, g} into one-to-one correspondence with {1, 2, 3, 4, 5, 6, 7}.

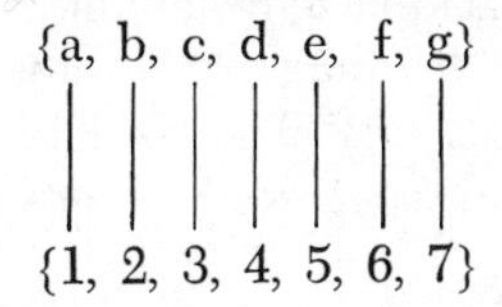

Once this is done, the student should be told that the last number used in counting the elements is the cardinal number of the set. Thus the given set has cardinal number 7. Many children will mentally perform these steps with very little help, but some will need careful guidance and many opportunities for practice.

Let us try another example. What are the mathematical prerequisites for learning how to add two numbers, each represented by two-digit numerals, using vertical form and expanded notation, a task that usually occurs in second grade? The arrangement of the work takes this form:

$$\begin{array}{r} 42 \\ +\,25 \\ \hline \end{array} = \begin{array}{r} 40 + 2 \\ 20 + 5 \\ \hline 60 + 7 = 67 \end{array}$$

In analyzing this task, we see that the student should be able to write numerals in expanded notation ($42 = 40 + 2$, $25 = 20 + 5$), find sums in basic addition facts ($2 + 5 = 7$), find simple sums ($40 + 20 = 60$), and write standard numerals when expanded notation is given ($60 + 7 = 67$). Furthermore, in order for the student to know why the ones and tens can be added separately (other than your simply telling him to add that way), he needs to understand the commutative and associative properties of addition. Examine these steps to see what is happening in the vertical algorithm:

$$\begin{aligned} 42 + 25 &= (40 + 2) + (20 + 5), \text{ the original problem} \\ &= (40 + 20) + (2 + 5), \text{ the way the addition is performed} \end{aligned}$$

This last step is justified by the commutative and associative properties of addition. Prerequisite learning for two-digit addition begins in first grade and should be reviewed in second grade before a relatively complex problem such as $\begin{array}{r} 42 \\ +\,25 \\ \hline \end{array}$ is introduced. To provide continuity in teaching, you, as teacher, should be aware of subsequent learning also. In this case, helping students learn addition requiring regrouping, e.g., $\begin{array}{r} 42 \\ +\,29 \\ \hline \end{array}$, will be an upcoming objective.

The task of finding the cardinal number of a set was analyzed by a descending task analysis—the end goal was stated and then prerequisite learning was identified. In the second example, a radiating task analysis was made—steps before and after the identified goal were considered. Another type of task analysis is an ascending approach which starts with the simplest behavior and builds up to the end goal.[4]

Teaching the Concept or Skill

Expository and Guided-discovery Methods. What is to be taught determines what type of instruction is appropriate. If the topic to be taught is verbal association or discrimination learning (Gagné's lower learning types), the method of instruction may appropriately be didactic, or expository. The vocabulary, symbolism, and notation used in mathematics cannot be discovered. Students must be told that "3" is the symbol for "three" and that $2 + 5 = \square$ is an open sentence. Furthermore, use of expository teaching is fitting after students have formed concepts and discovered principles and generalizations to review their discoveries and point out some ways they will be used later. If the lesson concerns a concept, principle, or problem-solving situation (Gagné's higher types of learning), the guided-discovery approach is recommended. In this approach, the teacher facilitates learning by asking pertinent questions and giving verbal cues. This type of instruction lends itself well to small group work and laboratory activities in which the students manipulate materials, interact with their peers, discuss strategies for solving problems, and verbalize their discoveries.

Inductive and Deductive Approaches. In addition to adjusting the type of instruction to the task to be taught, the early childhood educator should understand how the inductive method is used to teach most topics at the early childhood level. Although a few superior learners may have reached the formal-operational stage of intellectual development and be able to benefit from deductive exercises, for the most part early childhood experiences in mathematics follow the inductive strategy. Children need many examples to investigate, and with your questioning and making suggestions, they make generalizations and discover patterns. Many young children recognize a symbolic pattern before they can verbalize it. For example, a first-grade student may find $1 + 0 = 1$, $2 + 0 = 2$, $3 + 0 = 3, \ldots, 9 + 0 = 9$ and predict from the pattern that $849 + 0$ is 849, but he may not be able to verbalize, "Zero added to any number gives a sum equal to the original number." Provide the verbalization as a model for pattern description until the students can do so.

Components of a Mathematics Lesson. Usually a mathematics lesson involves a statement of objectives, list of prerequisite learning, list of materials, teaching procedure, and plan for evaluation. Objectives are often given in behavioral terms in which the expected performance of the student and the selected materials for the activity are related. Some objectives, however, can be designed to provide open-ended experience in a certain content. Prerequisite learning is determined when the task to be taught is analyzed. Suitable materials, which may be commercial, teacher-made, pupil-made, or ordinary, must be chosen. Some commercial manipulative aids are described subsequently in this chapter. The teaching procedure should answer the question, "What do I have to do to help students acquire the expected behavior?" and should include what you and your students will be doing during the activity. Also, questions you plan to ask to guide the students in the activity and to help them relate previous knowledge to the new learning should be planned. Evaluation may consist of your observing the performance of individual students as they participate in the activity while listening to their oral responses, your presenting students with a task similar to the ones done during the activity and observing their performance on it, or your giving them a written test.

The sample first-grade lesson that follows is directed to helping a child, small group, or entire class conceptualize fractional numbers.

Objective: Each student should write the fractional number represented by folded paper.

Prerequisite learning: Each child should be able to distinguish whether or not a shape is divided into parts of the same size and be able to determine the cardinal number of a set.

Materials: Each student should have a crayon, two strips of paper folded into nonequivalent parts, two unfolded strips, and four rectangular strips folded into equivalent parts and selected from among these: one strip in five parts with two parts shaded, another in six parts with five parts shaded, and the others in thirds, fourths, eighths, and tenths with none shaded.

Procedure: Give each student two strips of prefolded paper (the fifths and sixths), two strips of unfolded paper, and a crayon. Ask the students to find the strip of paper folded into five parts. Establish with the students that all five parts are the same size. Write 5 on the chalkboard. Now ask them how many parts of the strip of paper are shaded. Since two of the five parts are shaded, record 2 over the 5 on the chalkboard: $\frac{2}{5}$. Emphasize what the 2 and 5 mean. Tell the students this is a fractional number, is read as "two-fifths," and means two-fifths of the strip of paper is shaded. Repeat this procedure with the paper folded into sixths, recording and having students to record (on the back of the strip of paper) and read the notation. Now ask them to fold one of the unfolded strips of paper into two parts, each the same size, and to color one of the parts. Verbalize that one out of two equivalent parts is colored, record $\frac{1}{2}$ on the chalkboard, and emphasize that one-half of the region is colored. Have the students fold the other strip into four equivalent parts, color, for example, three of the four parts, and write the correct fractional number on the back of the paper.

Next give each student the other strips of paper (two prefolded into equivalent parts and two folded into nonequivalent parts). Not all children should have the same examples of the prefolded paper. Ask the students to look at their four strips and sort them into two sets, one having paper folded into parts the same size and the other having paper folded into nonequivalent parts. When they have done this, point out that fractional numbers can only be shown when all parts are equivalent. Then ask them to identify the number of parts in the equally sized strips, color the number of parts they wish, and then write on the back of the paper the fractional number represented. When students have finished, let each child present one of his two examples to the class. (Select students if time is limited.)

Evaluation: Observe children as they fold the paper and write the fractional numbers represented, and listen to their oral responses.

Practicing for Mastery

When a student has learned a concept, he needs ample opportunity for practice so that he can develop skill in the area. This step is particularly important in becoming skillful in arithmetic. Practice activities are generally given at the symbolic level and are often in the form of games, puzzles, and competitive exercises for individuals or groups of the class. Ideas for practice are given in subsequent chapters.

The following sample practice, or drill, activity on multiplication is appropriate for a small group of students in second or third grade.

Objective: The student should be able to have immediate recall of the basic multiplication facts.

Prerequisite learning: the concept of multiplication

Materials: multiplication flash cards, jack ball, cups on which are numerals corresponding to the products on the flash cards

Procedure: Introduce the game to the group of students. Designate one child as leader of the group. Explain that he should hold up flash cards, one at a time (with the answer on the back for him to check the answers of the players), and each player should read the card, tell the product, and then try to throw the jack ball into the cup corresponding to the product. If the player knows the answer, he gets one point; if he "rings" the correct cup, he gets another point. If the player does not know the product, the next student can try to answer the same problem. The game should continue until each student has several turns. The player with the most points wins the game. An alternative procedure would be to divide the class into groups and let the teams compete.

Some books that have numerous suggestions for practice activities and games are *Games for Individualizing Mathematics Learning* by Leonard M. Kennedy and Ruth L. Michon, Charles E. Merrill Publishing Company, 1300 Alum Creek Drive, Columbus, Ohio 43216; *Math Activities for Child Involvement* by Enoch Dumas, Allyn and Bacon, Inc., 470 Atlantic Avenue, Boston, Massachusetts 02210; and *Plus* by Mary E. Platts, Educational Service, Inc., P. O. Box 219, Stevensville, Michigan 49127.

Applying the Concept or Skill

Acquiring skill in areas of mathematics becomes especially meaningful to the student when he understands when to use it in his everyday life. In the classroom, the application step manifests itself generally in the form of story problems and problem-solving discovery situations, both of which are discussed in chapter 15. Problem-solving activities should also include the use of measurement skills and properties in geometry.

Instructional Materials

Instructional materials discussed in this section are manipulative aids for abstracting mathematical concepts, the pocket calculator, elementary mathematics textbooks, individualized mathematics programs, and programmed instruction.

Manipulative Aids

The learning psychology described in chapter 1 points to the need for using concrete materials to help young children conceptualize mathematical ideas. Because of the abundance of teaching aids available, it is important to exercise good judgment in choosing materials for the classroom. These are some criteria to consider in the selection of manipulative materials:

1. The material should foster concept or skill development and/or attitude improvement.
2. The material should be interesting to the child but not so appealing that he merely wants to play with it and not use it to make mathematical discoveries. You should allow students, however, to explore any manipulative aid before using it in a structured activity. If you do not let them have this time of free play, you will probably not realize the mathematical objectives of your activity.
3. The aid should serve to teach several mathematical concepts. Such a feature encourages the child not to tie any one concept to a particular aid and hence promotes mathematical abstractions.
4. The product should be usable by children of various learning styles—visual, aural, tactile, and kinesthetic.
5. The aid should serve as a learning device for children at various levels of concept formation or skill development.
6. The material should not require extensive time on your part and your students' part to learn how to use it. In this way, the concept being abstracted can be the focal point rather than the object being manipulated.
7. The material should withstand normal use by you and your students.
8. The material should not require excessive space for use or storage.
9. After the aid has been introduced and demonstrated, individual children or small groups should be able, without constant teacher supervision, to use the aid to discover mathematical relationships and make generalizations.
10. The commercial product should offer utility and attractiveness that a similar teacher-made product would not offer.
11. The level of difficulty of content and format should be appropriate for the intended age level.
12. The material should fit into the program at your school and complement the textbooks in use.
13. The cost of the material should not outweigh its curricular functions.

These criteria are reflected in the evaluation form on p. 37.

Attribute Materials. Attribute blocks, characterized by various attributes such as shape, color, and size are designed to foster the development of logical thought. These blocks are useful in preschool through grade 3 for teaching logical classification, set inclusion, union and intersection of sets, logical terms such as "not," "and," "or," "some," "all," and "none," number, and matrix arrays. You may make your own set from colored poster board, but the commercial blocks are more appealing to children and are not excessively priced. These blocks contribute in the affective domain as well as the cognitive domain. Students enjoy playing such games as "Detective" in which one child hides a block behind him and other students ask him logical questions about the shape, size, and color of the block to determine the correct block.

MATHEMATICS TEACHING AIDS EVALUATION FORM

Evaluator ______________________________

Teaching Aid ______________________________

Manufacturer ______________________________

Address ______________________________

Type of Product ______________________________
(manipulative materials, slides, games, etc.)

Cost $__________ Age Levels __________

Fosters concepts______, skills______, attitude______ (Check those which apply.)

Useful for individual children____, small groups____, teacher demonstration____
(Check those which apply.)

I. Rating
Rate the aid on each criterion listed. Use the scale 0 (poor), 1 (fair), 2 (good), and 3 (excellent). Then add the individual ratings to determine your overall rating of the aid: 0-12 (poor, 13-20 (fair), 21-28 (good), 29-36 (excellent).

Criteria	*Rating*
1. Concept or skill development and/or attitude improvement	____
2. Interest appeal to children	____
3. Multi-topic applicability	____
4. Usability by children with various learning styles	____
5. Provisions for use by children at various performance levels	____
6. Reasonableness of required preparation time	____
7. Durability	____
8. Storage space and area required for use	____
9. Practicality for use by individual children or small groups	____
10. Comparability to similar teacher-made product	____
11. Appropriateness of level of difficulty of content or format for intended age level	____
12. Complementation of other materials used in the program	____
Total Points	____
Overall Rating	____

II. Questions

1. What concepts, skills, and/or attitudes is the product likely to foster?

2. Are the objectives cited in (1) above appropriate for the early childhood curriculum?

3. Does this material lend itself to the integration of at least one other subject area with mathematics? Explain.

4. What aspect of the product will make children interested?

5. Compare the quality of the product with the cost to determine the desirability of purchase.

Additional remarks:

Other attribute materials include People pieces and Creature cards.* The People pieces have cartoon-like pictures of people which can be distinguished by the attributes of age, sex, color of clothing, and body type. Creature cards have various designs and encourage discriminatory thinking. Student activity cards can be purchased with the aids, but children should be encouraged to create their own cards to add to the commercial set. Plastic templates containing the shapes of the attribute blocks assist students in designing their own activity cards.

Cuisenaire Rods. These rods, developed by Belgian educator Georges Cuisenaire, are useful in preschool through grade 3 for teaching ordering, fractional numbers, even and odd numbers, prime and composite numbers, operations of whole numbers, patterns, and length in the metric system. The rods range in length from one centimeter to ten centimeters (one decimeter), each length being a different color. Cuisenaire squares and blocks can be combined with the rods to obtain a model for place value through the thousands.†

The rods form color families based on the primary colors of red, blue, and yellow and the nonchromatic colors of white and black. The red family (red, purple, and brown rods) and yellow family (yellow and orange rods) are based on doubling–the length of one rod is doubled to obtain the next rod in the family. The blue family, consisting of the green, dark-green, and blue rods, represents an arithmetic progression: the length of the green rod plus itself is the length of the dark-green rod, and the length of the dark-green rod plus the green rod length equals the length of the blue rod. The white and black rods form the remaining color family.[5]

Even though much of the early work with Cuisenaire rods involves the association of the numerals 1, 2, . . ., 10 with the rods, they are not limited to this correspondence. When fractional numbers are modeled, the unit rod will be some rod other than the white one; for example, if the yellow rod represents 1, then the white rod is associated with $\frac{1}{5}$; the red rod, with $\frac{2}{5}$; the green rod, with $\frac{3}{5}$; . . ., the orange rod, with 2. This feature encourages the abstraction of mathematical concepts. In a given discussion, it is important to clarify what numbers are represented by the rods.

Unifix Cubes. These cubes are useful in teaching the one-more-than concept, cardinal and ordinal number, place value, arithmetic operations, and measurement of length, area, and volume. The cubes are interlocking to enable the student to construct longer rods from single cubes, and a set consists of ten different colors although color is not significant to the numerical values represented.

Base-ten blocks. These wooden blocks can be used in kindergarten through third grade for developing the concepts of number, place value, and addition, subtraction,

* These materials are included in the set of attribute materials published by Webster Division, McGraw-Hill Book Company, New York, New York. The set of thirty-two attribute blocks published by this company consist of four colors, four shapes, and two sizes.

† The name Cuisenaire and a color sequence of the rods, squares, and cubes are trademarks of Cuisenaire Company of America, Inc., and used with its permission.

multiplication, and division and for teaching measurement of length, area, and volume. Recent sets with metric dimensions are useful in work with the metric system. A set of blocks consists of unit cubes (ones), "longs" (tens), "flats" (hundreds), and "blocks" (thousands). The tens, hundreds, and thousands blocks are scored to give the appearance of a unit-cube composition. This is a top view of the representation of 132:

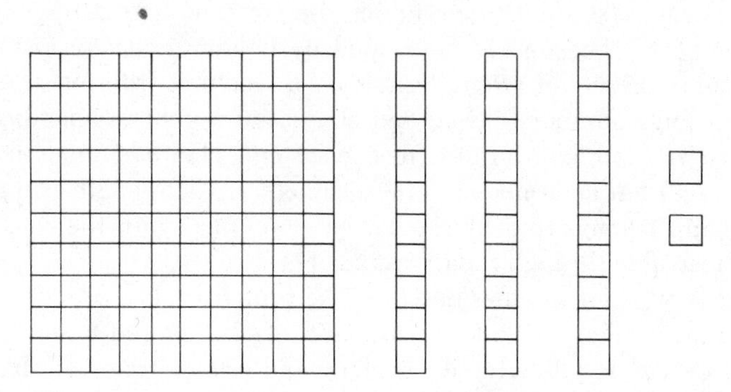

Abacus. An abacus can be used in kindergarten through grade three in activities on counting, place value, addition, subtraction, multiplication, and division. It is especially helpful in grades two and three in representing large numbers and modeling addition and subtraction algorithms for whole numbers. An abacus can be purchased commercially, but teacher-made ones are functional also. The following open-end abacus illustrates 26,045.

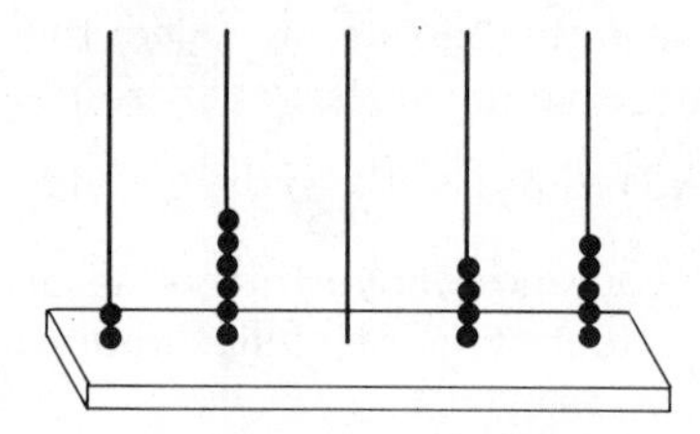

Other commercial varieties have loops and an upright partition; numbers are represented by bringing the needed counters from the back over the partition. Calculation on the abacus will be discussed in later chapters. Similar procedures apply to the Cranmer abacus, a calculation device used by blind and partially sighted students.

Geoboard. The geoboard is a device with pegs arranged in an array that allows students to abstract mathematical concepts and discover patterns by making shapes and designs with rubber bands. You should set standards on the use of the material, especially the rubber bands. The geoboard can be used at all levels of early childhood education and by large and small groups as well as by individual students. Although it is basically a geometry aid, it can be used to teach number patterns, fractional numbers, and the array approach to multiplication. The manipulation of the rubber bands helps students develop fine motor skills and eye-hand coordination. Congruence, similarity, symmetry, rotation, reflection, perimeter, and area are some of the geometry and measurement concepts that can be taught with the geoboard. This illustration shows a geoboard representation of two congruent right triangles:

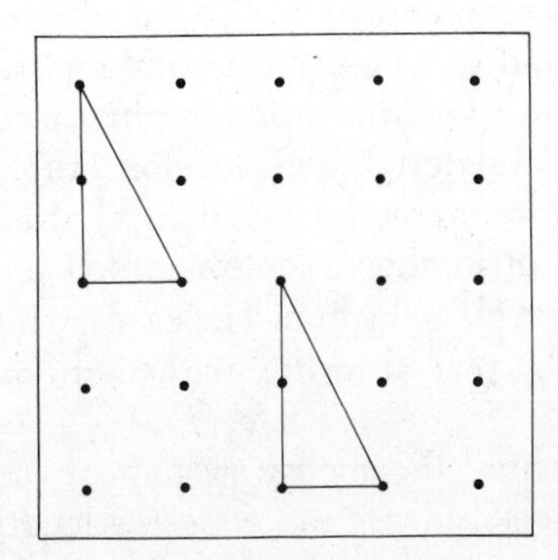

Dot paper can be used along with the geoboard and is especially helpful when recording answers found on the geoboard. Activity cards can be bought or generated by you and your pupils.

Students enjoy making their own geoboards. Individual five-nail by five-nail metric geoboards can be constructed by arranging nails on twenty-centimeter by twenty-centimeter sanded boards so that adjacent nails in both the horizontal and vertical directions are four centimeters apart. Have students leave two centimeters of space on the border so that several geoboards can be pushed together to form a large array.

Tangrams. Sets of tangrams, a seven-piece puzzle originating in China, consist of five triangular pieces, one square shape, and one parallelogram shape. Preschool, kindergarten and first-grade students can learn the names of the shapes of the pieces, discriminate among the sizes of the triangular pieces, and use the tangrams to solve puzzles. They enjoy making animal figures and letters with the tangram pieces. The instructional aid can be used to teach second and third graders congruence, similarity, symmetry, perimeter, and area. Tangrams are especially good in helping students see that a given plane region can be decomposed into various shapes. For example, two triangles and a square cover the given parallelogram shape.

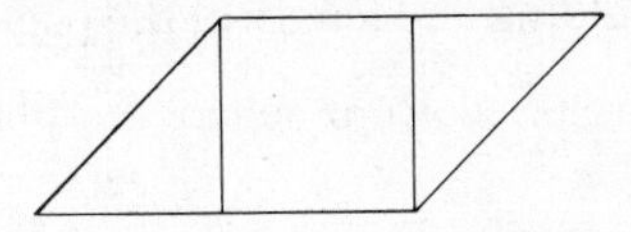

Experience such as this also helps children see relationships among the areas of geometric regions and discover formulas for area.

Puzzle cards on which students can place the tangram pieces are available commercially. At first let students solve the puzzles by placing the tangram pieces directly on the puzzle card, but gradually prod them to solve the puzzle to the side of the card as they look at the shape on the problem card. Some cards show only a small-scale version of the puzzle in which case the student is forced into a higher level of performance of solving the puzzle off the card. Encourage children to create their own puzzles and exchange them with their friends or carry them home for their parents to solve.

Mirror Cards. A set of mirror cards consists of unbreakable mirrors and groups of picture and pattern cards. Students place mirrors on the problem cards to reproduce given patterns. The aid can be used in kindergarten through third grade to foster a positive attitude toward mathematics, improve a child's self-concept through success in working with the material, and develop problem-solving skills. Encourage students to observe, make predictions, and check predictions as they work with the cards. Some concepts in mathematics for which mirror cards are servicable are reflection, symmetry, and classification. The cards, which can be used by individuals or small groups, can be placed in the mathematics center for students to use during their free time.

Metric Materials. Teaching the metric system of measurement begins in the early childhood years. When students can conserve length and deal with informal, nonstandard units of measurement, they are ready to measure with standard units. Classroom materials needed for teaching the metric system at the early childhood level are listed here:

1. meter stick divided into centimeters
2. centimeter rulers
3. meter tape measures
4. a balance scale
5. metric weights (1-, 2-, 5-, 10-, 20- and 50-gram weights)
6. bathroom scale calibrated in kilograms
7. centimeter cubes each weighing one gram (Interlocking cubes are functional for measuring length, area, and volume as well as weighing objects in grams.)
8. Celsius thermometer
9. centimeter grid paper
10. centimeter grid transparency (useful for measuring area especially of irregularly shaped regions)
11. a liter container
12. a milliliter graduate[6]
13. a trundle wheel (useful for measuring distances such as the length of the hall in the school building and length of the playground)

These materials can be purchased from school supply houses and publishers of manipulative materials.

The Minicomputer. A versatile, inexpensive instructional device that may be helpful with some children is the minicomputer designed by Frédérique Papy. Since it is color coded with Cuisenaire rods, it should not be used until children have gained some experience with Cuisenaire rods. In particular, the white (1), red (2), purple (4), and brown (8) rods should be kept visible for reference during Minicomputer activities. Children in the first grade may use the Minicomputer to model counting numbers and to add and subtract them; second-grade children may use it for addition and subtraction algorithms and for multiplication; children in third grade may model basic division facts and extend addition, subtraction, and multiplication.

A Minicomputer board can be constructed using squares of white, red, purple, and brown construction paper or poster board. The four squares can be taped or glued together and covered with self-adhesive clear plastic.

Brown(8)	Purple (4)
Red (2)	White (1)

Plastic chips or paper cut-outs can be used for counters. For numbers greater than or equal to 10, two or more boards are needed and place value is utilized. The board can be adapted for blind students by placing felt numerals on the squares or by pasting the appropriate number of strips on each square.

There is one general basic ground rule in modeling numbers on the Minicomputer: the smallest number of counters possible should be used to represent a single number. In the initial work, an appeal to the four Cuisenaire rods is helpful in understanding this ground rule. For example, 4 is shown with Cuisenaire rods as the single purple rod; similarly, 4 is indicated on the Minicomputer by placing a counter in the purple square.

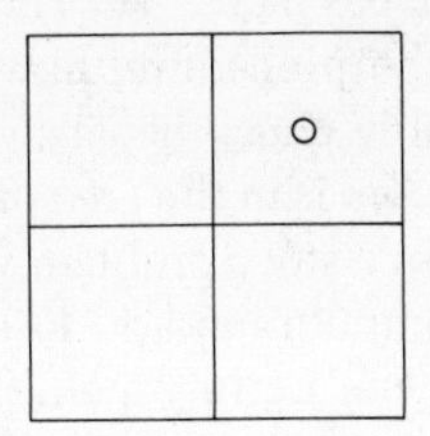

Performing operations of whole-number arithmetic on the Minicomputer will be illustrated later as the operations are discussed.

The Pocket Calculator

The pocket calculator was introduced by the Japanese in 1968. The size of a large paperback book, it cost about four hundred dollars. The cigarette pack-sized version was introduced by Bowmar Instrument Corporation in 1971 for about two hundred forty dollars. Texas Instruments soon presented their mini-calculator for one hundred dollars. Today we can buy similar calculators for less than ten dollars.[7]

Parents and teachers are often hesitant to let their children use calculators for fear they will depend on the calculator and never develop computational abilities. The calculator, however, offers many possibilities in early childhood education not as a replacement but as a supplement to any mathematics program. It can be treated as any other instructional aid.

Calculators serve several important instructional purposes. One is the development and extension of patterns in arithmetic. For example, you could have first-grade students use their skill in addition and subtraction to compare the sums $5 + 8$ and $(5 + 2) + (8 - 2)$ and other pairs of sums in which you increase one addend by a certain number and decrease the other addend by the same number. They should find each pair is equal. Then you could have the children predict whether $748 + 263$ and $(748 + 90) + (263 - 90)$ would be equal and then check their prediction by performing the computation on the calculator (since they have not yet learned procedures for doing these computations with paper and pencil). Ask questions such as "Do you think the pattern will always hold? Can you find an example when it does not hold? Can you make a generalization?" Students should indicate that the sum stays the same when a whole number is added to one addend and subtracted from the other addend. They should learn, however, to moderate, especially by the time they are in third grade, the blind acceptance of patterns with a healthy skepticism about whether the pattern is generally true. Gifted students should become cautious of the possible limited extent of patterns even earlier than third grade.

The calculator is also useful in the development of algorithms for arithmetic operations of whole numbers. For example, when teaching a third grader the process of dividing whole numbers by the distributive method, you can have him find the quotients

$$4\overline{)348} \qquad 4\overline{)300} \qquad 4\overline{)40} \qquad 4\overline{)8}$$

on the calculator and then ask, "Does $348 \div 4 = (300 \div 4) + (40 \div 4) + (8 \div 4)$?" Once the student understands the procedure, however, he should compute using pencil and paper and then possibly check his answers with the calculator. A third use of the calculator in the classroom is in the area of word problems. When students must perform the computations in a story problem with pencil and paper, they may become sidetracked with the computation and lose sight of the actual process involved in solving the given problem. Letting a child use the calculator for the computation allows him to focus on the process alone.

One problem in the use of calculators is that they give decimal quotients rather than whole-number quotients and remainders. Consequently, work with them should be limited when third graders explore division. Division at that level requires whole-number quotients, and when a remainder exists, it is listed separately as a whole number.[8]

Calculator games help develop skills of estimation and mental calculation. This area has long been an important part of instructional programs of blind students but has often been neglected in regular classrooms. One such estimation game for primary level involves presenting the students with several numeral cards from which they must choose the two they think represent numbers which give a sum closest to a given number (one that you specify). Children select the two numbers by estimating the sum or by mental calculation, and each student records on a sheet of paper his choice. Two students, who act as judges, compute on the calculator the answers of each child to determine which students have answers closest to the given number. Each of these students receives a point. The game continues for several more rounds. The winner is the child with the most points at the end of the activity.[9]

The calculator is well suited for problem-solving activities in a laboratory setting. Here is a problem card designed to help third-grade students conceptualize one million.

1. Write the numeral for one million. ______________
2. How would you spend a million dollars without giving any away to charity? Do the following to help you answer this question.
   a) Write down the names of some items you would like to buy, and estimate the the cost of each item.
   b) Use your calculator to add the costs of the items. Continue listing until you have spent a million dollars.[10]

Elementary Mathematics Textbooks

A good mathematics textbook is the instructional aid to which a teacher refers most often. Textbooks are usually written by elementary school teachers and mathematics educators and should reflect modern research findings and the best thinking of professional individuals concerned with mathematics education. Many effective teachers individualize instruction by using several textbooks from which they select material appropriate for groups of students or individual students. Pages from various textbooks can be placed in clear plastic covers or laminated for student use, or a teacher can modify certain exercises and construct work sheets or task cards. Although the textbook is clearly vital, it should not be the initial aid used when presenting new concepts because it only offers material which is pictorial and symbolic. Concepts should be introduced at the concrete level using suitable materials.

These criteria should be considered when selecting and evaluating a student textbook:

1. Mathematics content (Is attention given to arithmetic, geometry, and algebra?)
2. Presentation of content (Is attention given to the development of a concept using concrete or pictorial aids followed by meaningful "practice" activities? Are inductive patterns developed and generalizations provided for as aids for learning?)
3. Sequencing of content (Do the topics seem to be sequenced well developmentally and mathematically?)
4. Organization of content (Is the content organized around major concepts in mathematics? Does the textbook clarify the scope and purpose of individual lessons, chapters, and units?)
5. Mathematical precision in language (Are statements mathematically correct? For example, is "union of sets" used instead of "adding sets" when illustrating addition of cardinal numbers?)
6. Provision for individual differences (Are enrichment and remediation activities included?)

7. Adaptation of materials (Can the textbook be adapted to meet the needs of "normal" as well as "special" students?)
8. Vocabulary (Are names given to important ideas?)
9. Readability (Does the textbook accommodate the range of abilities and backgrounds of the children who will use it? Are directions clear and precise? Are examples heavily dependent on reading or are many nonverbal?)
10. Physical make-up (Are the cover and illustrations attractive? Is the book durable enough to withstand normal use? Is the size appropriate for young children?)
11. Appropriateness of illustrations (Are the illustrations fitting for the grade level being considered? Do the pictures depict males and females enjoying a variety of roles? Do they reflect the pluralistic, multi-ethnic nature of our society?)
12. Tests and other aids in the textbook (Are pretests and posttests given? Is adequate attention given to review? Are references and the glossary adequate in terms of scope and content?)

Features of the teacher's manual and offerings of the publisher should be considered also.

1. Does the manual include suggestions for optional activities and specific instructional procedures?
2. Are objectives stated for the activities?
3. Does the manual give a scope and sequence chart?
4. Does the teacher's edition include a reproduction of the pages in the student textbook?
5. Does the publisher offer other services such as manipulative aids, filmstrips, activity cards, and overhead-projector transparencies?

Individualized Programs

Individualized programs such as Individually Prescribed Instruction (IPI) and Program for Learning in Accordance with Needs (PLAN) are designed to channel the learner through a mathematics program commensurate with his rate of learning. These programs include behavioral objectives, diagnostic tests, lessons, and posttests. The teacher administers tests and prescribes appropriate activities in the program for the student. Most research indicates that students in these programs show no significant achievement gains over conventional programs.[11] Many companies, for example, Ginn and Houghton Mifflin, publish individualized programs.

Programmed Instruction

Evidence is inconclusive as to whether programmed instruction or conventional methods produce greater learning. Some studies have pointed to the fact that students differ on how much they benefit from programmed material.[12] Research does indicate that some mentally retarded children have profited from programmed learning as evidenced from the fact that their retention rate was increased, time required to attain skills was reduced, and negativism and hostility were reduced when they used programmed materials.[13]

The Laboratory Approach

In general the laboratory approach to learning involves the students in making discoveries in active problem-solving situations. In addition to being suitable for problem-solving activities, the laboratory setting can be used to reinforce and extend concepts and principles already introduced in class. In this approach, the classroom is organized into learning stations where students work on the same or related content with different materials at the same time or with the same materials at possibly different levels. The teacher is cast in the role of facilitator as she circulates among the groups. Manipulative materials such as filmstrips, records, and cassette tapes can be set up in the centers. Young children whose reading level may be such that they cannot read the directions for activities at each center will need to have the directions read to them orally or given on cassette tape. Even kindergarten classrooms can be successfully organized into learning-listening centers for instruction in mathematics.

An example of a laboratory activity on addition follows. It is geared to first-grade level and is designed to reinforce and extend ideas dealt with in class as the teacher introduced the concept of addition of whole numbers to the students.

Objective: The students will participate in four centers on addition and do the required work with 100 percent accuracy.
Prerequisite learning: addition at the concrete, pictorial, and symbolic levels
Materials: described at each center
Procedure: Give each student a record sheet:

Center 1	*Center 2*	*Center 3*	*Center 4*
1.	1.	1.	
2.	2.	2.	
3.	3.	3.	
4.	4.		
	5.		

Explain the tasks involved at each center. Allow each group to remain at each table from five to ten minutes, or let students rotate individually among the centers when a vacant space occurs. Center 3, however, requires group work.
Center 1. For each child at the center, arrange two small margarine bowls side by side with a designated number of objects (dried beans, plastic counters, etc.) in each bowl. The student determines the number of objects in each bowl (set), empties the contents of one bowl into the other, determines the number in the joined set, and writes the associated addition sentence on his record sheet. The student then rearranges the objects so that each of the two bowls

has the original number of objects in them. He then moves to a vacant position at this center and repeats the procedure. Each child should work at least four exercises at this table.

Center 2. Provide for each student 10 to 20 large paper clips and laminated index cards–one blank card, several cards each having a one-digit numeral on it, and one card for each of the symbols "+" and "=." Each child should make five open addition sentences with the cards, clip paper clips on each card to make a set associated with the numeral on the card, and write the sum on the blank card. An example of his work is suggested here:

Each time the student should write the completed sentence on his record sheet and wipe off the numeral on the blank card.

Center 3. Provide a walk-on number line and task cards, each with an example such as $4 + 1 = \square$. Each child selects three task cards. Students take turns holding up task cards in front of the other members of the group and walking on the number line to solve sentences. To solve $4 + 1 = \square$, a student begins at 0, takes four steps, pauses, takes one more step, stops, and says the number sentence $4 + 1 = 5$. He then writes the sentence on his record sheet.

Center 4. Have students cut out pictures from magazines or draw pictures to generate examples representing addition. Each child should join two sets and then write the associated number sentence for each example he develops. This open exercise is especially intended for advanced learners. Encourage them to examine open sentences such as $7 + 0 = \square$, $0 + 7 = \square$, $6 + 9 = \square$, and $9 + 6 = \square$ and to look for patterns among them.

Closing

Teaching concepts in mathematics proceeds through the steps of establishing readiness, presenting the concept, providing practice activities to promote mastery, and designing activities for applying the concept. Diagnosis and evaluation should be ongoing throughout this instructional sequence. The inductive approach is the usual one in presenting mathematical ideas at the early childhood level. The teacher may vary lessons from the telling, or expository, type to a guided-discovery approach. Discovery methods with a minimum of guidance are often effective with mathematically precocious students engaged in problem-solving work. Well-organized laboratory activities in which students discuss their work with their peers, make generalizations, and verbalize their findings are well suited to the application step of the instructional sequence. A laboratory setting is also useful for reinforcing a concept and supplying activities and games for practice.

Appropriate materials must be chosen for all levels of the instructional sequence. These may include manipulative teaching aids, elementary mathematics textbooks, programmed-instruction workbooks, or individualized programs. The pocket calculator also has its place in the early childhood classroom, especially in the areas of teaching properties of arithmetic operations, processes of computation, procedures for solving word problems, and estimation.

Notes

1. Richard W. Copeland, *How Children Learn Mathematics*, 2nd ed. (New York: Macmillan Publishing Co., Inc., 1974), p. 42.
2. Robert M. Gagné, *The Conditions of Learning*, 2nd ed. (New York: Holt, Rinehart and Winston, Inc., 1970), pp. 300–301.
3. Jerome S. Bruner, *The Process of Education* (New York: Vintage Books, 1960), p. 39.
4. Fredricka K. Reisman, *Diagnostic Teaching of Elementary School Mathematics* (Chicago: Rand McNally College Publishing Company, 1977), pp. 50–51.
5. "A Guide to Using Cuisenaire Rods," Filmstrip Notes, frame 29 (New Rochelle, New York: Cuisenaire Company of America, Inc., 1973).
6. Mary Richardson Miller and Toni Cresswell Richardson, *Making Metric Maneuvers* (Hayward, California: Activity Resources Company, 1974), p. 100.
7. "Calculator Lesson Plan," a booklet (Fullerton, California: Hunt-Wesson Foods, Inc., 1976), p. 1.
8. Max S. Bell, "Calculators in Elementary Schools? Some Tentative Guidelines and Questions Based on Classroom Experience," *The Arithmetic Teacher* 23 (November 1976): 502–3.
9. James V. Bruni and Helene J. Silverman, "Let's Do It! Taking Advantage of the Hand Calculator," *The Arithmetic Teacher* 23 (November 1976): 499.
10. "Calculator Lesson Plan," p. 5.
11. Marilyn N. Suydam and J. Fred Weaver, *Using Research: A Key to Elementary School Mathematics* (Columbus, Ohio: ERIC Information Analysis Center for Science, Mathematics and Environmental Education, 1975), p. 4–6.
12. Herbert F. Spitzer, *What Research Says to the Teacher, Teaching Elementary School Mathematics* (Washington, D.C.: National Education Association, Association of Classroom Teachers, 1970), pp. 25–26.
13. Leroy G. Callahan and Vincent J. Glennon, *Elementary School Mathematics: A Guide to Current Research* (Washington, D. C.: Association for Supervision and Curriculum Development, 1975), p. 72.

Assignment

1. Select two elementary mathematics textbook series that have recent publication dates, and evaluate two student books at a selected level among K through 3 using the list of criteria given in this chapter. Include a rating showing the extent to which the materials fit the criterion considered (0–2, no or little extent; 3–4, to some extent; 5–6, great extent) and a brief comment on each item. Also evaluate the accompanying teacher's editions.
2. Examine as many of the manipulative aids described in this chapter as you can, and evaluate each one using the form given in this chapter.
3. Study the "I Do and I Understand" booklet which describes the Nuffield mathematics project. (Nuffield materials are published in the United States by John Wiley and Sons.) Look over some of the other booklets in the series. Answer these questions:
   a) What is the philosophy of the program?
   b) How are the booklets organized?
   c) How can ideas of this program assist you in designing teaching and laboratory activities?
4. Examine an individualized program in mathematics, and try to interview two or more teachers who use such a program. Then answer these questions.
   a) What are the teachers' views on the program?
   b) Do the teachers think a special group of students—fast, average, or slow learners—is benefitted by such a program?
   c) Do the teachers use the individualized program along with an elementary mathematics textbook?
   d) What do you think may be advantages and disadvantages in using such a program in the early childhood years?
   e) How much does an individualized program cost? Evaluate the merits of the system in relation to its cost.
5. On pages 5–1 and 5–2 of *Using Research: A Key to Elementary School Mathematics* by Marilyn N. Suydam and J. Fred Weaver (see Bibliography for complete reference), read a summary of research on analyses of and vocabulary used in mathematics textbooks. Make a list of major points.
6. Consult the Educational Index in the periodical section of your college library to find at least one article on the use of attribute materials, the geoboard, tangrams, or mirror cards. Read the article. Then examine the teacher's manual which accompanies the teaching aid. How do ideas presented in the article and the teacher's manual compare? Does the article present a case in favor of or disfavor of the use of the material?

Laboratory Session

Preparation:

1. Study Chapter 3 of this textbook.
2. Examine the November, 1976, issue of *The Arithmetic Teacher* or consult other sources to generate a list of at least five pocket-calculator games that can be used for motivation, interest appeal, or instruction. Try to select possibilities that are suitable for various age and ability levels in the early childhood years. Duplicate at least one of the games for distribution to the class members.
3. Read at least one article on the laboratory approach to teaching mathematics. You may select one of the following articles in *The Arithmetic Teacher*. Make a note of principal ideas in the article.
   a) "What's *Your* Position on . . . the Role of Experience in the Learning of Mathematics?" by Edith E. Biggs, May 1971,
   b) "What's *Your* Position on . . . the Role of Experience in the Learning of Mathematics?" by Maurice L. Hartung, May 1971,
   c) "A Model for the Construction and Sequencing of Laboratory Activities" by Thomas R. Post, November 1974,
   d) "Success for All: An Adventure in Learning" by Eileen M. Rouda, January 1972,
   e) "The Miquon Mathematics Program" by Don and Lore Rasmussen, April 1962,
   f) "A Mathematics Laboratory—From Dream to Reality" by Patricia S. Davidson and Arlene W. Fair, February 1970,
   g) "Geometry Alive in Primary Classrooms" by Janet M. Black, February 1967,
   h) "The Mathematics Laboratory For The Elementary and Middle School" by Alan Barson, December 1971.

Objectives: Review the laboratory approach to teaching mathematics, participate in games for the calculator, and criticize a teaching activity.

Materials: Pocket calculators (each student should try to bring one to class), ideas collected by the students.

Procedure: Form groups of four or five and do the following:

1. Exchange ideas on the article you read about the laboratory approach to teaching mathematics. Take notes on suggestions you wish to use in your teaching.
2. Guide members of your group through your calculator game.
3. Discuss the following kindergarten lesson intended to teach the concept of cardinal number of a set. Does the activity reflect careful thinking? Is the prerequisite learning reasonable for a child in kindergarten? Is the concept of cardinal number actually taught in this lesson? Suggest improvements.

Objectives: Each student should identify the cardinal number of a set and generate a representative set for a given cardinal number.

Prerequisite learning: Rote counting, concept of a set, one-to-one correspondence, equivalent sets

Materials: Plastic counters, a die or large cube with faces like those on a die

Procedure: The teacher places a certain number of plastic counters on the table in front of the group and asks the students to select the same number of counters as she has. She repeats this procedure several times. She then tosses a die (a large teacher-made cube if the entire class is participating) and has students to place the same number of plastic counters on their desks as dots on the upturned face of the die. She selects children to tell the cardinal number of each set made. Next she distributes pegboards to the students, writes numerals on the chalkboard, and has students place the correct number of pegs in the pegboard each time.

Evaluation: The teacher observes students as they make the sets and listens to their oral responses.

Bibliography

"A Guide to Using Cuisenaire Rods," Filmstrip Notes. New Rochelle, New York: Cuisenaire Company of America, Inc., 1973.

Bell, Max S. "Calculators in Elementary Schools? Some Tentative Guidelines and Questions Based on Classroom Experience." *The Arithmetic Teacher* 23 (November 1976): 502–9.

Bruner, Jerome S. *The Process of Education.* New York: Vintage Books, 1960.

Bruni, James V., and Silverman, Helene J. "Let's Do It! Taking Advantage of the Hand Calculator." *The Arithmetic Teacher* 23 (November 1976): 494–501.

"Calculator Lesson Plan," a booklet. Fullerton, California: Hunt-Wesson Foods, Inc., 1976.

Callahan, Leroy G., and Glennon, Vincent J. *Elementary School Mathematics: A Guide to Current Research.* Washington, D. C.: Association for Supervision and Curriculum Development, 1975.

Caravella, Joseph. "Selecting a Minicalculator." *The Arithmetic Teacher* 23 (November 1976): 547–50.

Copeland, Richard W. *How Children Learn Mathematics,* 2nd ed. New York: Macmillan Publishing Co., Inc., 1974.

D'Augustine, Charles H. *Multiple Methods of Teaching Mathematics in the Elementary School,* 2nd ed. New York: Harper & Row, Publishers, 1973.

Gagné, Robert M. *The Conditions of Learning,* 2nd ed. New York: Holt, Rinehart and Winston, Inc., 1970.

Grossnickle, Foster E., and Reckzeh, John. *Discovering Meanings in Elementary School Mathematics,* 6th ed. New York: Holt, Rinehart and Winston, Inc., 1973.

Miller, Mary Richardson, and Richardson, Toni Cresswell. *Making Metric Maneuvers.* Hayward, California: Activity Resources Company, 1974.

Reisman, Fredricka K. *Diagnostic Teaching of Elementary School Mathematics.* Chicago: Rand McNally College Publishing Company, 1977.

Reys, Robert E. "Considerations for Teachers Using Manipulative Materials," in W. George Cathcart, ed. *The Mathematics Laboratory, Readings from the Arithmetic Teacher*. Reston, Virginia: National Council of Teachers of Mathematics, 1977, pp. 101–8.

Spitzer, Herbert F. *What Research Says to the Teacher, Teaching Elementary School Mathematics*. Washington, D. C.: National Education Association, Association of Classroom Teachers, 1970.

Suydam, Marilyn N., and Weaver, J. Fred. *Using Research: A Key to Elementary School Mathematics*. Columbus, Ohio: ERIC Information Analysis Center for Science, Mathematics and Environmental Education, 1975.

4

Diagnosis in the Cognitive Domain

Considerations given in this chapter for assessing students' thinking levels include Piagetian-task interviews, standardized diagnostic tests in mathematics, and teacher-constructed paper-and-pencil diagnostic instruments. These devices, along with day-to-day observations made by the teacher, the child's responses to oral and written questions, and evaluative comments of former teachers, should help a teacher identify strengths and weaknesses of individual students and make instructional decisions. Now that many classrooms are being mainstreamed and contain exceptional children, teachers must attend to a variety of needs and performance levels. Methods of diagnosis, then, both informal and formal, are an important consideration in a mathematics program.

Conducting Diagnostic Interviews

Interview procedures provide information that standardized tests do not. In an interview, the way a child organizes information is revealed as it occurs, and how he justifies his answers often gives further enlightenment on his level of thought. Also, the teacher can note behaviors (the attention level and the emotional or physical state of the child while he is performing the tasks) which may influence a child's performance level. The Piagetian interviews presented in the next section are appropriate for children from ages four through eight.

Some considerations for conducting a diagnostic interview follow. Items 1 through 5 apply to any diagnostic interview; 6 and 7 are additional points applicable to the interviews on Piagetian operations.

1. Set up materials for conducting the interview in a readily accessible area of the classroom. A parent, student teacher, practicum student, or other assistant can help administer the interviews if he or she is thoroughly familiar with the materials involved.
2. Read the directions clearly to the student.
3. Allow the student to think about his answer without your talking at that time.
4. Do not tell the student whether his answers are right or wrong. The purpose of the interview is to assess and not to teach a concept.
5. File student responses so that they are readily available.
6. On each Piagetian task, there is provision for a rating which refers to the stages of intellectual development given by Piaget. The stages applicable to the early childhood years are (a) preoperational (no understanding of the task), (b) transitional (some but not complete understanding of the task), and (c) concrete operational (complete understanding of the task).[1] Rate each child on each task as preoperational, transitional, or concrete operational according to his level of understanding.
7. On the tasks that involve multiple parts, you need not complete a task when a student cannot perform a given part of it. Rate the child preoperational or transitional on that task and proceed to the next Piagetian operation.

Piagetian-Task Interviews

Background for the Interviews

Often a child's reasoning is baffling to the adult mind. For example, if you twist a balloon from normal expanded shape to a new shape, a child may think the balloon in one case holds more air because "it is longer" or "it is fatter." Responses such as these typify the logic of children who are preoperational in their intellectual development.

In Piagetian theory, the years spent in kindergarten through second or third grade are those in which most children's thought becomes operational. Five areas of cognitive growth are classification, seriation, number, geometry, and measurement. A child develops in these areas in an interrelated way. When classifying, a child must attend to certain attributes and ignore irrelevant ones as he groups objects that are alike in some way. Classification experiences include sorting by size, color, shape, thickness, texture, and combinations of these. The first three Piagetian-task interviews given in this section deal with logical classification. Most children are able to classify by the time they are five or six years old. Seriation involves the ordering of objects and events, e.g., blocks from smallest to largest or games that one likes most to those one likes least. To seriate correctly, a student must make comparisons and detect differences. Most children can seriate the tasks in interviews 4 and 5 by the time they are in first grade.

According to Piaget, the ability to classify and seriate are vital to the understanding of number. Interviews 6 through 9 should help you as teacher assess a child's level of thinking pertaining to both cardinal and ordinal uses of number. Most children can conserve number by the time they are in kindergarten or first grade.

A child's conception of space is quite different from that of an adult. Perhaps you have observed a small child who thought other people could not see him when he covered his eyes and could not see them. Determining a child's level of thinking on certain geometric (spatial) notions is the purpose of interviews 10 and 11.

The last five tasks deal with measurement of quantity, length, liquid capacity, area, and time. Piagetian research indicates that conservation of quantity occurs by first or second grade; length, by second grade; area, by third grade; and time, by third or fourth grade. Most children do not conserve weight and volume until they are nine or ten years old.

A learning disabled child may do poorly on some of these tasks. Although the Piagetian interviews are not designed to diagnose learning disabilities, a child's performance on the tasks may lead you to suspect that a student has perceptual difficulties.

Some possible teaching implications follow each interview. Specific instructional activities are examined in later chapters.

The Piagetian Tasks

The materials used in these tasks require a minimum of expense. Since interest and emotional factors may influence a child's performance, materials with which the egocentric young child can identify have been chosen. Also, alternative materials have been suggested in circumstances involving certain special children.

1. Classification by one attribute at a time

Materials: triangular and square shapes in two sizes, each size and shape in two colors. (Substitute the attributes of texture or thickness for color for visually handicapped students.)

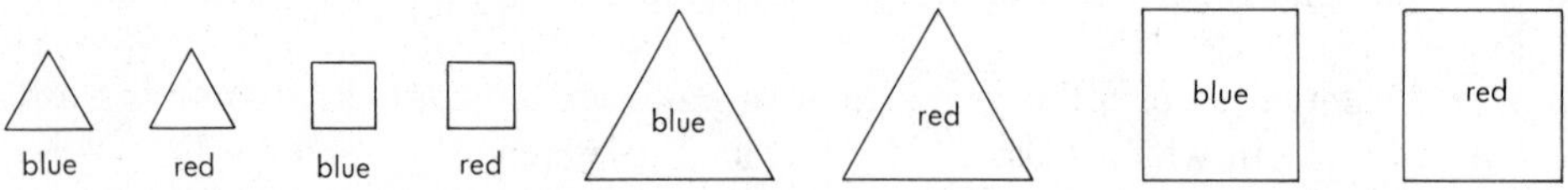

blue red blue red blue red blue red

Procedure: Mix up the objects in front of the student, and ask him to separate them into piles so that all of the objects in each pile are alike in some way.

Justification: Ask, "Why did you put the objects where you did?"

Child's comments: ______________________

Mix up the shapes again and ask the student to group them in another way (by another criterion). Have him justify the answer. Repeat the process another time to determine if he can sort by all three criteria—shape, color, and size.

Rating: ______________________

Rate the student at the preoperational level if he is unable to group the objects by shape, color, or size. A child at this level may randomly select the objects or make a house with the □ and △, but he exhibits no consistent thinking strategy. He may begin to sort by one criterion such as shape and then switch to another criterion such as color. A student at the transitional level may be able to form sets each of which has elements the same shape, size, or color, but he is unsure of his work and makes errors before making the classification. A child who is at the concrete operational level is able to classify the objects correctly without any trial and error.

Teaching implications: Children classify in mathematics, science, social studies, health, and reading as early as preschool and kindergarten. Young children need experiences in classifying plants and animals, good health habits and poor health habits, vowels and consonants, and so on. In mathematics students may participate in classification tasks such as sorting ordinary playing cards by color and kind; buttons by color, size, and pattern; and polygonal figures by number of sides. Even children who are concrete operational on the above Piagetian task can participate in activities such as these for purposes of reinforcement. Prompt these students to extend their knowledge of classification to more than one attribute at a time. Ask questions about the relationships among the sets, and encourage the children to compare and discuss various subsets they have made of a given set.

2. Classification by two attributes simultaneously

Materials: same as those needed for the first task

Procedure: Ask the student to place all the large, square shapes together.

Justification: Ask, "Why did you choose those shapes?"

Child's comments: ______________________________

Rating: ______________________________

Teaching implications: If a child does this task correctly, activities requiring sorting by two or three attributes are in order for purposes of reinforcement and extension. These students should organize the classification process using yarn loops and label cards for their sets. Terms such as "disjoint" and "overlapping" describing set relations, set operations of intersection and union, and logical terms such as "not," "and," and "or" can be introduced at this time. A child who does not work the Piagetian task correctly needs instructional and laboratory experiences involving multiple classifications. (See chapter 5.)

3. Set inclusion

Materials: pictures of six dogs and two cats (Adjust the materials if a student is blind or partially sighted. Use scraps of different fabric—six scraps of terrycloth and two of polyester, for example, and modify the questions accordingly.)

Procedure: Ask the child to look at the pictures of the dogs and cats. Say, "Are there more dogs or more animals in the pictures?"

Child's response: ______________________________

Justification: Ask, "Why do you think so?"

Rating: ______________________________

Teaching implications: This concept is important in the relationship of part to whole (significant in work with fractional numbers), joining sets to obtain another set (fundamental in addition and multiplication of cardinal numbers), and separating a set into subsets (basic to subtraction and division of cardinal numbers).

In instructional activities, ask questions to guide children's thinking: Are all the crayons blue? Are all the students in the classroom boys? Are all the boys in the classroom students? If all the boys went outside, would there be any students left inside? If all the students went outside, would any boys be left inside? Be sensitive to the fact that the concept of set inclusion is prerequisite learning to many future mathematical explorations and each child needs opportunities to develop the concept.

4. Seriation of events

Materials: five picture cards of a tree as it looks when it falls from an upright position to the ground

   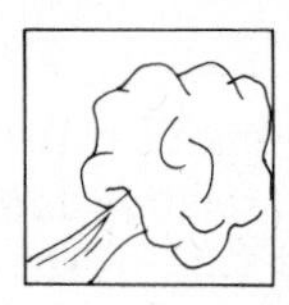 

Procedure: Ask the child to think how a tree looks when it is falling to the ground. Then ask him to look at the drawings on the cards and place the cards in the order that shows how the tree falls.

Justification: Ask, "Why did you place the pictures in that order?"

Child's comments: ______________________________

Rating: ______________________________

Teaching implications: Sequencing events is readiness for sequencing numbers and is important in understanding measurement of time. Reading and mathematics can be integrated by reading a story to the children and then having them sequence picture cards which represent events in the story.

5. Seriation of objects

Materials: eight pencils graduated in length

Procedure: Place the pencils in a pile in front of the child. Ask him to arrange them in a row from shortest to longest.

Justification: Say, "Why did you arrange them that way?"

Child's response: ______________________________

Rating: ______________________________

Teaching implications: For a student who does poorly on this task, provide activities in which he must order only two or three objects and gradually increase the number of objects to seriate. If a student does perform the Piagetian operation correctly, have him order two series simultaneously, e.g., ten paper cylinders (graduated in length) from shortest to longest or vice versa and ten straws (graduated in length) in the same order. Then reverse the order of one series, point to an object, and have the child designate the matching object in the other series. In chapter 5 some suggestions on teaching seriation are given.

6. Conservation of number

Materials: eighteen cut-outs of balls

Procedure: Place three balls in one row on the table, and directly below them form a second row of three balls.

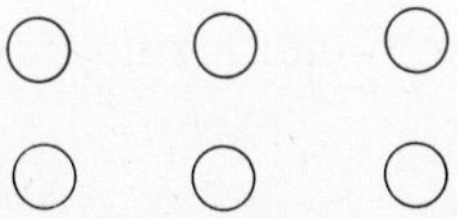

Say, "Are there more balls in this row (point to first row), are there more balls in this row (point to second row), or do the rows have the same number of balls?"

Child's response: ______________________________

Justification: Ask, "Why do you think so?"

Now spread out the second row to produce this appearance:

Say, "Are there more balls in this row (point to first row), are there more balls in this row (point to second row), or is the number of balls the same in each row?"

Child's response: ______________________________

Justification: Ask, "Why do you think so?"

Increase the number of balls in each row to nine and repeat the above tasks and questions.

Child's response when row 2 is lined up directly below row 1: ______________

____

Child's response when row 2 is spread out: ______________

____

Rating: ______________________________

A student who does not know the number is the same in the two rows in the arrangement

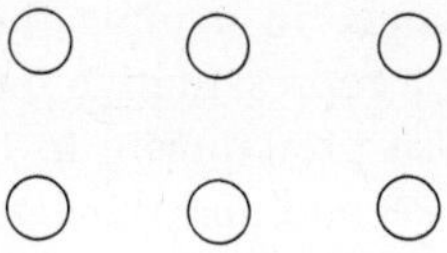

is preoperational and shows no readiness for number concepts. Some children understand that the rows in

O O O

O O O

have the same number, but when the number in each row is increased to nine, the considerable change in visual display causes them to revert to nonconservation responses. These students are transitional. Students who are concrete operational recognize the number property of a set and are not distracted by perceptual change, i.e., they conserve number.

You may gain further insight into a student's understanding of number by transforming the arrangement

O O O　　　　to　　　　O O O

O O O　　　　　　　　　OOOOO

in which the length of the two rows remains constant but the number in one row increases. Ask the child if the two rows have the same number of balls.

Teaching implications: To help a child understand number, arrange objects in sets in many different ways. For example, in a feltboard activity use configurations such as O and O O rather than only O O O, and spread out the elements
O O
O
sometimes and cluster them at other times. Question the students on whether number is changed when you move the objects of the sets.

7. Equivalence of sets

Materials: 15 to 20 plastic counters

Procedure: Pile the counters in front of the student and choose five of them to make "your set." Tell the child, "Take the same number of counters from the pile as I have."

Child's response: ______________________________

Justification: Say, "Why do you think your set has the same number of counters as mine?"

Rating: ______________________________

Teaching implications: A student who works this task demonstrates understanding of equivalent sets. He is ready to learn that all sets that can be placed into one-to-one correspondence have the same number of elements and hence the same number name. He needs experience in naming the number when given a representative set and in constructing a representative set when given the number name. Work with equivalent sets is also prerequisite to the study of multiplication and division of cardinal numbers in second and third grades.

8. Ordinal number

Materials: a picture of five "stick" people facing the left. (Cover the figures with felt if you administer this interview to a visually handicapped student.)

Procedure: Show the pictures to the student and tell him the people in the picture are lined up to go to a movie. Ask him to point to the fourth person in the row.

Child's response: ______________________________

Justification: Say, "Why do you think that person is the fourth one?"

Rating: ______________________________

Teaching implications: A child who performs this Piagetian operation can be provided reinforcement activities dealing with ordinal number and can extend his knowledge of ordering to larger numbers. If he knows the order for 1 through 9, help him extend the pattern to the decade 10 through 19 and higher decades. If he cannot identify the correct position in the Piagetian task, design appropriate teaching activities (see chapter 5) and use the terms first, second, and so on in situations that arise during the school day.

9. Ordinal number in two dimensions

Materials: 25 plastic counters

Procedure: Form a 5 by 5 array with the counters.

Ask the student to touch the third counter in the second row.

Child's response: ______________________________

Justification: Say, "Why do you think that counter is the third one in the second row?"

Rating: ______________________________

Teaching implications: Students who understand ordinality and can identify a position in two dimensions are ready to study coordinate geometry, which usually enters the curriculum in second grade. Also, multiplication using an array approach is fitting at this time.

10. Horizontal-vertical reference system

Materials: for a "happy face," a large yellow circular felt cut-out with the mouth glued on, two small circular felt cut-outs for the eyes

Procedure: Tell the student this is a "happy face" without the eyes. Give him the two small circular felt pieces and ask him to place them on the happy face.

Child's respone: ____________________

Justification: Say, "Why do you think the eyes go there?"

Rating: ____________________

To determine if a student has a vertical reference system, draw a picture of a house with a sloping roof and ask him to draw a picture of a television antenna or a chimney on the roof. Children who have not internalized a vertical reference system often draw the chimney or antenna perpendicular to the roof.

Teaching implications: In his art work, a three- or four-year-old child often arranges people's eyes in nonhorizontal positions or one eye may even be outside the head. Help students refine their thinking by providing magazine pictures or pictures of class members and asking them if they have drawn the eyes, ears, and so on in the same place as shown in the pictures. Also, when working with the number line, draw it in the vertical position sometimes.

Lavatelli suggests some structured activities to help a child coordinate horizontal and vertical positions into a single frame of reference.

a) Have students draw the water level in a jar when the jar is in a tilted position. Students who have not internalized a horizontal-vertical reference system often draw the water level parallel to the bottom of the jar. To help move the student's reference plane away from the bottom of the jar to the table top or floor, gradually move a brightly colored cardboard from the top of the bottle to the table top, asking the child to draw the water level as the cardboard is lowered.

b) Position a slanted board near a wall of the classroom so that one end is on the floor and the other is on a concrete block. Place brightly colored strips of paper on the floor (horizontal) and on the wall (vertical) near the board. Choose a student to walk part of the way up the board and stop. Ask other students to observe the child, the slanted board, and the horizontal and vertical strips and to draw a picture of what they see.

c) Ask children to draw a mountain on paper and color a horizontal strip in one color at the bottom of their paper and a vertical strip along the left or right side in another color. Instruct the students to place toothpicks the same color as the vertical strip on the mountain to represent people.[2]

11. Topological and Euclidean geometry

Materials: crayon, ball, topological forms

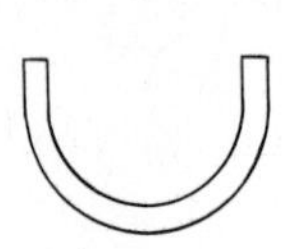

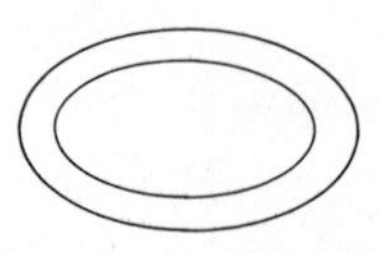

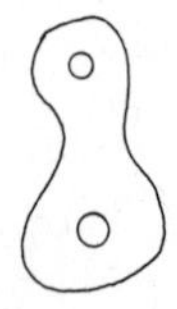

cut out of poster board, Euclidean shapes

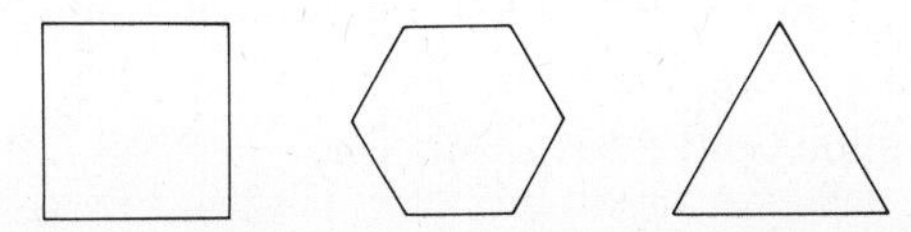

cut out of poster board, a poster of pictures of all these materials (or mount felt pieces representing these objects on the poster), a manila folder to serve as a screen

Procedure: Stand the manila folder up and place the objects behind it so that you can see them but the child cannot. Put the poster of pictures by the child. Ask him to reach around the folder with one hand, feel one of the objects, and then point to it on the poster with the other hand. Continue until he has exhausted the objects.

Child's response: ______________________________

Justification: Say, "Why do you think that is the picture of the object you are holding?" as he feels the object.

Rating: ______________________________

Teaching implications: Two- and three-year-olds recognize the familiar objects (crayon and ball) only. Most children who are four years old can make distinctions among the topological forms, and between the ages of four and six or seven they can distinguish among the Euclidean shapes.[3] This situation suggests that the teaching sequence in geometry begin with topological ideas (points inside or outside a closed curve, open and closed curves) followed by recognition and representation of Euclidean shapes.

12. Conservation of continuous quantity

Materials: two balls (the same size) of modeling clay

Procedure: Ask the student to observe the two balls of clay, and say, "These balls have the same amount of clay. Do you agree?" Now flatten one of the balls into the shape of a hamburger patty.

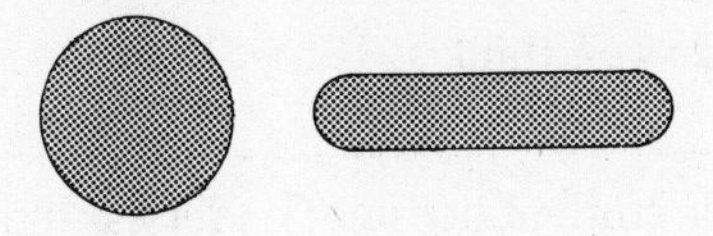

Ask the student, "Does the hamburger-patty shape have more clay, does the ball have more clay, or do they have the same amount of clay?"

Child's response: ______________________________

Justification: Ask, "Why do you think so?"

Rating: ______________________________

Teaching implications: As children use modeling clay in play activities, reinforce their understanding of quantity if they do conserve quantity and question those children who do not perform the conservation task correctly on the amount of clay in each shape as they form many shapes from a single piece. Encourage them to discuss what they are doing.

13. Conservation of length

Materials: two strips of paper, each about eight inches long

Procedure: Place one strip on top of the other to establish with the child they are the same length. Then line up the strips this way:

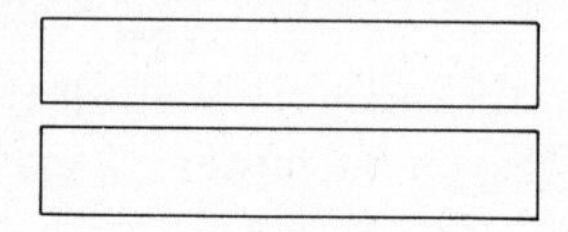

Ask, "Are the strips the same length?"

Child's response: YES NO

Move the bottom strip to produce this arrangement:

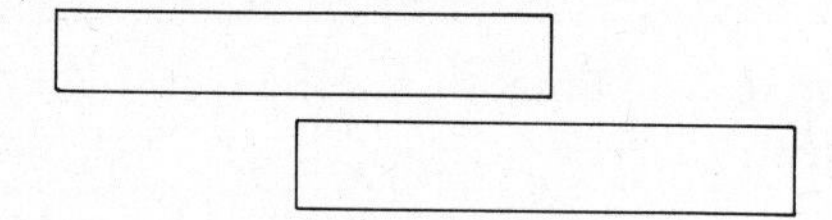

Say, "Are the two strips the same length?"

Child's response: YES NO

Justification: Ask, "Why do you think so?"

Now fold the bottom strip to make an L-shape.

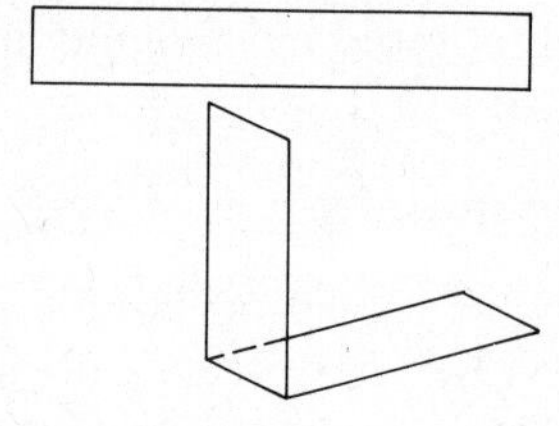

Say, "Are the two strips the same length?"

Child's response: YES NO

Justification: Ask, "Why do you think so?"

Rating: ______________________________

Teaching implications: If a student does not understand that the length of the strips remains invariant regardless of their positioning, he is not ready to measure with rulers. In order to use a ruler in a meaningful way, a child must know it stays the same length regardless of its position. When working with the number line, a student who does not conserve length may think that the number pictured on the second number line is greater than the number represented on the first number line.

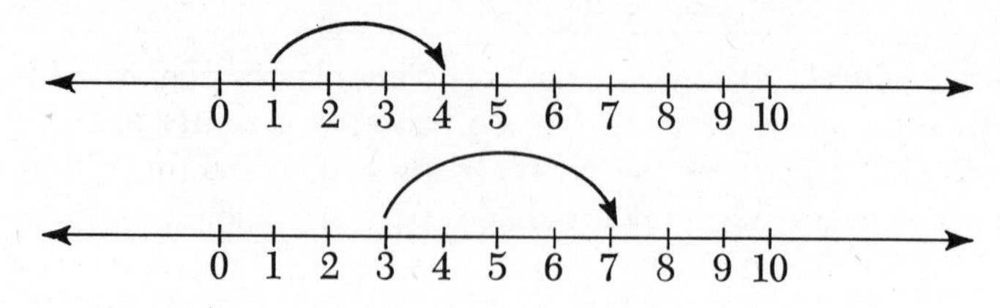

A student who conserves length is ready to begin informal activities with measurement of length.

14. Conservation of liquid capacity

Materials: two identical short jars, one taller jar with a diameter less than that of the short jars, food coloring, water to fill the small jars

Procedure: Fill the two short jars with water and add some food coloring to each. Ask the student to observe that the two jars are the same size and contain the same amount of water. Now, ask him to watch as you pour the water from one of the short jars to the taller jar.

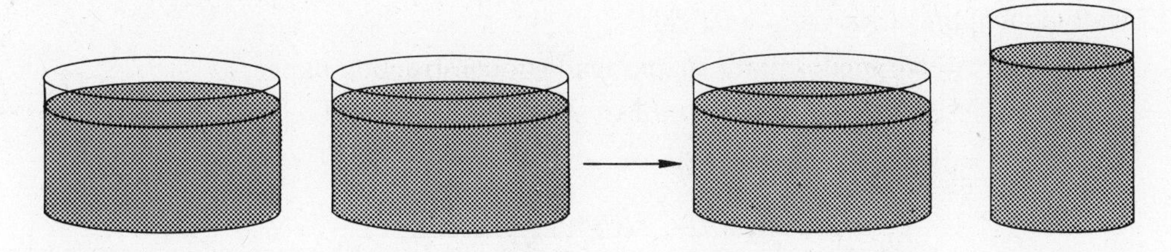

Say, "Does the tall jar have more water, does the short jar have more water, or do they have the same amount of water?"

Child's response: ______________________________

Justification: Ask, "Why do you think that?"

Rating: ______________________________

Teaching implications: Plan cooking activities and activities in which sand is poured from one container to another of a different size, question students on their work, and encourage them to verbalize as they work. This kind of experience is good for preschool, kindergarten, and first-grade levels.

15. Conservation of area

Materials: two sheets of gray construction paper, sixteen toy cars (or sixteen cubical blocks)

Procedure: Show the two pieces of construction paper to the student, and tell him they represent parking lots. Place one sheet on top of the other and note with the child they are the same size. Place two cars on each piece of construction paper, clustering them on one sheet and scattering them out on the other. Ask the child, "Are the cars on the two lots taking up the same amount of parking space?"

Child's response: YES NO

Justification: Ask, "Why do you think so?"
Increase the number of cars (cubes) on each sheet of paper to eight.

Say, "Is the space taken up by the cars on the lots the same?"

Child's response: YES NO

Justification: Ask, "Why do you think so?"

Rating: ______________________________

If a child just counts the cars and says the parking space taken up by the cars on each lot is the same, he may be conserving number (how many) and not area (amount of surface covered). To determine whether this situation does exist, try stacking the cars (cubes) on one lot and ask if the parking space taken up on the lots is the same.

An alternative to the above interview follows. It can be modified and used for instructional purposes.

Materials: twelve small square shapes made of construction paper

Procedure: Arrange squares on a table to make these two rectangular shapes:

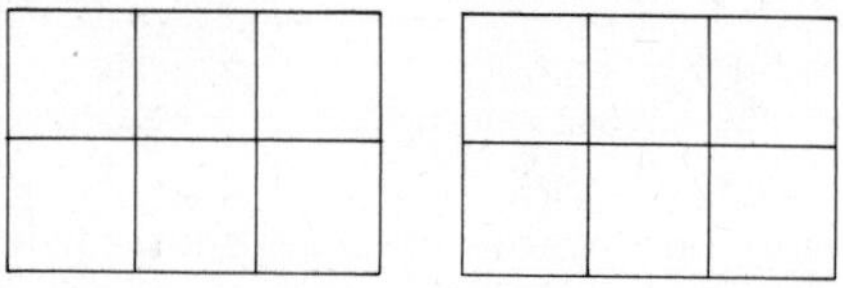

Note with the child the two shapes look exactly alike. Next, stack the squares in the second rectangular shape to produce this arrangement:

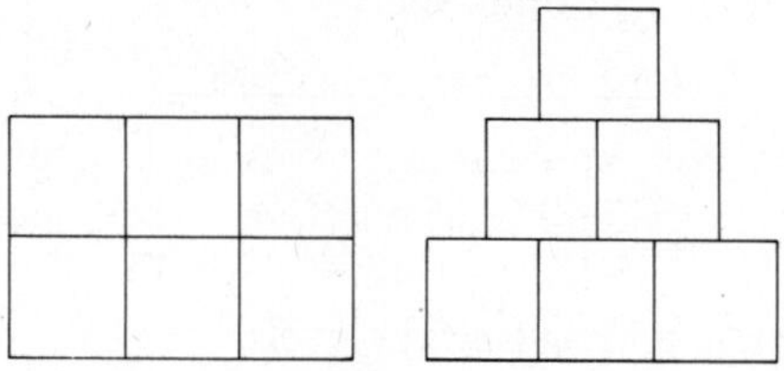

Say, "Does this shape (point to the first one) cover more of the table top, does this shape (point to the second one) cover more of the top of the table, or do they cover the same amount of the table top?"

Child's response: ______________________________

Rating: ______________________________

Teaching implications: It is dubious that a student would understand the assigning of a number to a region determined by a geometric figure if he thinks the amount of surface covered changes when the region is cut up and altered in appearance. This consideration should be made before formal instruction on area occurs.

16. Conservation of time

Materials: two pictures of children cut out of a magazine and backed with tagboard

Procedure: Tell the student the children in the picture are going to take a walk and he is to tell when for them to start and stop their walk. Line the pictures up side by side and when he tells you to start, move them at the same rate. Stop them side by side when he tells you to stop.

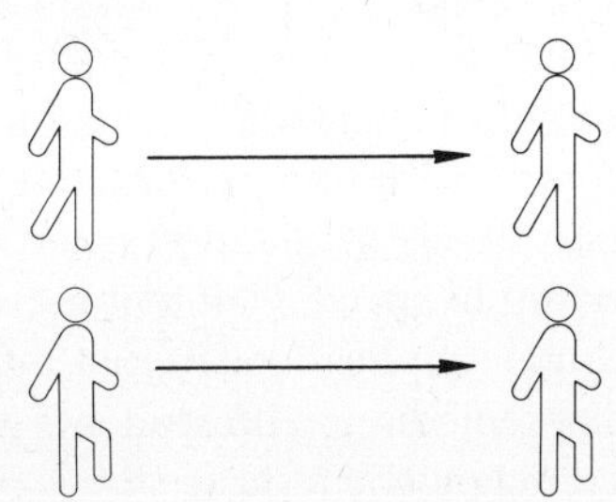

Say, "Did it take the children the same amount of time to take their walk?"

Child's response: YES NO

Justification: Ask, "Why do you think so?"

Repeat the above experiment but move one of the pictures at a faster rate this time so that they are lined up when they start but not when they stop.

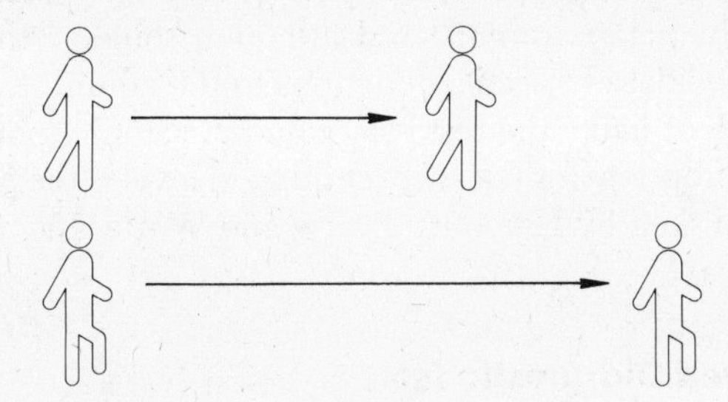

Say, "Did the children walk the same amount of time?"

Child's response: YES NO

Justification: Ask, "Why do you think so?"

Rating: ______________________________

Teaching implications: Piaget claims that ". . . one does not find that the young child has a concept of time which is radically independent of speed."[4] Certainly, if a student does not understand that speed can vary while the lapse of time is the same, he will likely have difficulty with the fact that the minute hand makes a complete revolution while the hour hand advances from one numeral to the next!

Reisman found that children in first grade could learn the skill of telling time on a clockface but not until third grade could students work correctly the Piagetian conservation-of-time task.[5] Although teaching time on the clock may begin in first grade, teachers should be aware that many students might not fully conceptualize measurement of time until third grade or later. Teaching the skill of telling time is itself a worthy goal, especially for retarded youngsters. Thus instructional goals must, as usual, take into account the particular needs of the student.

Standardized Diagnostic Tests

In many elementary schools, standardized tests are administered near the beginning of the school year. These tests may assess achievement in mathematics or serve as diagnostic instruments. Achievement tests survey a student's attainment of the concepts and skills generally taught in elementary school mathematics and allow each child's performance to be compared to the performance of the persons of the standardized sample. The primary focus of diagnostic tests, on the other hand, is the student's level of understanding in specific content areas. The results of diagnostic tests, then, have direct classroom implications in that a teacher has a basis for planning instruction and providing remediation for students who need it.

Two major standardized diagnostic tests are the KeyMath Diagnostic Arithmetic Test and the Basic Educational Skills Inventory (BESI). The KeyMath test is particularly useful in preschool through sixth grade. It is designed to be individually administered and is organized into three areas of mathematics—content, operations, and applications. Content is subdivided into numeration, fractions, geometry, and symbols. An easel kit containing all the materials for giving the test permits convenient administration.[6] The BESI instrument, restricted to basic skills in the lower grades, is to be individually administered although some of the activities can be given in groups. A booklet of specific directions to the administrator and associated booklets for the student make the test easy to administer.[7] Another diagnostic test offering material appropriate for the early childhood level is the Stanford Diagnostic Arithmetic Test published by Harcourt, Brace, and World, Inc. in 1966.

Teacher-Constructed Diagnostic Tests

If a student is experiencing a learning difficulty in a certain area of mathematics content as evidenced by his performance on a standardized test or in class activities, you can design your own informal tests to determine the level at which the child's understanding breaks down. These tests can be administered to a group of students or an individual student in an interview.

Test items usually begin at the most elementary level allowing students to use concrete and pictorial aids and advance to the symbolic level. Sequencing of questions, however, is usually not so rigid that a student who cannot work a lower-level task will necessarily not be able to perform a succeeding task. The test items can be given to the students on duplicated sheets. For a child in kindergarten or first grade, however, it is preferable to cut up the duplicated sheets and present the questions to the child so that he sees only one at a time. Such an arrangement holds his attention better and reduces the chances of his being overwhelmed by the test. Elementary mathematics textbooks are a good source of ideas for test items.

The following sample tests can be used for diagnostic purposes and as pretests to mathematical topics as they are introduced during the school year. The indicated grade level is approximate.

Counting and Cardinal Number—Kindergarten, First Grade

1. Copy these numerals: 5, 2, 9, 0, 6.
2. Write the numerals the teacher calls out. (The teacher reads 4, 7, 8, 12, and 27.)
3. Write the numerals in sequence for the counting numbers from 1 through 10.
4. Show a one-to-one correspondence between these sets. (Either concrete objects or pictures can be used.)

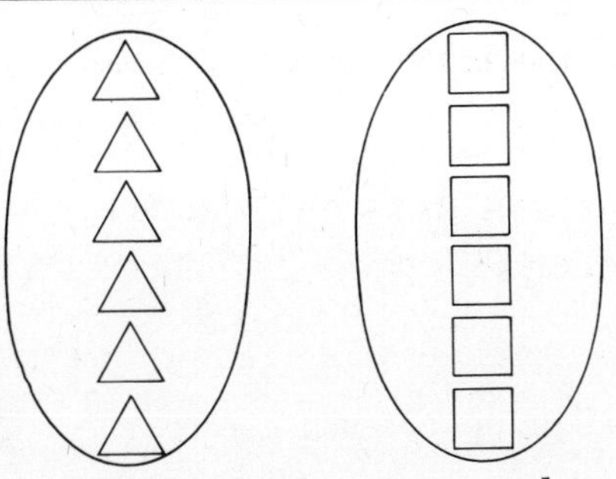

5. Place an "X" by the set in each pair that has more members.

a)

b)

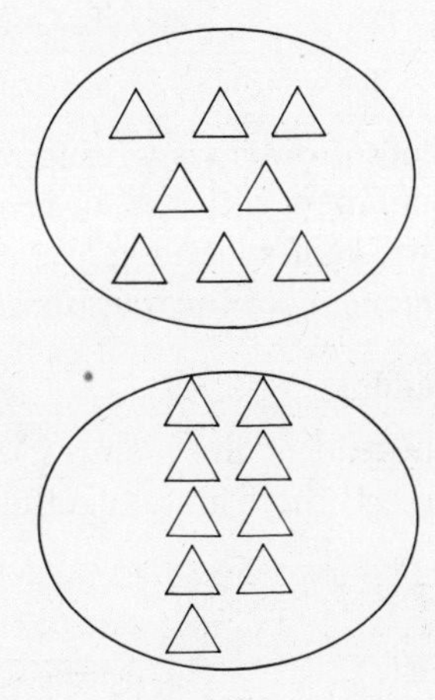

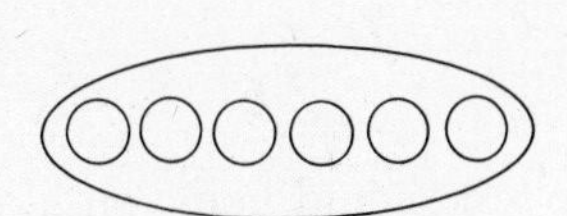

6. Write "Yes" by each set equivalent to

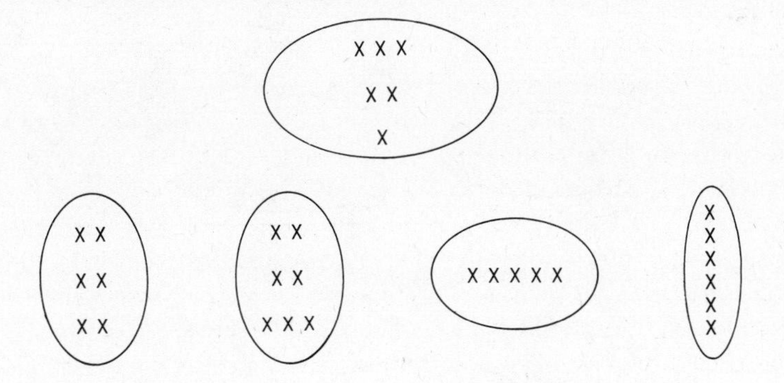

7. Write the numeral that tells the number of members in each set.

a)

b)

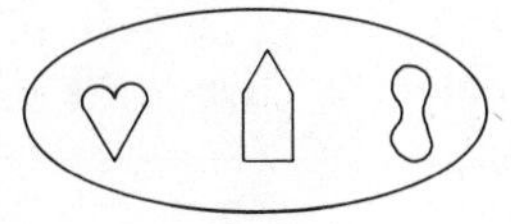

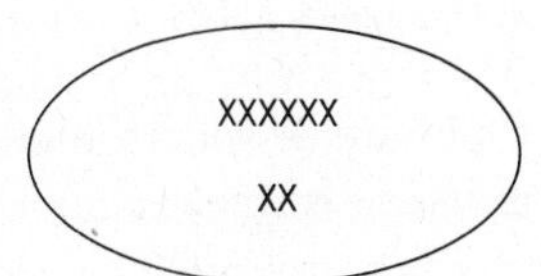

8. Draw a picture of a set that has the number of members given.

   a) 5 b) 9

9. Draw a ring around all groups of seven objects.

X X X
X
X X

X
X
X
X
X
X

X X X
X
X X X

X X X X
X X X X

The test should help you identify if the student's problem is writing numerals, counting, recognizing equivalent sets, associating numbers with sets, or recognizing sets that have a given number property. Notice that Gagné's learning hierarchy and Bruner's levels of learning are evidenced in the sequencing of the questions.

Addition—First Grade

1. (On this item, you may use concrete objects instead of the pictures.) In each blank, write the numeral that tells how many objects there are altogether.

   a) ____ b) ____

   c) ____

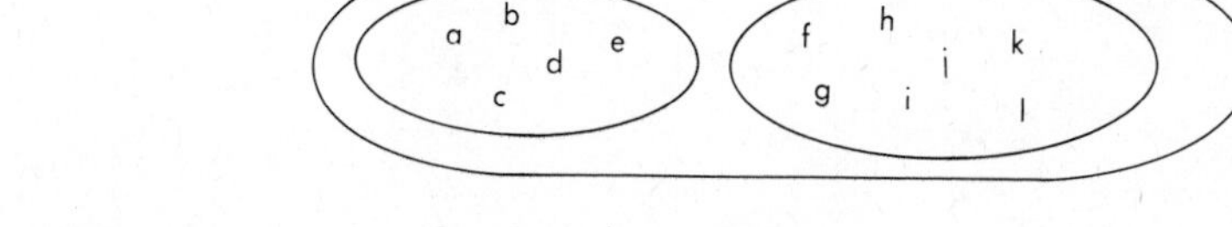

2. Fill in the frames with the correct numeral.

   a) b) c)

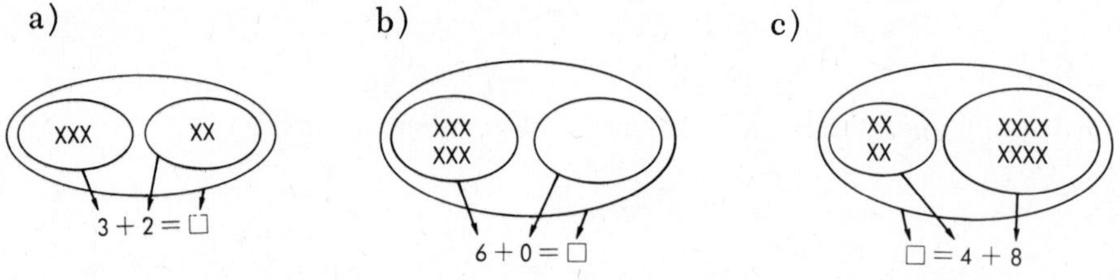

3. Draw a picture of sets to match each addition sentence. Then write the correct numeral in the frame.

a)

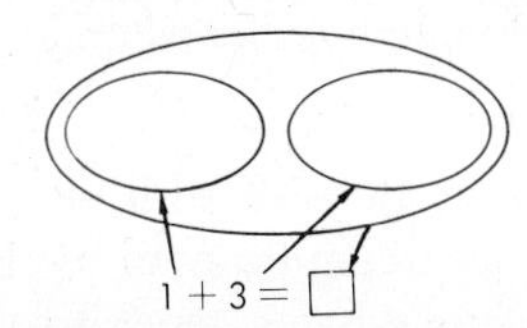

b)

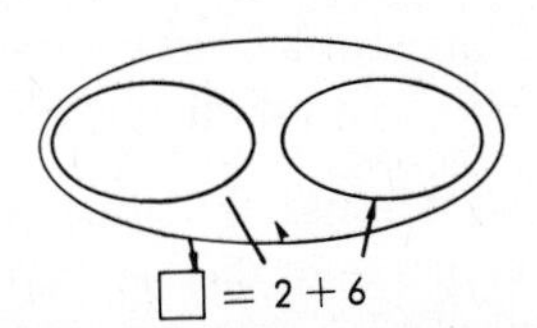

c)

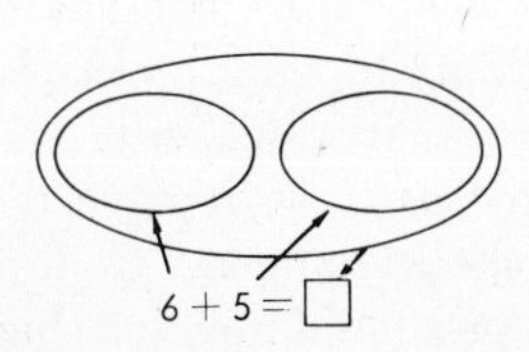

4. Write the sum.

a) $4 + 5 =$ ____ b) $9 + 0 =$ ____ c) ____ $= 7 + 3$

d) $\begin{array}{r} 2 \\ +7 \\ \hline \end{array}$ e) $9 + 8 =$ ____ f) ____ $= 6 + 8$

5. Write the missing numeral in each blank.

a) $7 + 6 =$ ____ $+ 7$ b) $4 +$ ____ $= 9 + 4$
c) $3 + (5 + 4) = (3 +$ ____ $) + 4$
d) ____ $+ (6 + 3) = (4 + 6) + 3$

6. (Use bundles of straws when presenting parts (a) and (b). Use the pictorial model for (c) and (d).) Write the numeral that tells the number of objects altogether.

a) ____ b) ____

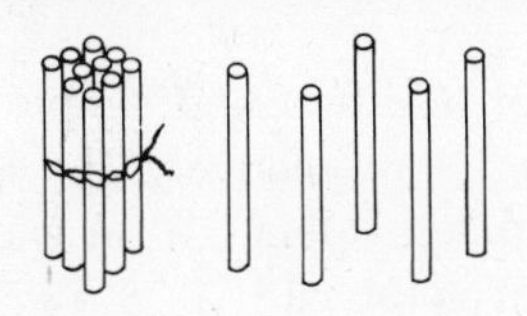

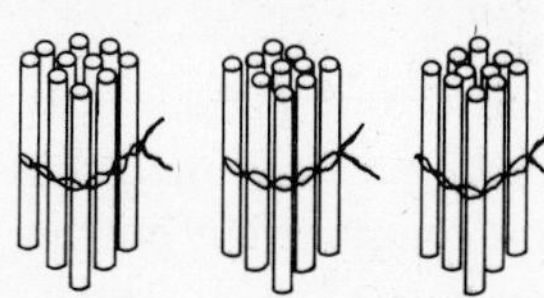

c) ____ d) ____

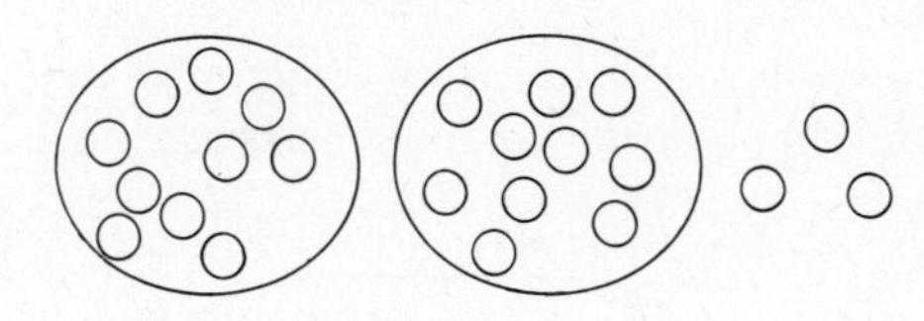

XXXXX
XXXXX

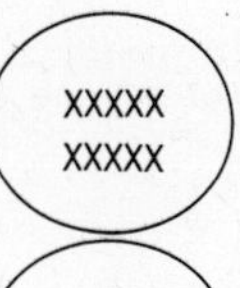

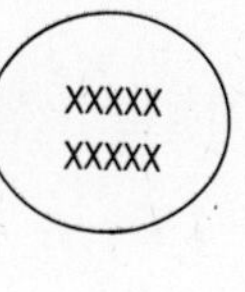

XXXXX
XXXXX

7. Write the correct numeral in each blank.

a) $10 + 9 =$ ____ b) $42 =$ ____ $+ 2$ c) $10 + 30 =$ ____
d) $50 + 60 =$ ____

8. (Read this problem to the child.) If you have 3 yellow balloons and 5 blue balloons, how many balloons do you have? ________

Items 1 through 5 of this test deal with the basic facts of addition sequenced from the concrete to the symbolic level. If a child has difficulty with the basic facts at the symbolic level (question 4) but can find the sum when a concrete or pictorial model is given (items 1 and 2), he probably has conceptualized addition but needs practice at the symbolic level. Capitalize on his ability to work problems such as 1 and 2 of the test. Give him representative sets and have him write and read the corresponding addition sentence; during the same activity, remove the physical model and ask the student to again read the symbolic sentence. Continue in this fashion for many of the basic addition facts. Then present him with a sheet of problems similar to problem 4. Involve the child in board games or competitive games in class which allow him to practice the basic facts. Consult chapter 7 for specific instructional ideas on the basic addition facts.

Notice that problems 2(c), 3(b), and 4(c) and (f) are written in the less common form $\square = a + b$. Many students who are able to solve open sentences of the form $a + b = \square$ cannot solve $\square = a + b$; i.e., they cannot use the symmetric property of equality (if $x = y$, then $y = x$). Weaver investigated the performance of first, second, and third graders on solving open sentences and found that they did better on open-sentence forms $a + b = \square$, $a + \square = c$, $a - b = \square$, and $a - \square = c$ than on equations symmetric to these ($\square = a + b$, $c = a + \square$, $\square = a - b$, and $c = a - \square$).[8] He suggested that a child's knowing how to solve $8 + 4 = \square$ and $9 - 7 = \square$, for example, does not necessarily mean he can solve $\square = 8 + 4$ and $\square = 9 - 7$. Students need a balance of these forms in instructional activities in order to cope with all of them successfully.[9]

A child who does poorly on addition properties—identity property (2(b) and 4(b)), commutativity (5(a) and (b)), and associativity (5(c) and (d))—needs instruction that fosters the discovery of patterns and formulation of generalizations. (See chapter 7.)

Problems 6 and 7 deal with numeration facts of addition and simple sums. These instructional tasks are discussed in chapters 6 and 9, respectively. If a student does well on computation but not on the verbal problem (item 8), he needs instruction in solving story problems, a topic considered in chapter 15.

Subtraction—First Grade

1. (Use concrete materials for this question. Remove the objects as suggested by the arrow.) In each blank write the numeral that tells how many objects are left.

a) ____ b) ____ c) ____

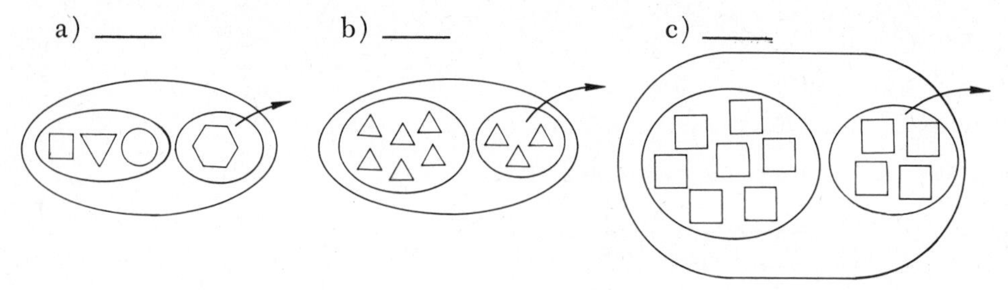

2. In each blank write the numeral that tells how many objects are left.

   a) ____ b) ____ c) ____

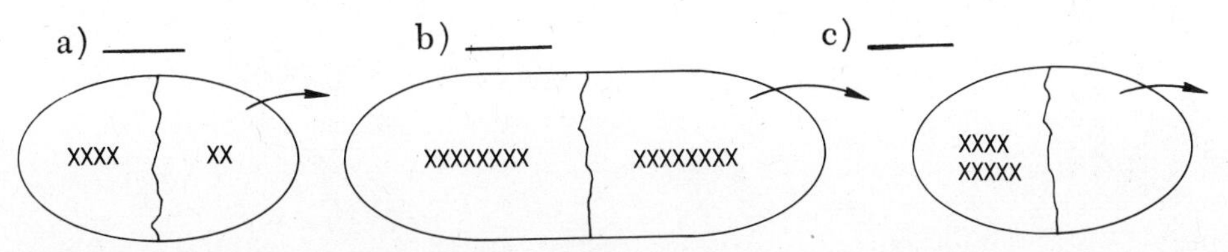

3. Fill in the blank with the correct numeral.

   a) $9 - 2 =$ ____ b) ____ $= 5 - 4$ c) $14 - 8 =$ ____ d) $8 -$ ____ $= 3$

   e) ____ $- 4 = 13$ f) $10 - (2 + 1) =$ ____ g) $8 - (3 - 1) =$ ____

4. Write the missing addend in each frame.

   a) $2 + \square = 6$ b) $\square + 1 = 10$ c) $8 + \square = 15$

5. Subtract.

   a) $\begin{array}{r} 10 \\ -\ 4 \\ \hline \end{array}$ b) $\begin{array}{r} 17 \\ -\ 9 \\ \hline \end{array}$

6. Write the numeral telling the number of cents left. ("D" represents dime here.)

   a) ____ b) ____

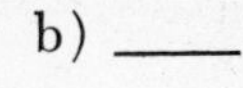

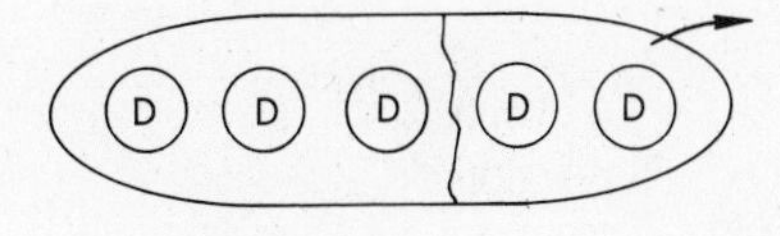

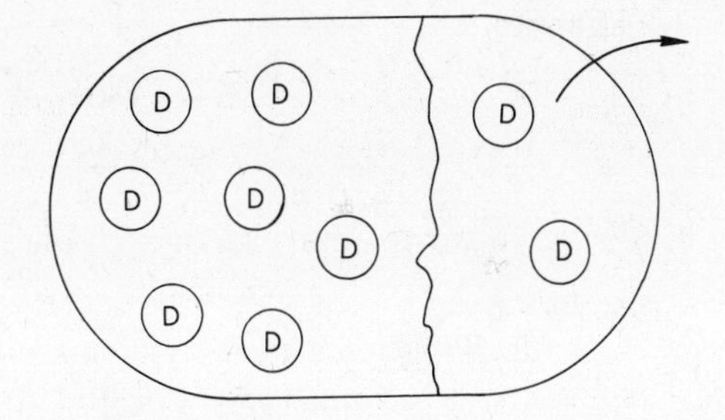

7. Subtract.

   a) $90 - 30 =$ ____ b) $\begin{array}{r} 120 \\ -\ 50 \\ \hline \end{array}$

8. If you had 15 toy cars and lost 6 of them, how many would you have left? ____

Some common trouble spots are represented in items 3(f) and (g) (notation with parentheses) and 4 (missing addend). If a student has trouble with expressions having parentheses, stress the fact that computation inside the parentheses must be done first and encourage him to record the "hidden" difference, e.g.,

$$7 - \underbrace{(5 - 1)}_{4} = \square \longrightarrow 7 - 4 = \boxed{3} \longrightarrow 7 - (5 - 1) = \boxed{3}$$

Missing addend sentences help a student relate subtraction to addition. A child who does poorly on problem 4 needs experience in finding the missing set and the missing addend. When he can do this task, have him rewrite the addition sentence as a subtraction sentence.

$\square + 4 = 7$ $\qquad$ $7 - 4 = \square$

Simple differences such as those in questions 6 and 7 are usually taught in first grade and can be modeled at the concrete level with an abacus, money, or Cuisenaire rods.

Subtraction—Third Grade

1. Write two subtraction sentences represented in the picture.

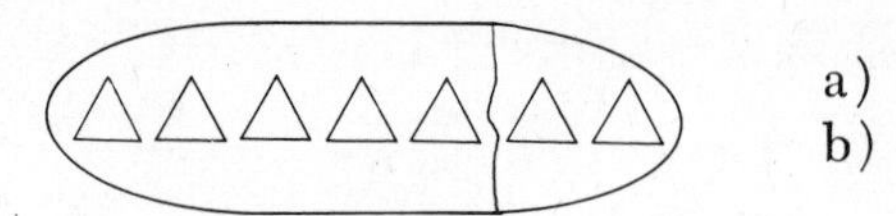

a)
b)

2. Complete each open sentence.

a) $16 - 9 = \square$ $\qquad$ b) $\square = 12 - 8$ $\qquad$ c) $9 - \square = 5$

d) $\square - 4 = 13$ $\qquad$ e) $50 - 20 = \square$ $\qquad$ f) $1700 - 900 = \square$

3. Compute.

a) $\begin{array}{r} 15 \\ -\ 7 \\ \hline \end{array}$ $\qquad$ b) $\begin{array}{r} 180 \\ -\ 90 \\ \hline \end{array}$ $\qquad$ c) $\begin{array}{r} 1400 \\ -\ 800 \\ \hline \end{array}$

4. Write the correct numeral in each blank.

a) $70 + 8 = \_\_\_\_ + 8$
$\quad = \_\_\_\_ + 18$

b) $49 = 40 + \_\_\_\_$
$\quad = 30 + \_\_\_\_$

5. Fill in the blanks.

a) $\begin{array}{r} 27 = 20 + 7 \\ -\ \underline{3} = \underline{\qquad 3} \end{array}$

$\_\_ + \_\_ = \_\_$

b) $\begin{array}{r} 38 = 30 + 8 \\ -\ \underline{23} = \underline{20 + 3} \end{array}$

$\_\_ + \_\_ = \_\_$

c) $\begin{array}{r} 43 = 40 + 3 = 30 + 13 \\ -\ \underline{25} = \underline{20 + 5} = \underline{20 + \ 5} \end{array}$

$\_\_ + \_\_ = \_\_$

6. Subtract.

a) $\begin{array}{r} 38 \\ -\ 5 \\ \hline \end{array}$ $\qquad$ b) $\begin{array}{r} 76 \\ -24 \\ \hline \end{array}$ $\qquad$ c) $\begin{array}{r} 83 \\ -27 \\ \hline \end{array}$ $\qquad$ d) $\begin{array}{r} 352 \\ -\ 36 \\ \hline \end{array}$

7. Jim had 19 pieces of bubblegum but gave 12 pieces to his mother. How many pieces of bubblegum did Jim have left? _____

Success on question 1 indicates that a student can associate the separation of a set with a subtraction sentence—an idea basic to the conceptualization of subtraction. Better performance on problem 2 than 1 might suggest that the student is learning by rote rather than with meaning. Such a student needs to associate concrete and pictorial models with symbolic sentences.

Children should be able to compute differences such as those in 2(e) and (f) and 3(b) and (c) before being introduced to the subtraction algorithm. Have students who cannot compute these differences model them with a teaching aid such as an abacus. (See chapter 9.) If a student cannot perform the regrouping in 4(a) and (b), he is not ready to compute with the subtraction algorithm in which a regrouping, or "borrowing," step is required (problems 5(c) and 6(c) and (d)). Solving in a systematic way word problems such as problem 7 deserves serious attention in third grade.

Multiplication—Third Grade

1. (Present each situation to the student at the concrete level.) Look at the sets and answer the questions.

a)

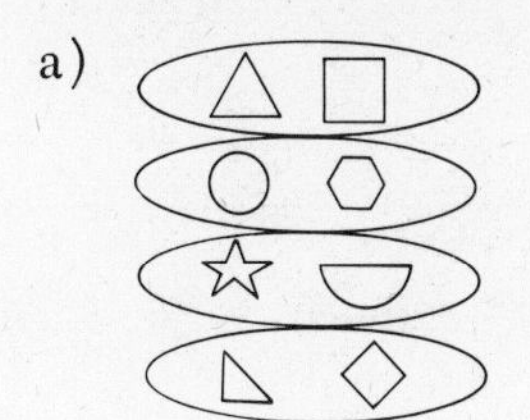

4 sets of 2
How many are there in all?

b) O O O O

O O O O

O O O O

3 rows, 4 columns
How many are there in all?

c)

2 cups, 3 saucers
In how many ways can the cups be paired with the saucers?

2. Look at the pictures and complete the sentences.

a)

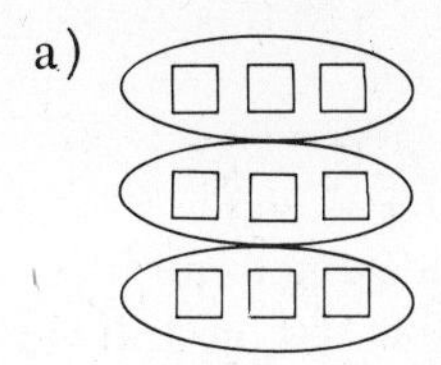

$3 \times 3 =$ _____

b) . . . . . .

. . . . . .

. . . . . .

$3 \times 6 =$ _____

c)

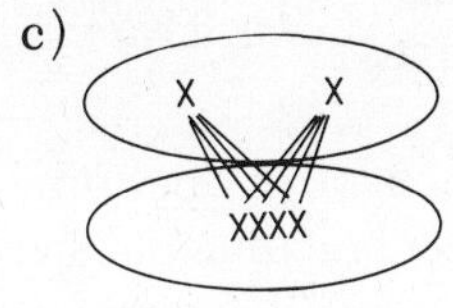

$2 \times 4 =$ _____

3. a) Draw a ring around four sets of three.

   △△△△△△△△△△△△△△△△

   b) Draw a ring around the part of the array that shows 5×2.

   ....

   ....

   ....

   ....

   ....

   ....

4. Fill in the blank with the correct numeral.

   a) $3 \times 4 =$ ____ b) $8 \times 1 =$ ____ c) ____ $= 7 \times 4$
   d) $9 \times 0 =$ ____ e) $6 \times 8 = 8 \times$ ____
   f) $1 \times (5 \times$ ____ $) = (1 \times 5) \times 6$
   g) $8 \times (7 \times$ ____ $) = 8 \times (2 \times 7)$
   h) $6 \times 12 = 6 \times ($ ____ $+ 2) = 60 +$ ____ $=$ ____

5. Write the correct symbol ($\times$ or $+$) inside each triangle.

   a) $7 \triangle 1 = 7$ b) $6 \triangle 4 = 10$ c) $5 \triangle 0 = 0$
   d) $4 \triangle 8 = 32$

6. Write the correct numeral in the blank.
   a) Five cars have ____ wheels.
   b) Four dimes are worth ____ cents.
   c) Eight dimes are worth ____ nickels.

7. a) I am thinking of a number. When it is multiplied by 2, the product is 12. What is the number? ____
   b) I am thinking of a number. It is the product of 4 and 5. What is the number?
   ____

8. Solve each problem.

   a) $\begin{array}{r} 23 \\ \times\ 3 \\ \hline \end{array}$ b) $\begin{array}{r} 64 \\ \times\ 3 \\ \hline \end{array}$

Students who perform well on items 4 and 5 but not on 1 through 3 exhibit skill in computation but show a lack of understanding of the meaning of multiplication. This situation may be due to a second-grade experience focused on drill without adequate conceptualization of the operation of multiplication. Have these students represent the facts they know with arrays, sets of equivalent-disjoint-sets models, or Cartesian-product models (see chapter 8) and then find situations in everyday life to which their models can be applied.

Students who do poorly on problems on the properties of multiplication—the property of zero in multiplication (4(d) and 5(c)), the associative property (4(f)), the commutative property when an extra factor is present (4(g)), and the distributive property (4(h))—need instruction in these patterns. (See chapter 8.) Problem 8 represents content usually taught for the first time in third grade. If this test is given early in the school year, most third-grade children will not be able to do these computations.

Closing

Some tools for diagnostic teaching considered in this chapter are Piagetian-task interviews, standardized diagnostic tests, and teacher-constructed tests. These can help a teacher identify needs and abilities of individual students and specific trouble spots in mathematics content for one child, groups of children, or the entire class. Diagnostic information in the hands of a skillful, sensitive teacher should result in an environment that fosters the learning of mathematics.

Notes

1. Richard W. Copeland, *How Children Learn Mathematics,* 2nd ed. (New York: Macmillan Publishing Company, Inc., 1974), p. 84.
2. Celia Stendler Lavatelli, *Piaget's Theory Applied to an Early Childhood Curriculum,* A Center for Media Development, Inc. Book (Boston: American Science and Engineering, Inc., 1970), p. 125.
3. Richard W. Copeland, *Math Activities for Children: A Diagnostic and Developmental Approach* (Columbus, Ohio: Charles E. Merrill Publishing Company, 1979), p. 95.
4. Jean Piaget, "Time Perception in Children," in J. T. Fraser, ed., *The Voices of Time* (New York: George Braziller, 1966), p. 208.
5. Fredricka K. Reisman, *A Guide to the Diagnostic Teaching of Arithmetic* (Columbus, Ohio: Charles E. Merrill Publishing Company, 1972), p. 142.
6. Austin J. Connolly, William Nachtman, and E. Milo Pritchett, *KeyMath Diagnostic Arithmetic Test Manual* (Circle Pines, Minn.: American Guidance Service, Inc., 1971), pp. 1–2.
7. Gary Adamson, Morris Shrago, and Glen Van Etten, *Basic Education Skills Inventory, Math Levels A & B* (Lenexa, Kansas: Select-Ed, Inc., 1972), pp. 2–3. (Pamphlet.)
8. J. Fred Weaver, "The Symmetric Property of the Equality Relation and Young Children's Ability to Solve Open Addition and Subtraction Sentences," *Journal for Research in Mathematics Education* 4 (January 1973): 53.
9. Weaver, "The Symmetric Property," p. 56.

Assignment

1. a) Administer the Piagetian interviews to at least ten children. Try to interview some students in kindergarten or first grade and some who are in third or fourth grade.
   b) Keep a record of each child's responses.
   c) Share the results with the child's teacher, and compare the student's performance on the Piagetian tasks with his performance in mathematics.
   d) On which tasks did you rate the youngest children preoperational? transitional? concrete operational? Do the same for the oldest children. Which concepts seem to be learned earliest? latest? Are the results consistent with the age levels suggested at the beginning of this chapter at which most children can perform these tasks?
   e) If possible, compare the performance on the Piagetian tasks of students who use only the mathematics textbook and that of students who use supplementary material such as concrete teaching aids, cassette tapes, and games along with the textbook. Does there seem to be a relationship between success on the Piagetian tasks and the kind of school experience a child has had?
   f) Write a report summarizing your findings in (b)-(e), and turn it in to your instructor.
2. Administer the KeyMath Diagnostic Arithmetic Test or some other standardized mathematics diagnostic test to a child. Does the test offer any suggestions for remediating the student's errors?

Laboratory Session

Preparation:

1. Study chapter 4 of this textbook.
2. Select two of the following areas of mathematics content and write a diagnostic test for each area at the indicated grade level:
   a) addition of whole numbers—second grade
   b) division of whole numbers—third grade
   c) fractional numbers—first, second, or third grade
   d) geometry—kindergarten or first, second, or third grade
   e) measurement—first, second, or third grade

   Use current mathematics textbooks and the sample teacher-constructed diagnostic tests in this chapter to guide you.

Objective: Write a diagnostic test for a selected area of mathematics content.

Materials: Several K-3 mathematics textbook series

Procedure: In groups of three or four, students should discuss the diagnostic tests they have constructed, giving attention to the sequencing of questions and problems with which young children may have difficulty.

Bibliography

Adamson, Gary; Shrago, Morris; and Van Etten, Glen. *Basic Education Skills Inventory, Math Levels A & B.* Lenexa, Kansas: Select-Ed, Inc., 1972. (Pamphlet.)

Connolly, Austin J.; Nachtman, William; and Pritchett, E. Milo. *KeyMath Diagnostic Arithmetic Test Manual.* Circle Pines, Minnesota: American Guidance Service, Inc., 1971.

Copeland, Richard W. *Math Activities for Children: A Diagnostic and Developmental Approach.* Columbus, Ohio: Charles E. Merrill Publishing Company, 1979.

———. *How Children Learn Mathematics,* 2nd ed. New York: Macmillan Publishing Company, Inc., 1974.

Lavatelli, Celia Stendler. *Piaget's Theory Applied to an Early Childhood Curriculum.* A Center for Media Development, Inc. Book. Boston: American Science and Engineering, Inc., 1970.

Piaget, Jean. "Time Perception in Children," in J. T. Fraser, ed., *The Voices of Time.* New York: George Braziller, 1966.

Reisman, Fredricka K. *A Guide to the Diagnostic Teaching of Arithmetic.* Columbus, Ohio: Charles E. Merrill Publishing Company, 1972.

Weaver, J. Fred. "The Symmetric Property of the Equality Relation and Young Children's Ability to Solve Open Addition and Subtraction Sentences." *Journal for Research in Mathematics Education* 4 (January 1973): 45–56.

5

Prenumber and Number Concepts and Skills

Preschool children and kindergartners are developmentally ready for beginning experiences with number, geometry, and measurement. Early work in geometry and measurement is discussed in chapters 12 and 13, respectively. Prenumber and number topics are considered in this chapter. The prenumber and number concepts and skills given in the following list are suitable for beginning experiences in mathematics. Much of this learning can be reinforced and applied in other subject areas as well as play activities.

1. observing and describing objects and groups of objects in the environment
2. comparing objects and using quantitative terms such as "heavier than" and "longer than"
3. generating sets and identifying subsets
4. classifying objects
5. ordering objects and events
6. putting sets into one-to-one correspondence
7. identifying and generating equivalent sets and determining if one set has more or fewer members than another set
8. copying and discovering patterns
9. reading and writing numerals
10. counting
11. determining the cardinal number of a set
12. identifying the ordinal number of an element in a series

13. uniting, or joining, sets
14. separating sets
15. using logical words of "not," "and," "or," "all," "some," and "none"

Observing and Describing Objects and Groups of Objects in the Environment

One of the first experiences a teacher should provide for a young child is ample opportunity to observe and describe objects and sets of objects in his environment. These may be items such as a toy, desk, brick, picture in a magazine, set of dishes, or group of birds. Suggestions for instructional possibilities follow:

1. Ask children to describe objects in the room. Encourage them to note differences and similarities among objects.
2. Have children bring their favorite toys or other interesting objects from home and describe them in class. Group objects that are alike in some way, and ask the children to tell how the objects are similar.
3. Enclose felt cut-outs of items such as flowers, birds, and geometric shapes in yarn on a feltboard, and let children describe the sets.
4. Select pictures in magazines for children to describe. Ask questions such as "Where have you seen these before?" "How are they used?" "What color are they?" and, if you think some children are ready for the number concept, "How many do you see here?"

Comparing Objects

Children must learn to detect differences and similarities among objects. Comparing items in length, height, and weight provides readiness for measurement and is a part of ordering objects correctly. Ask students to first compare two objects that are grossly different in size, and gradually increase the number to be compared and decrease the differential in size of the objects. Introduce terms such as thicker, thinner, longer, shorter, larger, and smaller. After having the children compare concrete objects, provide pictorial experience for them to sharpen their discrimination skills. For example, have them select from a set of shapes the shape that looks exactly like a given one. In the following example, a child should match the given square with the correct shape in the row below it.

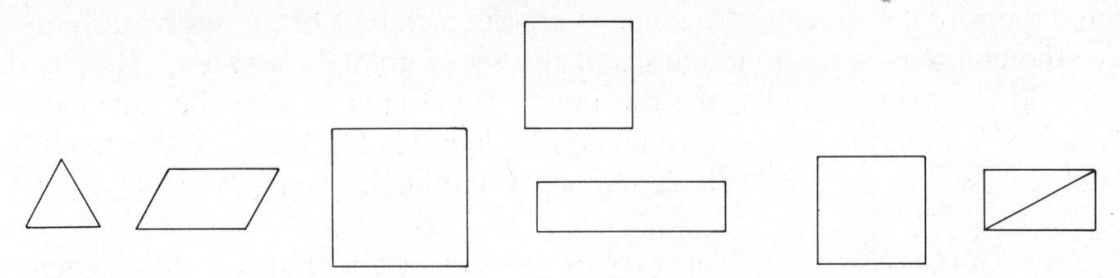

Guide children to see likenesses and differences among letters and numerals such as "b" and "d," "6" and "9," and "3" and "E."

Generating Sets and Identifying Subsets

Work with sets is basic to classification, number, and number operations. A set is a collection of objects called its elements, or members. Children learn about sets and elements of sets by working with many examples. In beginning experiences, students usually represent sets by placing a loop around concrete objects or drawing a ring around objects in a picture. Later the symbolism using the braces, e.g., $\{a, b, c\}$, is introduced. A subset, symbolized as "$\subseteq$," of a set can be a set of one or more of the set's elements or a set having no members. The pattern for counting all the subsets of a given set is an interesting one and is appropriate for superior learners at the second- or third-grade level to explore. The study of subsets furnishes the basis for joining and separating sets which in turn provide a foundation for the operations of addition, subtraction, multiplication, and division of whole numbers.

There are three special types of sets: the universal set, a singular set, and an empty set.* The universal set is the set of all elements in a particular discussion. A singular set contains exactly one member. Children usually develop the concept of a set having one element after they can logically group two or more objects together. An empty set is one that contains no members. Students enjoy examples such as "the set of all live elephants in the room" and "the set of all children who weigh a million pounds." To introduce the empty set, have three or four children stand inside a hula hoop and ask one at a time to step outside the hoop until no child is left inside the hoop. Point out that the hoop has no child inside and this set is an example of an *empty* set. Repeat the procedure by withdrawing objects from a box or having children take out objects enclosed in a loop on the feltboard. Represent the empty set on the chalkboard by drawing a closed curve with no elements inside. In second or third grade, the symbolism $\{\quad\}$ is introduced to name an empty set. Teaching the empty set should help a child conceptualize the cardinal number 0.

You may use the terms "set," "member of a set," and "subset" in everyday language before introducing them in structured activities. For example, ask, "Will the set of all children get ready for lunch?" "Who are members of the subset of children sitting at this table?" "Can you find the subset of red crayons in your box of crayons?" In a beginning structured activity, distribute strips of yarn and several kinds of plastic animals to each child. Ask the children to tell what the set is, and instruct them to place a loop around all the dogs in the set. Tell them the set of dogs is a *subset* of the set of animals. Continue with "Use your yarn strips to place a loop around the subset of horses. What colors are the members of the subset of horses?" Have the children form other subsets of the set of animals, and ask, "How is this animal different from (similar to) this one?" "What kind of animal is a member of this set?" "Would you like to have an animal like this for a pet?" "Which animals would you find on a farm?" "Which animals would be in a zoo?" "Which kinds have

* The terms *universal set* and *singular set* are generally not used with young children.

you seen before?" If the children understand number, include questions such as "Which subset has the fewest (most) members?" and "How many different subsets have you found?"

In another activity, present the children with a set of felt shapes on the feltboard. Ask a child to place a loop around a subset of the shapes and describe the subset. A child showing this work has selected a subset of blue objects. (B stands for blue; R, red; G, green; and Y, yellow.)

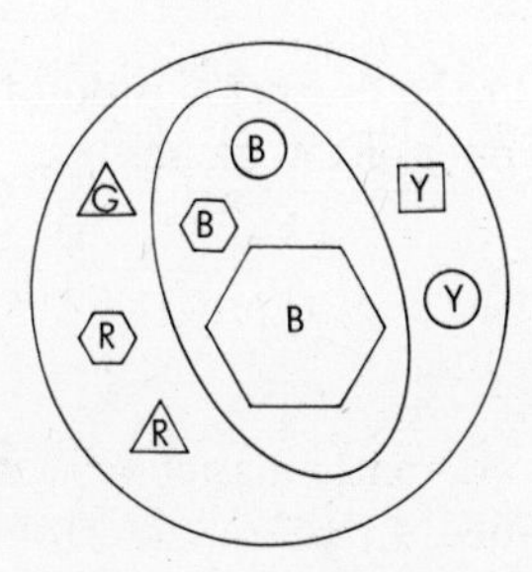

Guide children to generate subsets consisting of members of the same color, size, or shape. Be sure to include subsets having one member, e.g., the subset of all green objects or the subset of all large objects in the preceding example, and subsets with no members such as the subset of all pink objects in the example. After number is introduced, include directives such as "Show me a subset with five members. Name the members. Show me a subset with more than five members. How many members are in your subset? Show me a subset having fewer than five members. How many elements did you select?"

Young children naturally group objects that are alike in some way. Gradually introduce sets having dissimilar members, e.g.,

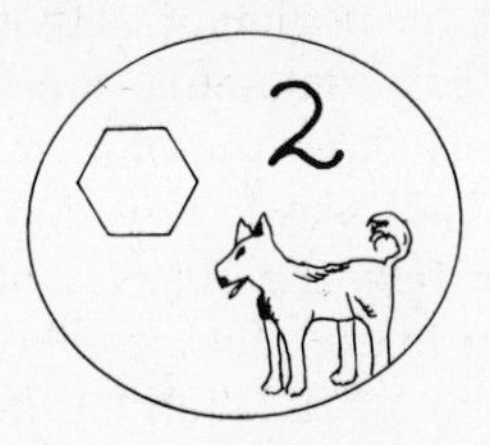

Emphasize that sets can be collections of any kind of objects.

Design experiences that involve the children themselves as members of sets. The set of boys and the set of girls are two possible subsets of a set of preschool children. Guide children to form others such as the set of children on the rug, the set of children wearing sandals, and the set of children who understand subsets.

A game which can be used for practice or for evaluation purposes is "Pin the Tail on Fluffy." Sketch and cut out a picture of Fluffy, the cat, on tagboard. Prepare several tails with label cards specifying shape, color, or size on them, and have a set of attribute blocks available for the group. Children should be blindfolded one at a time as they have a turn to select and pin a tail on Fluffy. Each child must generate the subset of attribute blocks specified on the tail he pinned on Fluffy.

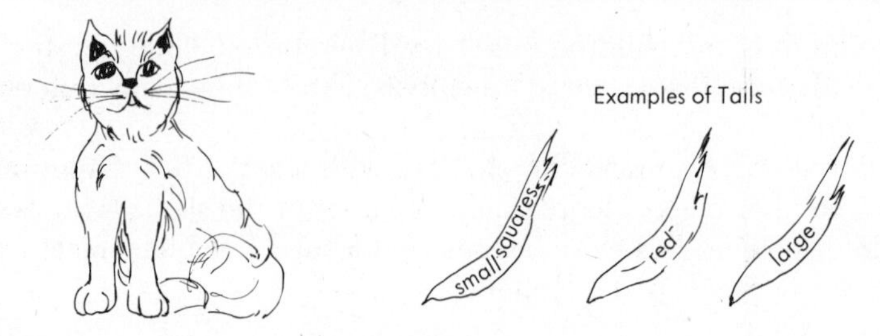

The player who comes closest to the correct position for the tail and who demonstrates his set correctly is the winner of the game.

Classifying Objects

In order to develop logically and mathematically, a child must learn to classify. Classification is an ongoing process that ranges from a "sorting" exercise in its earliest form to classification of rational, real, and complex numbers in later years. When classifying, a young child takes an assortment of objects, decides on a criterion, and generates sets so that all of the objects in each set are alike in some way. Notice that this process requires a strategy on the part of the child and is more difficult than finding all the red objects, all the circular blocks, or all the plastic horses of the preceding instructional task.

Commercial attribute materials can be used to give children early experiences in classifying. They can sort these materials by one attribute such as shape, two attributes, e.g., size and color, or three attributes such as size, color, and shape. The activity cards that can be purchased with the manipulative aids contain many classification tasks. You can also construct your own activities and materials.

These are some suggestions for beginning activities in classification:

1. Give each child a paper bag containing three or four of each of various objects such as rocks, buttons (large ones for small children), plastic counters, large paper clips, and small plastic toys. Materials may vary from bag to bag. Ask the children to take the objects out of their bags and make sets so that all of the members in each set are alike in some way. Have each child tell why he sorted the way he did, and let the children suggest other ways to classify the objects. Make a chart on the chalkboard of the attributes determined for each bag of materials.
2. Give children various colored and textured strips of paper of at least three different lengths. Ask them to sort the strips so that all of them in each set are the same length.
3. Distribute three paper plates and three each of plastic knives, forks, and spoons to each child. Ask the children to group the eating utensils and place them on the paper plates. Some will place the three knives in one plate, the forks in another plate, and the spoons in the other, but some may sort by place setting and group a knife, fork, and spoon with each plate.
4. Have the children play the same-different game. Make a partitioned sorting box with "Same" labeled on one side and "Different" on the other, giving

an example of each. Design several cards so that some have two designs that are alike and some have two designs that are different.

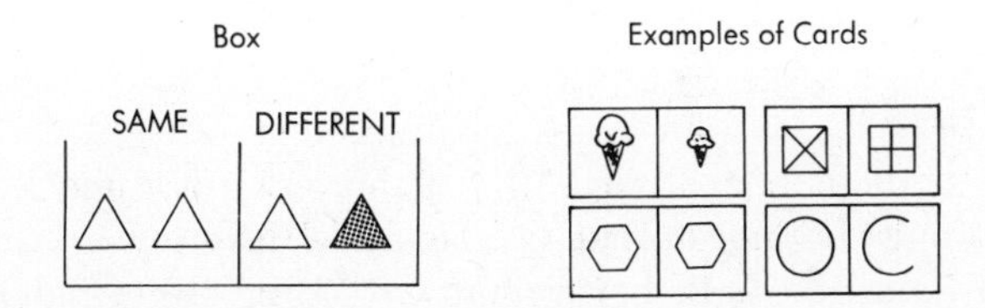

The child places cards having the same designs on the "Same" side of the box and those with different designs on the "Different" side. As children participate in the activity, ask, "How are these two pictures different? What does this picture represent? Why did you put this card on this side of the box?"[1] When a child learns how to place sets into one-to-one correspondence and understands equivalence of sets, include cards such as these:

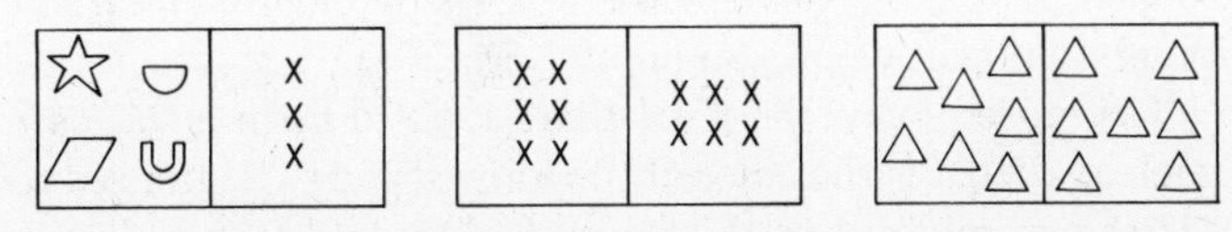

The child determines if the two sets on each card have the same number of elements and places it in the sorting box accordingly.

5. Cut out pictures of people, food, cars, airplanes, musical instruments, and so on from magazines and glue them on cards. Have children group cards together that have objects similar in some way.
6. Present magazine pictures of situations representing good health habits and poor health habits. Guide discussion of each picture, and let the children suggest whether to place the picture in the "Good Health Habit" or "Poor Health Habit" set.

Ordering Objects and Events

Experience in seriating objects helps a child order numbers in a meaningful way, and ordering events provides readiness for measurement of time. Some ideas for instruction follow:

1. Have children order objects such as straws by length, measuring spoons by capacity, and cylinders by height or diameter. Guide those who have difficulty with the task by statements such as "Try this," or "Do you think this will help?" Introduce quantitative terms such as longest, shortest, most, least, largest, smallest.

   Guide children to discover the transitive property. Ask them to order three cylinders from shortest to tallest (for our discussion, we will label the cylinders A, B, and C),

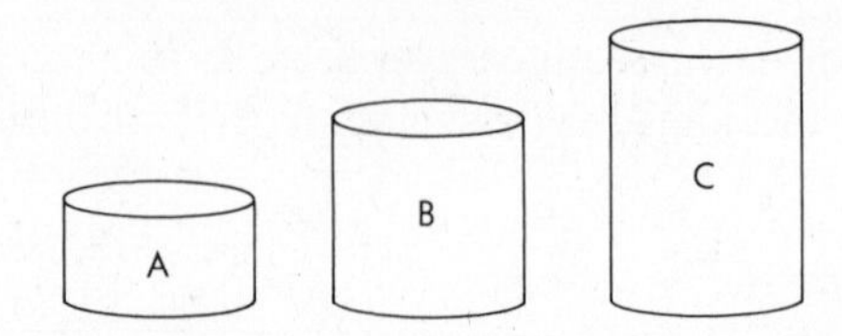

and help them understand that "If A is shorter than B and B is shorter than C, then A must be shorter than C." To guide them toward this end, ask, "Is A shorter than B?" "Is B shorter than C?" "How are cylinders A and C related?" If a child does not conclude A is also shorter than C, remove cylinder B and repeat, "How are A and C related?" Return B to the middle and ask if A is still shorter than C. The child should eventually conclude that A must be shorter than C regardless of whether an object is inserted in the middle or not. The transitive property is important in later work with comparisons of numbers.

2. Arrange objects on a large felt tree or beanstalk. Ask which object is highest, lowest, next to the highest, in the middle, above a given object, below a given object, and so on. Reverse the process by having children describe the location of an object when you point to it.
3. Have children sequence the Cuisenaire rods to form a staircase. Ask questions such as "What is the color of the longest rod?" "Is the red rod the shortest?" "Is the purple rod in the middle?" "Is the black rod longer than the yellow one?" "Is the light green rod shorter than the purple rod?"
4. Make a set of four or five picture cards showing a sequence of steps that occur as a kitten unravels a ball of yarn. Ask the child to arrange the pictures in the order the event takes place. The cards in random order may look like this:

5. On a stencil for duplication, draw pictures of certain occurrences from a story. Read the children the story, and ask them to cut out and order the pictures on their handouts according to events in the story.

Putting Sets into One-to-One Correspondence

The skill of one-to-one correspondence is important in determining if sets are equivalent which in turn is prerequisite to forming the concept of cardinal number. As a readiness experience, you might give each child a glass of juice. Emphasize that each child has exactly one glass of juice. In beginning activities on one-to-one correspondence, have children physically match complementary sets such as cups with saucers, flowers with vases (one flower in each vase), horses with riders, and birthday hats with children. Other work with complementary sets may include having children form a row of balls below a row of seals you have placed on the feltboard so that each seal has exactly one ball and putting triangles on top of squares you have placed on the feltboard to form houses.

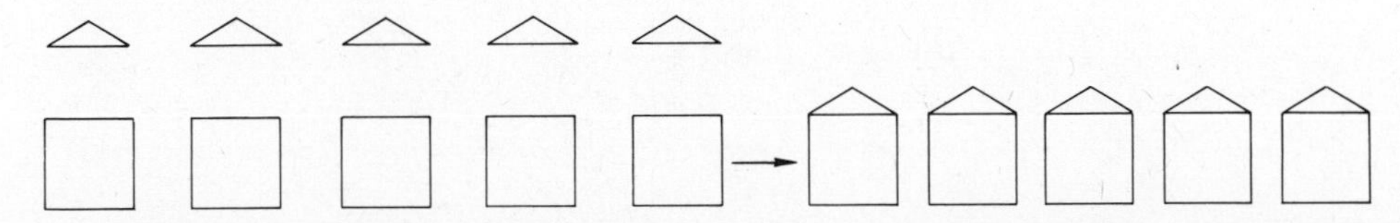

Children also need experience in demonstrating a one-to-one correspondence between two noncomplementary sets. According to Piaget, children should match noncomplementary sets after they can perform the easier task of matching complementary sets such as those in the preceding examples.[2] Have children match balls with balls or squares with circles by connecting matching elements with felt or ribbon strips.

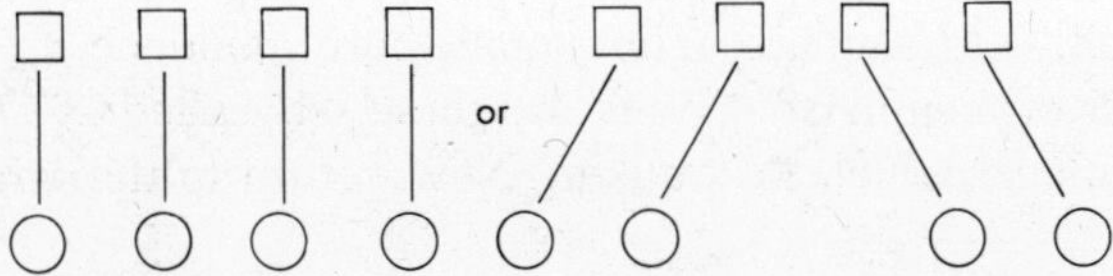

Guide them to exhibit other possible one-to-one correspondences such as

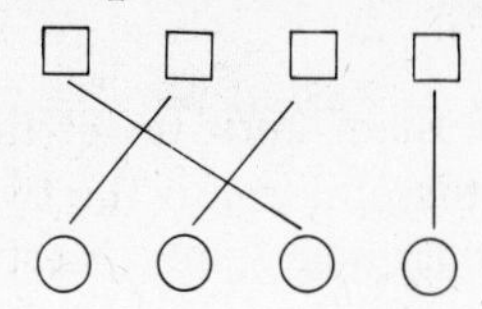

Challenge gifted learners to find all the possible one-to-one correspondences between two sets when the sets have one, two, three, or four elements and to make a generalization of the pattern.

Identifying and Generating Equivalent Sets

To understand number, children must recognize when sets are equivalent. They should have ample opportunity to compare sets using terminology of "more than," "fewer than," and "as many as." In beginning activities, help children understand that sets can be equivalent even though the elements may be different in nature and arrangement. To help toward this end, form a row of cut-outs of one color on the feltboard, and have a child place the same number of objects in a different color below the one you made. Then spread out the shapes in your row.

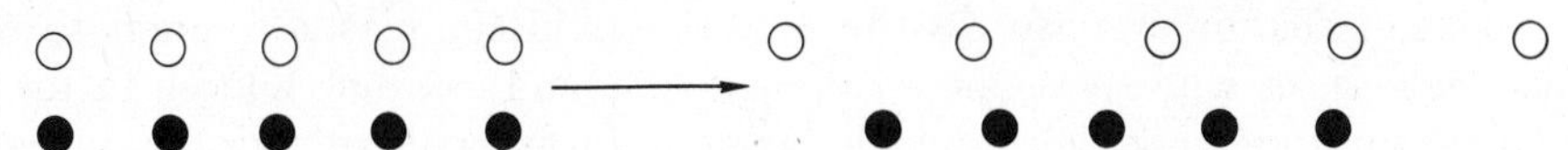

Ask if the sets in the second arrangement have the same number of objects. If a child does not understand that they do, move your cut-outs back to a position directly above his objects. If necessary, have him connect pairs of elements to show a one-to-one correspondence. Form other configurations with the top row of cut-outs, e.g.,

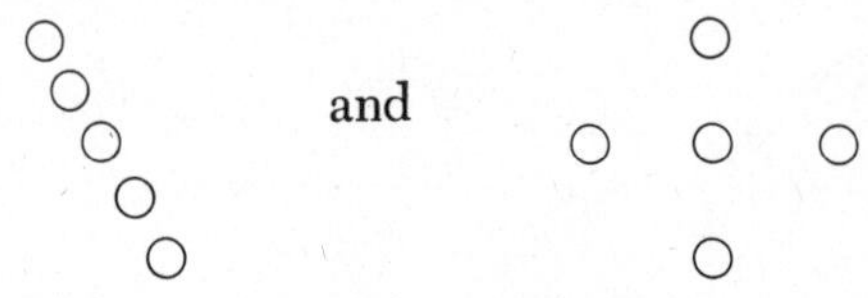

Each time ask if the two sets can be matched one-to-one. Tell the children that sets which can be placed into one-to-one correspondence are called *equivalent* sets. Emphasize that all of the sets you have demonstrated on the feltboard were equivalent although the arrangements were different. Also, illustrate equivalent sets having unrelated members, e.g., a set of a bird, circle, and square and another set of the numeral 5, a triangle, and the letter *a*. Ask questions such as "Can the sets you see on the feltboard be place in one-to-one correspondence?" "Do the sets have the same number of members?" "What are these sets called?" "Do equivalent sets always have members which look alike?" Now, return to the arrangement

and teach the concepts of "more than" and "fewer than." Increase the number of cut-outs in the top row, and ask the children if the two rows are equivalent. Select a child to attempt to demonstrate a one-to-one correspondence. His work may look like this:

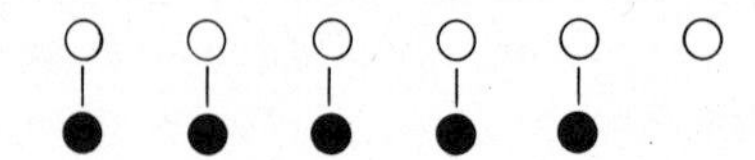

This time the top row has an element left unmatched. Tell the children this means that the top row has *more* members than the bottom row and the bottom row has *fewer* members than the top one.

Conduct another activity in which you form a half-circle of chairs (have fewer chairs than children) and have the children sit down in front of the chairs. Ask, "Is there exactly one chair for each student?" "Is the set of children equivalent to the set of chairs?" "If each child sat in a chair, would there be any empty chairs?" "Would there be a seat for every child?" "Are there more chairs or fewer chairs than children?" "How can you tell?"

As an evaluation exercise, distribute twelve to fifteen plastic counters to each child. Make a set with six counters, for example, and ask the children to form a set equivalent to your set, a set which has more members than your set, and a set which has fewer members than your set. Encourage them to discuss differences among their responses.

Iconic exercises are also important. Examples can be found in elementary mathematics textbooks or you may design your own. These are some sample exercises:

1) Draw a set equivalent to the given one.

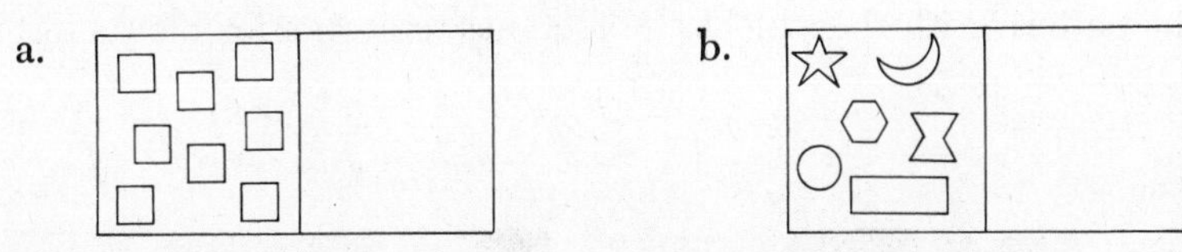

2) Draw a set that has more members than the given one.

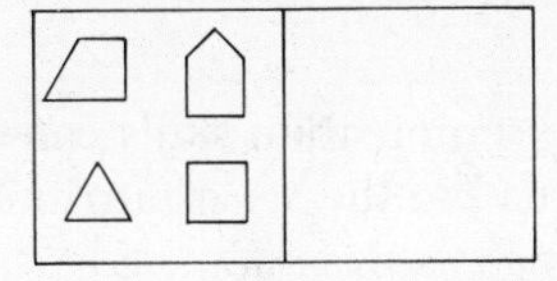

3) Circle the set that has fewer members.

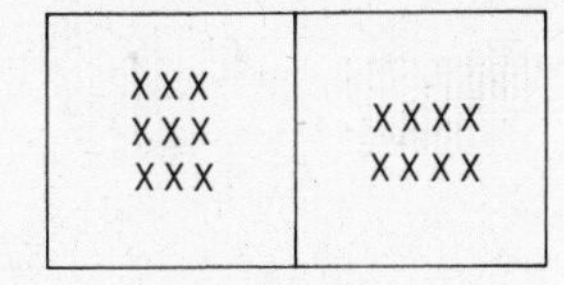

Copying and Discovering Patterns

Working with patterns helps children sharpen their perceptual skills and increases their awareness of order, shapes, and esthetics.[3] Students are frequently expected to copy problems from the chalkboard, overhead projector, textbook, and task cards. Teachers should not assume that all children can copy and recognize patterns with no systematic instruction. A child who has excessive difficulty with this task, however, may have a perceptual-oriented learning disability. Children having such a disability should be required to do little, if any, copying.

Some suggestions for work with patterns follow:

1. On the feltboard make a design such as

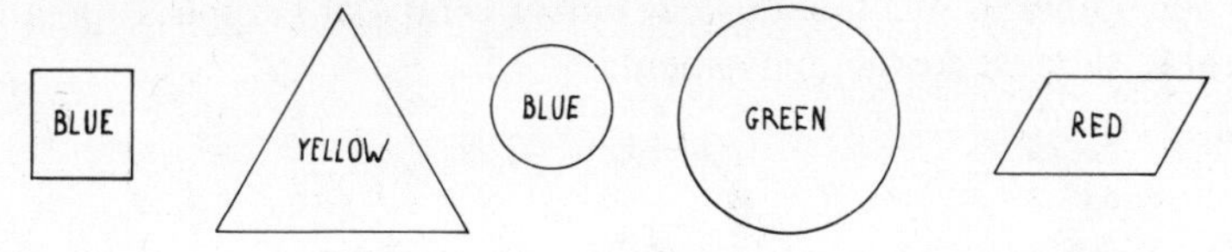

in which the children must attend to the attributes of shape, color, and size. Ask them to use attribute blocks to copy the design in the same order and then in reverse order.

2. Start a pattern such as

on the overhead projector and ask children to draw what they see, try to discover a pattern, and extend the pattern for five more shapes.
3. Make a color pattern with cubical blocks and have children copy and complete the pattern with their blocks. Sometimes omit interior blocks instead of ending blocks, e.g.,

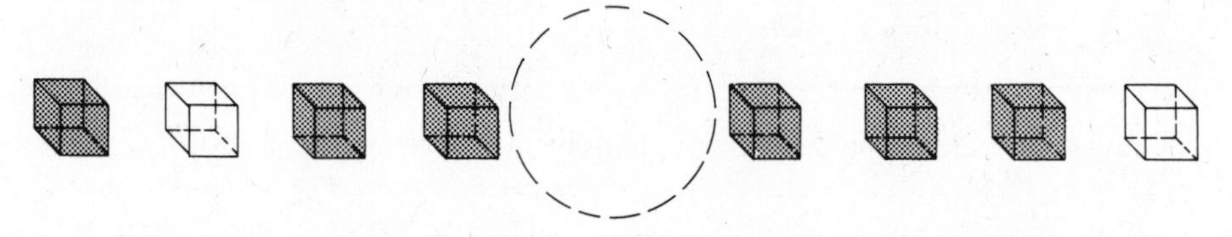

4. Refinement of a child's discrimination skills can also be fostered by using commercial Creature cards (in the set of attribute materials) or by creating your own task cards such as this one:

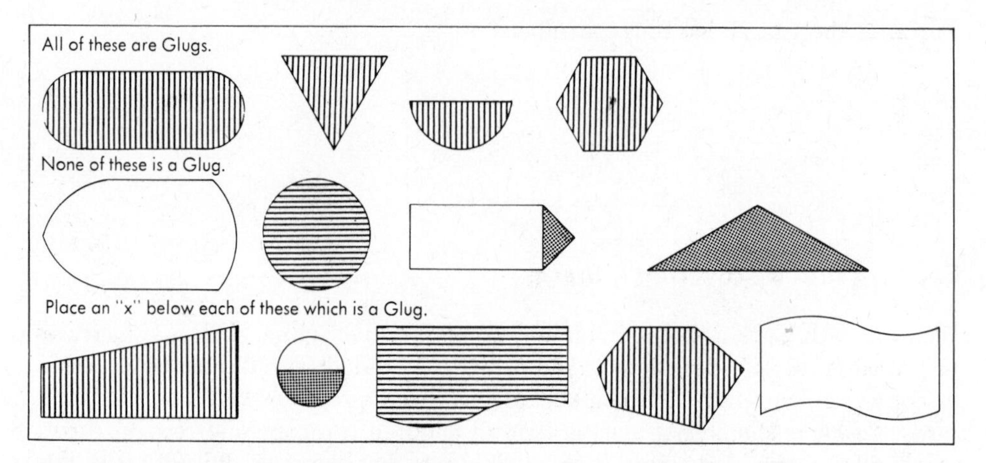

Reading and Writing Numerals

An early experience in kindergarten and preschool is writing numerals. Many children learn best when as many senses as possible are involved, so provide tactile, kinesthetic, and visual experience with numerals. Have children move their fingers along sandpaper numerals or numerals cut out of scraps of carpet. Tape an arrow on the numeral to suggest correct movement.

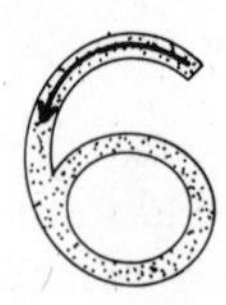

Pronounce and have children say the number names as they work with the numerals. Many children enjoy playing a game in which one child makes numerals with his finger on the back of a friend and the friend tells what the numeral is. Children can also construct numerals with popsicle sticks (to make straight marks) and yarn or ribbon strips (to make curves).

Children need many opportunities to copy numerals. A typical textbook exercise provides the numeral and spaces for practice.

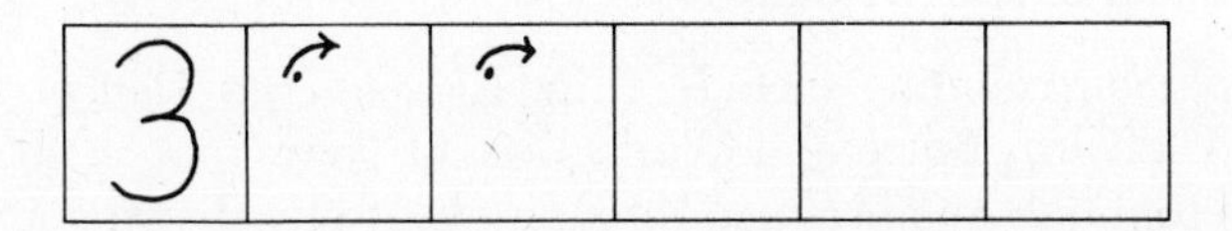

Other practice experience can be provided by calling out numerals in random order for children to record and presenting the children with numerals in random order to identify. Such discrimination learning is essential for success in subsequent work in mathematics.

Children also need to match the written number words to the numerical symbols 0 through 9. Make a self-correcting puzzle with the words zero, one, . . ., nine printed on the base and the symbols 0, 1, . . ., 9 written on the puzzle pieces.

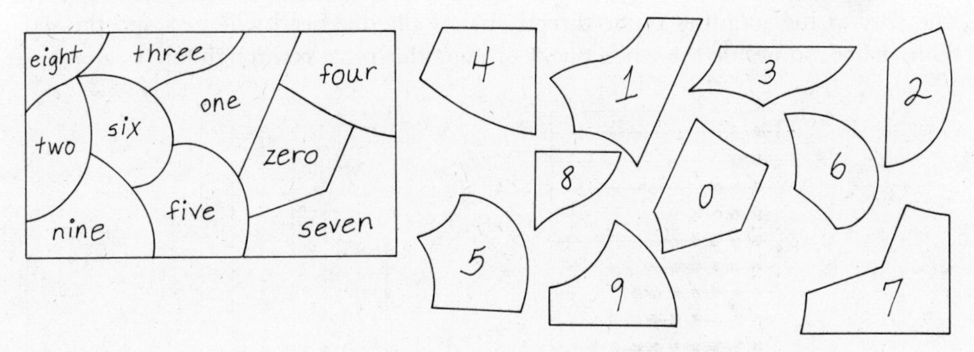

Children match the word with the numerical symbol to solve the puzzle.

Counting

By the time they are three or four years old, most children can say by rote the words for counting from one through ten. This type of learning is essential for work with cardinal and ordinal number even though it may have little meaning associated with it. For generations, songs and rhymes such as "One, Two, Buckle My Shoe" have helped young children learn to count. You might use such traditional material or you might wish to create your own counting songs or poems. Those which involve body action provide an extra dimension. Consider this example:

Count the Kittens

Two little kittens (everyone raises two fingers) were walking (walk the fingers of the right hand) down the street (count 1, 2)
One little kitten (raise one finger) they happened to meet
How many (gesture with opening arms and hands as if questioning) kittens can you count on the street? (count 1, 2, 3)

Three little kittens (raise three fingers) were walking down the street (count 1, 2, 3)
One little kitten (raise one finger) they happened to meet
How many (gesture with arms) kittens can you count on the street? (count 1, 2, 3, 4)
(Continue to the desired number.)

Children's books on counting include Fritz Eichenberg's *Dancing in the Moon* published by Harcourt, Brace and World, Inc. of New York in 1955 (This book introduces the counting numbers one through twenty by showing a variety of animals engaged in many activities.) and *Counting Rhymes* (a picturesque book containing sixteen rhymes on counting and number) published by Golden Press Inc. of New York in 1960.

For children to be able to count rationally, they must understand the concept of "one more than," the foundation of the counting process. To teach this concept, you might present the children with one bead on a counting frame or one white (unit) Cuisenaire rod. Tell them this set represents one. Have them find other sets with one object. Then introduce two as one more than one. Place two beads on the wire of the counting frame directly below the one-bead wire or stand the red Cuisenaire rod beside the white one. Continue this process through ten.

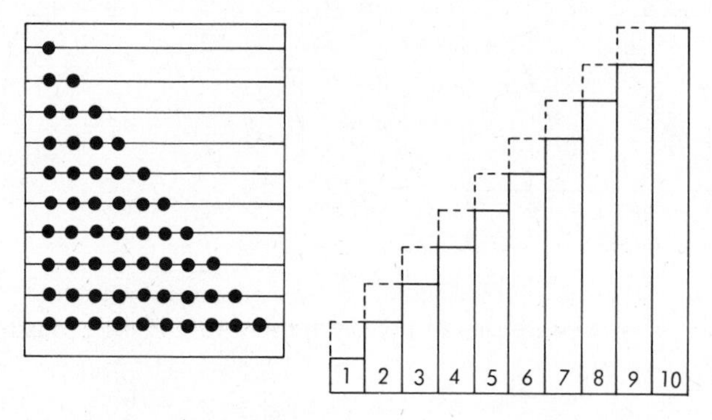

The dashed outline of the white Cuisenaire rod in the staircase suggests that you put the 1 rod in these positions to emphasize that each rod is one unit longer than its predecessor. Systematic instruction such as this should help the children generalize that every counting number has a successor which can be found by adding one to the original number.

Involve the children in many oral exercises in counting: "Count these blocks as I place them on the table. Count your crayons. Count the blue balls in your set. How many red ones do you have? Let us count together all the girls in this picture as I point to them." Let children play a game such as "Counting the Claps" in which they take turns clapping as other children count the claps.

Determining the Cardinal Number of a Set

This task was analyzed in chapter 3. We saw there that readiness includes the ability to order elements, count, and put sets in one-to-one correspondence. Furthermore, a child must be told that the last counting number used in making the one-to-one correspondence between the given set and a subset of the set of counting numbers tells how many elements the set has.

An alternative to the counting procedure just given for teaching a student how to find the cardinal number of a set is to present him with several examples of sets with one element, two elements, and so on, tell him the cardinal number of each of these sets, and then ask him to find other sets that have the same number. This inductive approach works especially well when students are learning the cardinal number of sets with five members or less and when students have limited experience in counting. As they gain experience in counting, they can use a counting technique to determine the cardinal number of sets which have more than five elements. As you design activities, include some requiring the children to tell the cardinal number when given the set as in

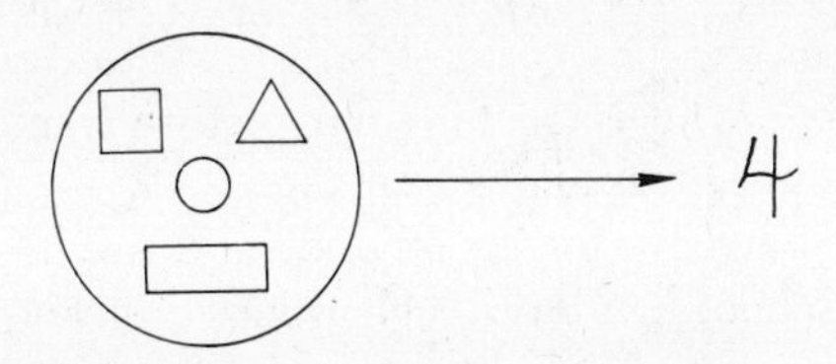

and some in which the child generates a representative set when given the number as in

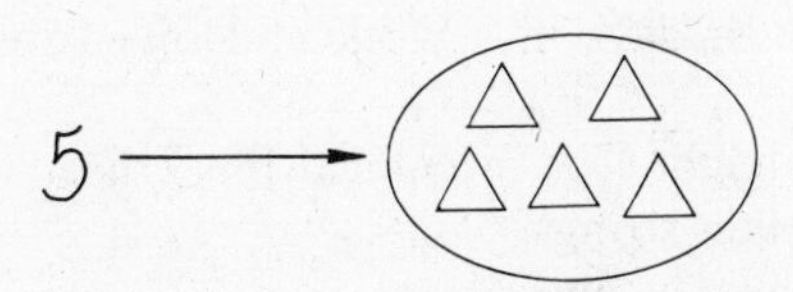

In order to conceptualize number, a child needs experiences in finding the cardinal number of many diverse sets. He learns through these examples and proper teacher questioning that many sets have the same number name. For example, all sets equivalent to $\{1, 2, 3\}$ have a common property, and this common property is what "threeness" is. Number is the concept formed by working with many examples of equivalent sets, and numerals are representations of that concept in spoken words or written symbols, e.g., 3 or three. The teacher should try to help the child make a distinction between the abstraction of number and a numeral which communicates that concept but should not insist on rigid use of the two terms.

Some ideas for teaching number follow. Most of them deal with cardinal numbers 0 through 10, but they can be adapted for use with numbers less than 10 for beginning activities.

·1. Give each child six containers such as margarine bowls labeled 0 through 5 and fifteen discrete objects such as buttons. Tell the children they will use all of the buttons in this activity. Let them identify the numerals on their bowls. Then tell them to put one button in the "1" bowl. Demonstrate the procedure, saying "one" as you put one button in the "1" bowl. Continue with the "2" bowl, but withdraw your assistance when the children can do the work alone. They should have used all of the buttons if they have correctly made all of their sets. If no child inquires about the empty bowl, you might ask, "Is there one bowl different from the others? What is the numeral on it? How many buttons are in that bowl?" Encourage the children to compare their sets, e.g., have them point to their "5" sets and tell how they are alike (same number of buttons) and how they are different (buttons of various colors or sizes).

2. Highlight one number each day. To generate sets with a specified number, children can use ordinary objects in the room, but other considerations include:
   a) A person has one mouth, nose, neck, tongue, and head; most dogs have one tail; a jar has one mouth; a car has one steering wheel.
   b) A person has two eyes, ears, and so on; a shirt has two sleeves; socks and shoes come in twos; a bicycle has two wheels.
   c) Many stools have three legs; a tricycle has three wheels; most traffic lights come in threes.
   d) Cars have four wheels; rectangular picture frames have four sides and four corners; most forks have four tines.
   e) Most people have five fingers on one hand and five toes on one foot; children go to school five days each week; a nickel has the same value as five pennies.
   f) A cubical block has six sides; insects have six legs; most guitars have six strings.
   g) There are seven days in a week and seven red stripes on the United States flag.
   h) Spiders have eight legs; a cubical block has eight corners; most mandolins have eight strings.
   i) A baseball team has nine players.
   j) Most people have ten fingers and ten toes; a dime is equivalent in value to ten pennies.
   k) There are eleven players on a football team.
   l) A year has twelve months; eggs are packed in cartons of twelves; a clockface shows twelve hours.
3. Give each child ten strips of yarn or ribbon, numeral cards for 1-10, and a paper bag containing one straw, two buttons, three crayons, . . ., ten empty spools. Objects may vary from bag to bag. Ask each child to sort the materials in his bag so that he has objects which are alike in each set, enclose each set with a yarn loop, and place the matching numeral by the set.
4. Have children draw their own sets on cards and pin them on the correct numerals on a large number line hanging on a wall or from the ceiling.
5. Place a set of Cuisenaire rods in a paper bag and distribute a set to each group of three children. Tell them the white rod represents 1 in this activity. Have each group make a staircase with the rods representing 1 through 10. Then instruct the children to take turns reaching in the bag for a rod, feeling it, saying the numeral associated with it, taking the rod out of the bag, and checking with the rods in the staircase to see if they have named correctly.

6. Write the numerals 0 through 10 on paper cups. Select various children to put the correct number of popsicle sticks or straws in each cup.
7. Design a functional, bulletin board display in which children identify the cardinal number of a set and make a set when given the cardinal number. For instance, construct dot-pattern cards, each having from 0 to 10 dots, and two sets of numeral cards for 0 through 10. Draw a rabbit with one floppy ear and a giraffe head and neck (with no spots), and place them on the bulletin board. Cut out at least ten black circular shapes from construction paper to use as the spots for the giraffe. Place a dot-pattern card on the rabbit and a numeral card on the giraffe face. Ask children to go to the bulletin board and place on the floppy ear of the rabbit the numeral card which matches the dot-pattern card and pin spots on the giraffe neck as specified by the numeral card on the giraffe's face.

8. Make a deck of forty cards with four cards in each "book" which consists of the nonphonetic numeral, the number word, a pattern of dots, and a set of objects, e.g.,

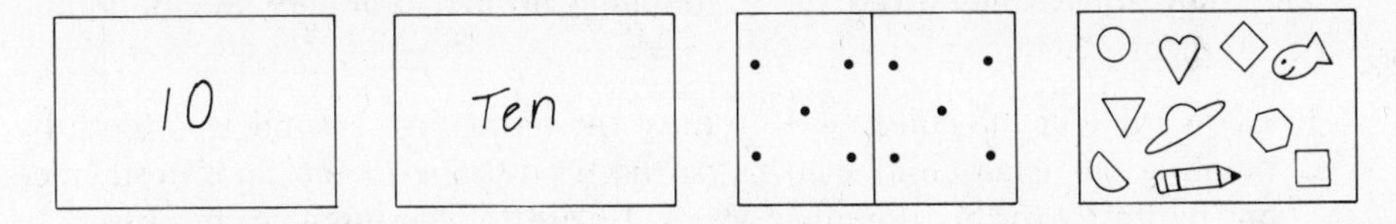

Similar "books" are made for 1–9. Ask an individual child to group the cards into ten sets of four members, or let two or three children play a game having the same rules as the "Go Fishing" card game in which they make "books" of four. If the game goes too slowly for young children when four cards are required for a "book," alter the rules and allow any two cards associated with the same number to constitute a "book." The player with the most "books" wins.

Teachers should be cognizant of the tendency of many children to continue to count to find the cardinal number of a set. In beginning experiences, counting to find the number of elements is desirable, but a child must learn to quickly identify the cardinal number of a set having up to five or six elements. Beyond that, some form of counting will continue to be utilized. To assist a young child in recognizing quickly the cardinal number of small-number sets, provide pattern arrangements such as  and

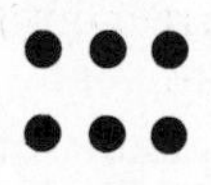

using dominoes, teacher-constructed domino cards, or ordinary playing cards for students to quickly tell the associated cardinal number. Hold up a large domino pattern card for a brief time for children to record the associated numeral. When they can respond quickly and correctly, review the equivalence of dot-pattern sets and sets with elements in various configurations, e.g.,

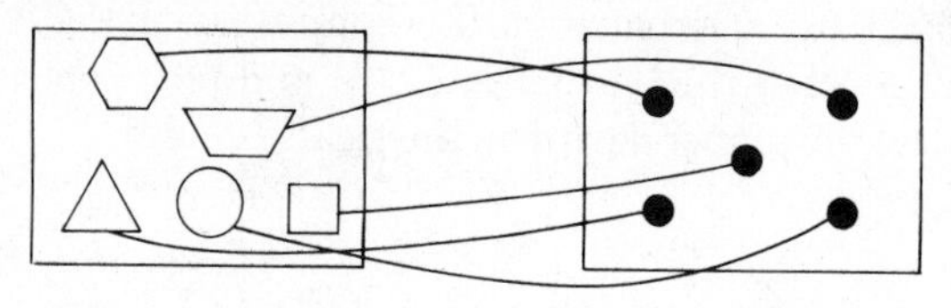

so they can determine at a glance the cardinal number of any small-number set.

Identifying the Ordinal Number of an Element in a Series

Ordinal number refers to the position of an element in a series whereas cardinal number tells how many elements there are in a set. Young children need experiences in both uses of number.† After a child can order events and objects, count, and write and identify numerals, he is ready to study ordinal number.

These are some suggestions for introducing ordinal numbers and providing practice in them:

1. Form a line of ten children. Introduce the terms first, second, . . ., tenth by pointing out each child who holds those positions in the line. Pronounce and write the ordinal-number terms. Relate the terminology to objects in the classroom. For example, ask, "Which table is the fifth one?" "Who is sitting at the first table?" "What is on the third desk in this row?" Include questions that involve both cardinal and ordinal numbers, e.g., "How many children are sitting at the second table?"
2. Reinforce the ordinal-number concept in everyday experiences and other subject areas: "Who went first to lunch today?" "Will the children at the fourth table go sit on the rug?" "What is the third letter of the alphabet?" "Who was the first President of the United States?" Incidental teaching greatly reduces the number of structured activities required for teaching ordinal use of number.

† Number can also be used for identification purposes. Telephone and car-tag numbers are examples of this use of number.

3. Many children learn ordinal words through poems and songs. Here is an example that involves both cardinal and ordinal uses of number:

4. Involve the children in active games for practice. Have them form a line, and lead or ask a child to lead a game with directives such as "Let's see the third child raise both arms high, the fifth and sixth children shake hands, and the eighth child bend over."
5. Give children a duplicated handout of pictures (in random order) representing occurrences from a story. Read the story to them and have them write by the pictures the ordinal-number names (first, second, and so on) that correspond to the sequence in the story.
6. Have children place on their notebook paper a "d" on the third line, an "i" on the fourth line, an "o" on the first line, an "r" on the second line, an "n" on the fifth line, an "l" on the seventh line, and an "a" on the sixth line. They will have spelled "ordinal" if they have placed the letters correctly.
7. Provide connect-the-dots sheets for children to work. These pages can be copied on sanded wood with carpenter's staples serving as dots. Attach a piece of yarn by the "1" in the picture and let the child put the yarn through the staples to complete the connect-the-dots picture. This manipulation promotes the development of eye-hand coordination in addition to giving the student practice in ordering numbers.

8. Demonstrate cardinal and ordinal uses of number on the number line.

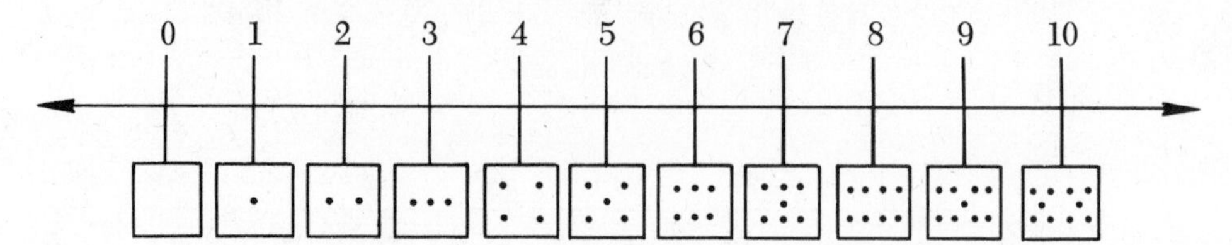

9. Children should learn to identify the ordinal number of an element in a row, column, and array and find an object when given its ordinality. Give the children picture cards or make worksheets showing rows, columns, and arrays of objects.

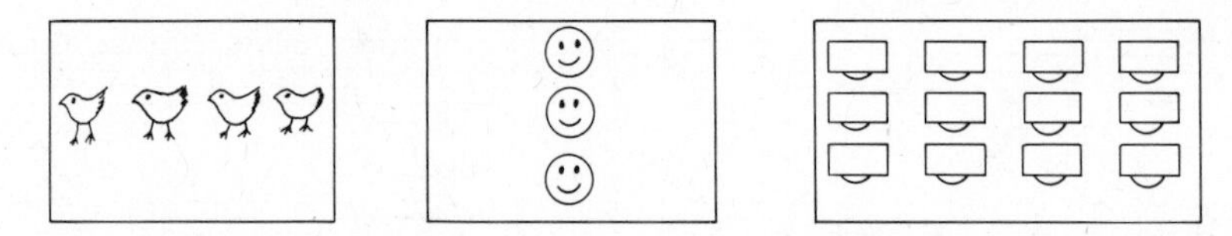

Using the first card, review the terms first, second, third, and fourth. Then ask the children to color the third bird blue, the second red, and so on. The second card can be used similarly. On the third card is a picture of rows and columns of desks. Draw the arrangement on the chalkboard or an overhead-projector transparency. Ask how many desks there are in the first (second, third) row. Suggest the rows by moving your hand horizontally. Next have the children tell the number of desks in each column, and gesture with a vertical stroke of your hand. Ask different children to point to the third row, first row, second column, and fourth column. If they can do this task, point to a specific desk and have them identify its position by using ordinal-number language, e.g., the desk in the second row and third column. Continue the activity by having them mark the desks in the positions you specify and also describe the position of desks to which you point or one of the children points.

Uniting, or Joining, Sets

Joining sets provides readiness for addition and multiplication of cardinal numbers. Children in kindergarten and first grade usually depict sets of discrete objects and their union by arrangements such as these:

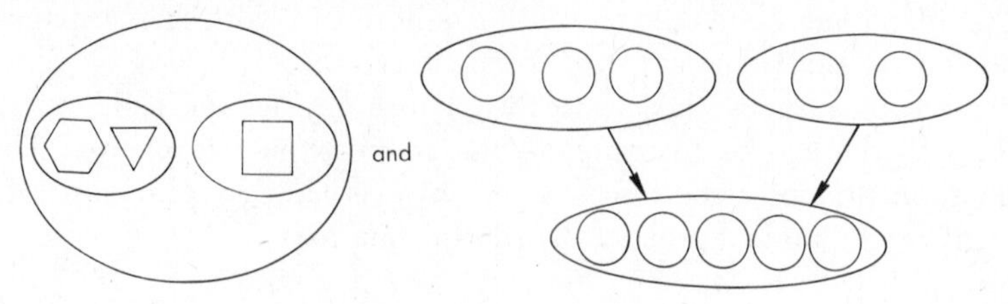

They should understand the union of disjoint sets in that it furnishes a model for addition and multiplication before beginning to unite overlapping sets. The symbol "∪" for set union usually does not occur until third grade or later.

Attribute blocks, plastic chips, and a counting frame can be used to teach children to join sets of discrete objects. The children themselves can also be used as members of sets. Have them stand in hula-hoops or chalk or rope circles on the floor, and select children to name the members of each original set and the joined set.

Children also need experience in uniting sets of continuous objects. Cuisenaire rods can be used for this purpose. Have each child select two rods, place them end to end to form a train, and place the rod that exactly matches the length of the train alongside the train. The work of a child who selects the red and green rods is suggested here:

red	green
yellow	

Separating Sets

Separating a set into disjoint subsets is prerequisite to studying the subtraction and division of cardinal numbers. To prepare children for the inverse relationship of addition and subtraction and also multiplication and division, you as teacher should design some experiences in which children "put together" and "take apart" sets in the same activity.

Attribute blocks are useful in activities dealing with set separation in that they can be partitioned into subsets, each of which contains members of the same size, shape, or color. Buttons, crayons, plastic counters, and similar objects are also serviceable in this instructional task. You should also use everyday happenings as examples of set separation: a child brings a set of cookies to school and distributes them among the class members; you pass out a set of materials to the children; and children in the class separate into groups to work on a mathematics activity.

As children separate sets, encourage them to compare their responses. Call their attention to the number of subsets each student forms and the number of elements in each subset. Have them consider whether or not the set can be separated into two equivalent subsets, three equivalent subsets, and so on. One worthwhile instructional activity involves having the students separate a set of plastic counters into exactly two subsets. Note different responses of the children and record the results. If they overlook some possibilities, coax them to seek other ways to separate the set. A set of four counters can be separated into two subsets in these ways:

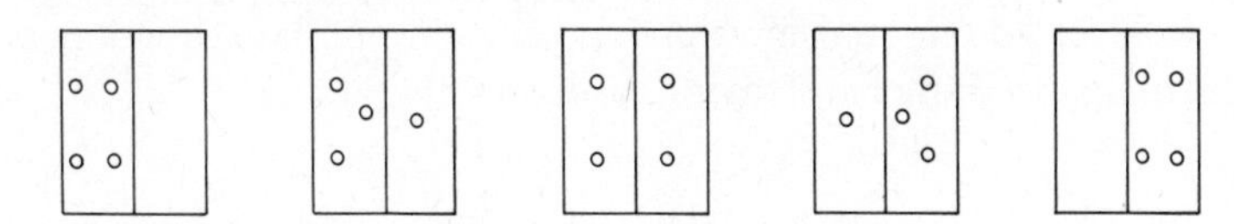

This experience should help children later rename a sum using several different addends.

Using Logical Words

Logical terms of "not," "and," "or," "all," "some," and "none" should be incorporated into activities on sets, subsets, and classification. As children work with attribute blocks, stress logical terminology: *some* of the yellow blocks are small and some are large; some of the square blocks are green; *none* of the square blocks is circular; *all* of the blocks are made of wood. When a child selects all of the blue blocks, point out that those he did not choose are *not* blue. Ask questions dealing with two or three attributes: "Which is the set of blue *and* circular blocks?" "Which block is blue, circular, and large?" "Which are small *or* square?" Many young children will not be able to grasp the concept of "or," but you should use it in informal situations whenever you can.

Be alert to opportunities to reinforce the use of logical words. These are some considerations: "Did all of the children wear raincoats today? Some, but not all, of us ride buses to school. None of us attends school on Saturdays. How many of you brought the materials for your science experiment today? Some, but not all, of you did. Who has not been absent this year? Stephen and Joanna can be group leaders today. We should eat a green or yellow vegetable every day. Which is the correct symbol for 5◯7—less than ($<$) or greater than ($>$)?"

Closing

The prenumber and number concepts and skills discussed in this chapter are generally appropriate for children of ages two through five. These ideas are reviewed and extended and teacher expectations may be upgraded in grades one, two, and three. Teachers should help children unify these experiences and apply them to new learning situations. For example, a child must deal with one-to-one correspondence, ordering elements of a set, and counting in order to find the cardinal number of set. In turn, finding the cardinal number of a set together with joining sets, recognizing numerals and operation signs, and finding representative sets enter into the tasks of adding and multiplying cardinal numbers. Beginners separate sets when learning to subtract and divide cardinal numbers.

Notes

1. Mary Baratta Lorton, *Workjobs* (Menlo Park, California: Addison-Wesley Publishing Company, 1972), pp. 82–83.
2. Richard W. Copeland, *How Children Learn Mathematics*, 2nd ed. (New York: Macmillan Publishing Company, Inc., 1974), p. 85.
3. Leonard M. Kennedy, *Guiding Children to Mathematical Discovery*, 2nd ed. (Belmont, California: Wadsworth Publishing Company, Inc., 1975), p. 75.

Assignment

1. Study a set of attribute materials and accompanying activity cards and teacher's commentary. Write a guided-discovery lesson plan using these materials for one of the topics:
   a) finding subsets
   b) classifying
   c) discovering patterns
   d) determining if sets are equivalent (concepts of equivalent, more than, fewer than)
   e) joining sets
   f) separating sets
2. Design a set of attribute materials different from the commercial ones you have examined in problem 1.
3. Make a list of ordinary materials appropriate for teaching preschool and kindergarten children prenumber and number concepts and skills. Write a brief activity for each concept for which your materials are suitable.
4. Examine a set of Parquetry blocks and materials. For what topics in this chapter are these materials helpful?
5. Talk to two or three kindergarten or preschool teachers about the prenumber and number ideas with which their children have the most difficulty and the least difficulty. Try to help a child who is having trouble with some prenumber or number aspect of the curriculum. List activities that seem to be beneficial to the child.
6. Create a laboratory activity dealing with cardinal number, ordinal number, or both cardinal and ordinal numbers. Provide for at least five centers.
7. Read and list teaching ideas in these articles of *The Arithmetic Teacher:*
   a) "Simple Materials for Teaching Early Number Concepts to Trainable-level Mentally Retarded Pupils" by Jenny R. Armstrong and Harold Schmidt, February 1972, pp. 149–53,
   b) "Sorting, Classifying, and Logic" by Douglas E. Cruikshank, November 1974, pp. 588–98,
   c) "Teacher-made Materials for Teaching Number and Counting" by Helene Silverman, October 1972, pp. 431–33.

Laboratory Session

Preparation: Study chapter 5 of this textbook.

Objectives: Each student should

1. use attribute blocks to investigate equivalent and nonequivalent sets, union, interesction, and complementation of sets,
2. use the inductive approach to write a generalization for finding the number of subsets of a given set,
3. relate set operations to addition and subtraction of whole numbers, and
4. relate objectives 1–3 to teaching children mathematics.

Materials: For each group, a set of attribute blocks, 5 or 6 yarn strips

Procedure: Students should work in groups of two or three on these problems:

1. a) Illustrate two equivalent sets with the blocks. Verify that the sets are equivalent by demonstrating a one-to-one correspondence between them with the strips of yarn.
   b) Use sets of attribute blocks and the yarn strips to show how you would help a preschool or kindergarten child determine when a set had more or fewer members than another set.
2. Let A be the set of all large circular blocks and B be the set of all square blocks.
   a) Do A and B have any blocks in commons? (i.e., Do they intersect?)
   b) Place a loop around A, another around B, and another around $A \cup B$. How many elements are in A, in B, and in $A \cup B$?
   c) For what important concept in the early childhood mathematics curriculum do (a) and (b) provide readiness?
3. Let A be the set of all green blocks and B be the set of all small green blocks. Place loops around A and around B.
   a) How are A and B related?
   b) How many elements are in A; in B?
   c) If the elements of B are removed, how many elements are left in A?
   d) Parts (a)-(c) serve as readiness for what concept in the mathematics curriculum?
4. a) Let A be the set of all small, blue blocks and B be the set of all small, yellow blocks. Place loops around A, B, $A \cup B$, and $B \cup A$. Are $A \cup B$ and $B \cup A$ equal? What property of addition do you think is based on this property of sets?
   b) Let A be the set of all small, circular blocks; B be the set of all small, square blocks; and C be the set of all large, blue, triangular blocks. Place loops around A, B, C, $B \cup C$, $A \cup B$, $A \cup (B \cup C)$, and $(A \cup B) \cup C$. Does $A \cup (B \cup C) = (A \cup B) \cup C$? What property of addition do you think is based on this property of sets?
5. It was mentioned in this chapter that it is worthwhile for certain gifted young students to explore the pattern for generating all the subsets of a given set. See if you can discover the pattern by considering these examples.
   a) Form a set having one attribute block. How many subsets does this set have? Describe the subsets.
   b) Form a set having two attribute blocks. How many subsets does this set have? Describe them.

c) Form a three-element set of blocks. How many subsets does this set have? Describe each subset.
d) If a set has no element (empty set), how many subsets does it have? What is the subset?
e) Summarize the results of (a)-(d), extend the pattern, and make a generalization in this table.

Number of elements in the set	0	1	2	3	4	. . .	n	. . .
Number of subsets of the set								

Bibliography

Copeland, Richard W. *How Children Learn Mathematics,* 2nd ed. New York: Macmillan Publishing Company, Inc., 1974.

Jungst, Dale G. *Elementary Mathematics Methods: Laboratory Manual.* Boston: Allyn and Bacon, Inc., 1975.

Kennedy, Leonard M. *Guiding Children to Mathematical Discovery,* 2nd ed. Belmont, California: Wadsworth Publishing Company, Inc., 1975.

Lorton, Mary Baratta. *Workjobs.* Menlo Park, California: Addison-Wesley Publishing Company, 1972.

Teacher's Guide for Attribute Games and Problems. New York: Webster Division, McGraw-Hill Book Company, 1974.

Numeration

In addition to learning what number means, children must learn to write numerals that represent numbers. A numeration system is a system for naming numbers. This chapter focuses on materials and techniques for helping students understand our own numeration system, the Hindu-Arabic system. The Hindus are given credit for inventing it, and the Arabs spread its use to the Western world. Keeping the following features in mind should assist you in guiding your students to develop a clear understanding of the Hindu-Arabic system of numeration:

1. It is a base-ten (decimal) system. Grouping is based on tens, e.g., 28 means 2 groups of ten and 8 ones.
2. It is a place-value (positional) system. The position of each digit influences the total value of the number represented. For example, 54 means 5 tens + 4 ones but 45 is 4 tens + 5 ones. In a whole number, the extreme right position represents ones, and each position to the left of the ones place represents ten times the value of the position to its immediate right. The place-value pattern is suggested here:

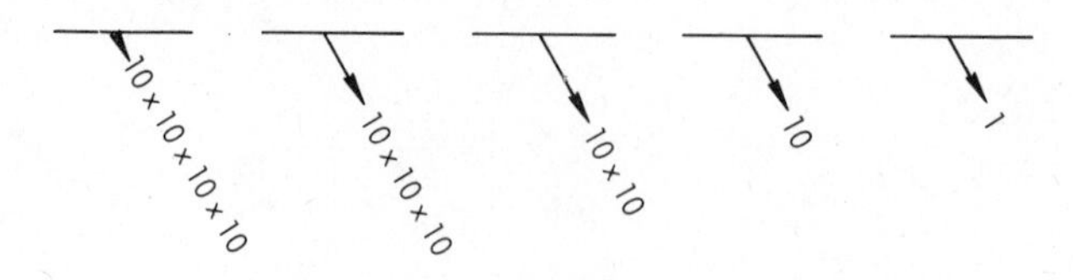

3. There is a zero symbol to indicate the absence of a group of ten. For instance, 805 means 8 hundreds, no tens, 5 ones. The ten symbols 0 through 9 serve to represent any number.
4. Our system is partially additive. Each digit (face value) is multiplied by the corresponding place value, and the resulting product is the total value of the digit. Addition is explicitly shown when the standard numeral 263 is written in expanded notation: $263 = 2 \times 100 + 6 \times 10 + 3 \times 1$. Pay special attention to the vocabulary of face, place, and total value and standard and expanded form of numerals.
5. Our system has a multiplicative feature as can be seen in the expanded form in the preceding example.

Place Value

The topic of place value is usually introduced in first grade and reinforced and extended in succeeding grades. It is basic to our numeration system and to the whole-number algorithms which occur in second and third grades. Several commercial structured materials such as base-ten blocks; Unifix cubes; Cuisenaire rods, squares, and blocks; and an abacus are helpful in teaching place value. These were described in chapter 3.

Everyday, teacher-made materials including a pocket chart, rectangular strips and squares, beansticks, plastic counters, and money are serviceable also. A pocket chart can be made by stapling manila envelopes to poster board and labeling the envelopes as Ones, Tens, and Hundreds. (Extend the chart as place value is expanded.)

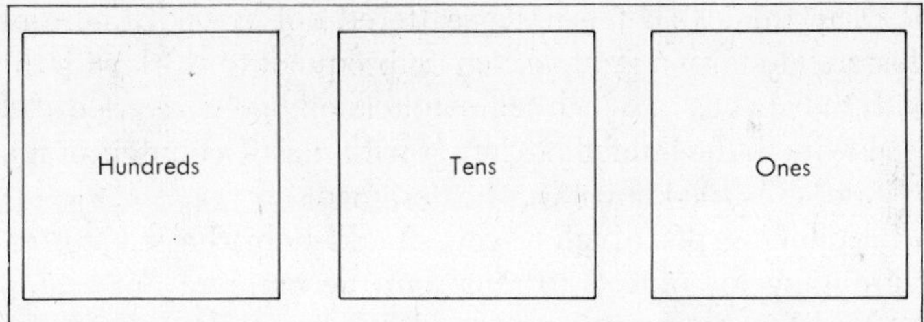

Rectangular strips and squares are a paper model of the unit, long, and flat base-ten blocks. Small squares, strips of ten squares, and large squares consisting of one-hundred units can be cut from grid paper to represent ones, tens, and hundreds. This is a model of 125:

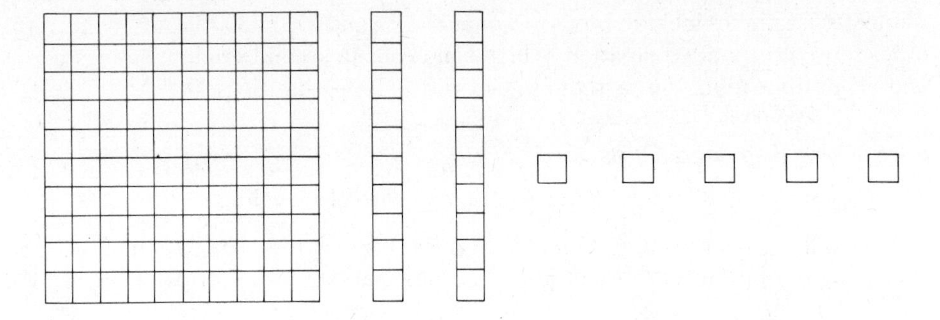

Using grid paper of square-centimeter units allows convenient coordination of the metric system and place value. Beansticks, which have proven successful in many classrooms, can be constructed with dried beans, popsicle sticks, and glue. Glue one bean on a stick for "1," two beans for "2," . . ., ten beans for "10." To form a hundreds stick, place ten of the tens sticks side by side, securing them in place by gluing two popsicle sticks across the back.[1] This picture shows sticks representing a hundred, ten, and four:

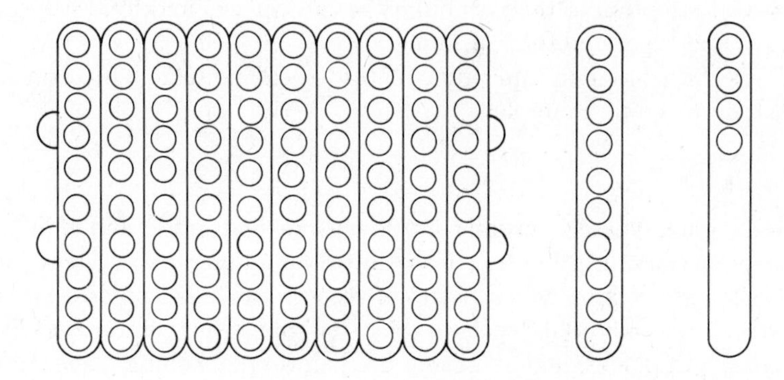

Bundles of straws or popsicle sticks in a pocket chart, base-ten blocks, rectangular strips and squares, beansticks, or some similar concrete aid in which ten ones can literally be seen in a group of ten should be used as the initial device for presenting place value. Also, a prerequisite to understanding work with these materials is the ability to conserve discrete quantity. If a student lacks this understanding, he may think that ten units scattered out are not the same quantity as ten units clustered to form a group of ten. Subsequent to work with this type of aid, activities with the abacus in which ten counters on one wire are exchanged for one counter on the wire to the immediate left or with plastic counters or money in which equivalent values must be known can be designed.

When introducing place value, you should help the student realize the advantage of grouping the objects to represent the number. For example, the Xs in XXXXXXXXXXXX must be counted to determine there are 12, but if they are grouped by tens as in (XXXXXXXXXX) XX, it is easy to see there are 12 objects. Students need to be able to change from the ungrouped form to the grouped form and vice versa. In first grade, children usually study the basic facts of addition in which the sum is ten or less before beginning work with place value. Notation for addition should be used in symbolic exercises with expanded notation and standard numerals. Examples include 2 tens + 8 ones = ____ and 372 = 300 + ____ + ____. Another form of expanded notation, which can occur in second grade when a student learns multiplication, is illustrated in $643 = 6 \times 100 + 4 \times 10 + 3 \times 1$.

Introducing Tens and Ones

These are some ideas for beginning experiences with place value:

1. Make a pocket chart of tens and ones. Extend the chart to hundreds and thousands later. You will also need nineteen straws or strips of paper long

enough to extend out the top of the pockets and one rubber band. Have the students count orally from one through nine with you as one child writes the numerals on the chalkboard and you place the corresponding number of straws in the ones pocket. Tell the students they are going to look at 10 through 19 in a new way. Make a chart of tens and ones on the chalkboard to correspond with the tens and ones on the pocket chart. Add a straw to the nine straws in the ones pocket, withdraw all ten, wrap a rubber band around them, and place the bundle of ten in the tens pocket. Tell the children that ten ones represent the same number as one ten. Record 1 in the tens column and 0 in the ones column on the chalkboard and then record the standard numeral 10 below the numerals 1 through 9 the student has written on the chalkboard. Put another straw in the ones pocket and point out that now there is 1 group of ten and 1 one. Record 11 accordingly on the chalkboard. Continue in this way through 19. The following picture shows 15 on the pocket chart and the chalkboard chart:

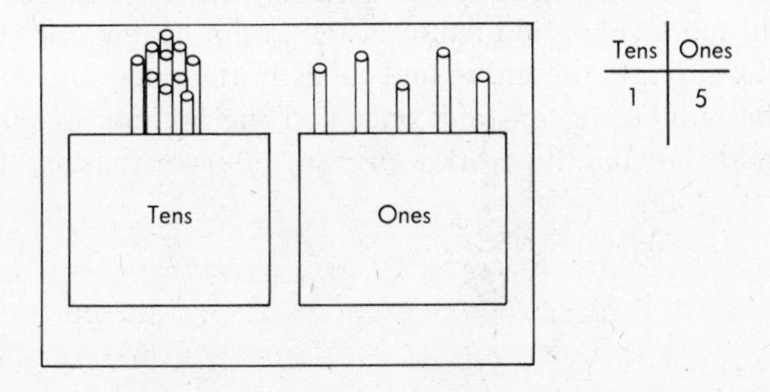

Tens	Ones
1	5

2. Present each student with 20, 30, or 40 plastic counters, five or six yarn loops, and numeral cards for making [1][0] at least four times and for making [2][0], [3][0], and [4][0]. Ask the children to make as many subsets of ten as they can, enclose the subsets of ten in the yarn loops, and place the correct numeral cards by each subset of ten and the entire set of counters. A student having thirty counters may show his work this way:

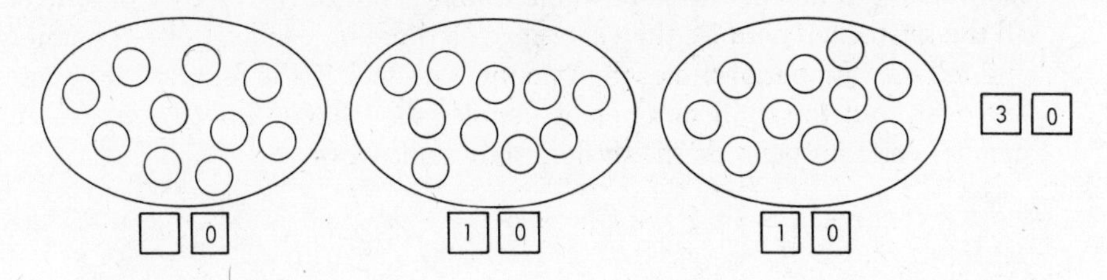

3. Give each student a multiple of ten dried beans and nine or ten of the tens beansticks. Instruct the children to make all the sets of ten they can with the beans and place a tens beanstick by each set of ten. Ask each student to tell the number of tens and the number of beans he has. For example, the child showing the following work may say, "I have 3 tens. That's the same as 30."

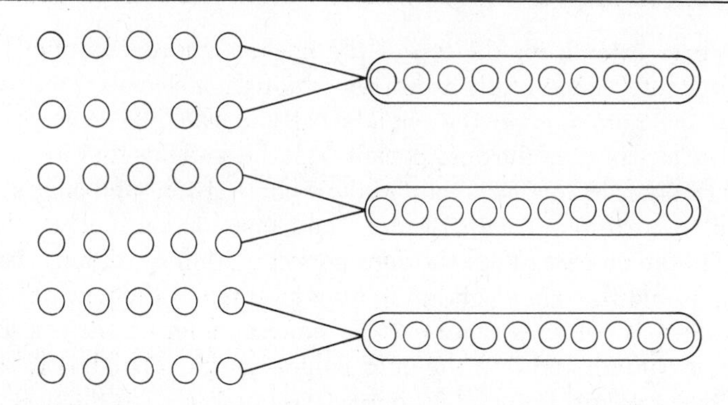

After students can deal with the multiples of ten, give them beans to group by tens and ones.

4. To each child, distribute between twenty and fifty popsicle sticks, five plastic-bag ties, and tens and ones place-value cups made by stapling together and labeling plastic drinking cups. Instruct each student to make all the sets of ten he can, tie together each group of ten, and place sets of ten in the tens cup and the individual sticks in the ones cup. Ask each student to tell the number of tens and ones and the number of popsicle sticks he has. A child showing this work would say he has 2 tens and 6 ones, or 26:

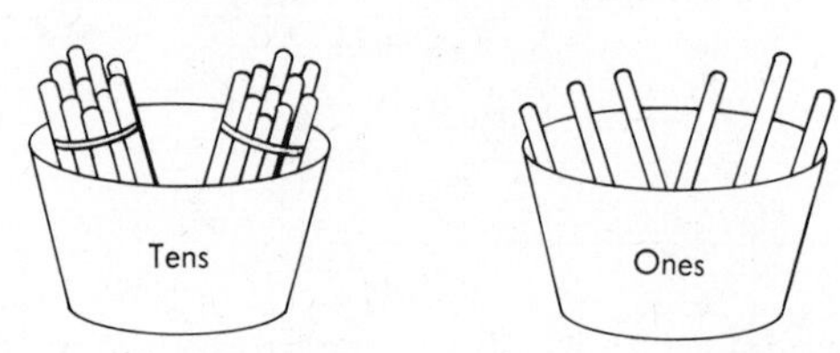

A similar activity can be done using interlocking plastic centimeter cubes or Unifix cubes. Students make sets of ten by hooking cubes together.

5. Distribute to each child between twenty and forty cut-outs of pennies and four or five cut-outs of dimes. (Play money can be obtained from most educational supply houses.) Ask the students to count out ten pennies. Tell them that a dime has the same value as ten pennies. Instruct them to make all the sets of ten pennies they can, place a dime by each set of ten pennies, and tell the number of dimes their pennies are worth, the number of pennies left over, and the total number of pennies they have. For two sets of ten pennies and four pennies, the student shows this work

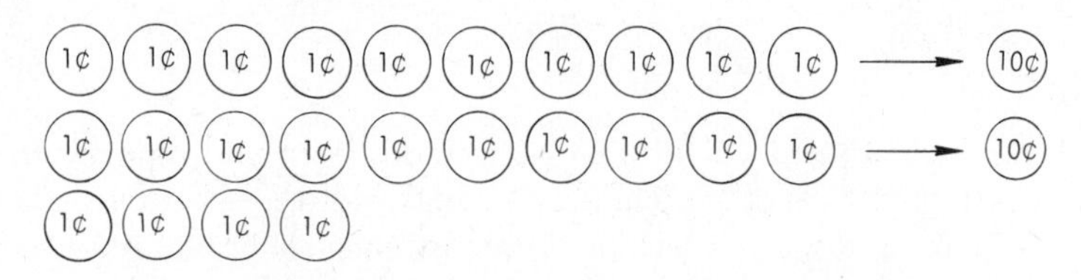

and says, "My 24 pennies are the same as 2 dimes and 4 pennies."

6. Give each child 25 to 30 white Cuisenaire rods and 3 or 4 orange rods. For activities dealing with 100s and 1000s, use the Cuisenaire squares and cubes.

Tell the students each white rod is one unit. Ask them to find the number of white rods required to match the length of an orange rod. Reinforce the fact that 10 ones = 1 ten. Give the children a worksheet or task cards with problems such as the following:

a) Represent 23 with the Cuisenaire rods.
b) Show 1 ten and 9 ones with the rods.
c) Fill in the blanks: 64 = ____ tens, ____ ones.

Encourage the students to check each other's work and create their own problems.

You may coordinate the metric system of measurement with the teaching of place value. For example, one white Cuisenaire rod is one centimeter long and the orange one is ten centimeters, or one decimeter, long. Furthermore, students may line up decimeter rods alongside a meter stick to determine there are ten decimeters in one meter. Thus a model for ones, tens, and hundreds is readily found in measurement of length.

7. After students have concrete experience with place value, have them group objects in a picture and write the numeral by each picture. An example may look like this:

Help students understand that the number of objects in each set can be found by never counting beyond ten: count to ten, group, count to ten again, group again, and continue in this way until all the objects are exhausted.

8. Concepts often seem less formidable when they appear in story situations. Create a story involving funny characters such as a Tens Clown and a Ones Clown. Tell the students that the Tens Clown must always line up to the left of the Ones Clown as they face us in a parade and that each clown holds a certain number of balloons. Have the students count the balloons and tell the number the clowns represent. For example,

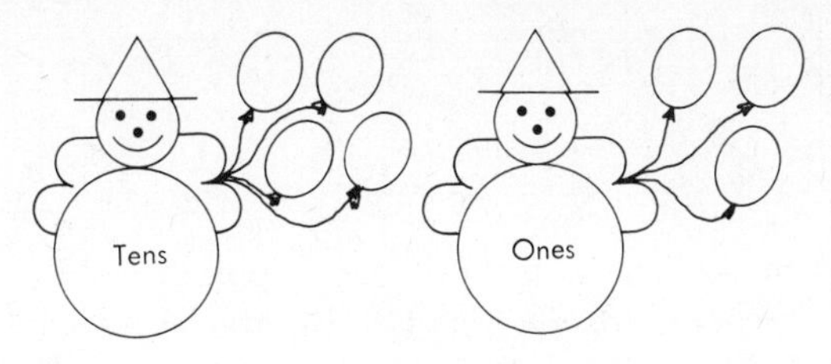

means 4 tens and 3 ones, or 43. This idea can be extended to hundreds and thousands.

9. Have students compare the repetitive pattern in decades of numerals.

   1 2 3 4 5 6 7 8 9 10
   11 12 13 14 15 16 17 18 19 20
   .
   .
   .
   91 92 93 94 95 96 97 98 99 100

   Provide exercises in which children name the number or numbers occurring before, after, or between given numbers, e.g.:

   ____ 39 ____ and ____ 70 ____ 73.

10. Give students symbolic exercises in expanded notation.

    63 = ____ tens, 3 ones
    42 = 4 tens, ____ ones
    57 = ____ tens, ____ ones
    ____ = 2 tens, 9 ones
    ____ = 8 tens, 0 ones

Introducing Hundreds

When students understand place value in the numerals representing counting numbers through 99 and have worked with addition and subtraction, they should be introduced to hundreds. This task usually begins in first grade. Children should learn that one hundred is ten tens, or a hundred ones. Base-ten blocks, beansticks, money, and other place-value devices can be used to introduce this level of numeration. For example, you may represent one hundred with beansticks and ask the children to tell the number of tens and ones contained in one hundred. Then record 100 on the chalkboard. Represent and record 200, 300, . . ., 900. After students understand counting by hundreds, present them with a certain number of hundreds, tens, and ones and have them tell the number of each represented. For the example

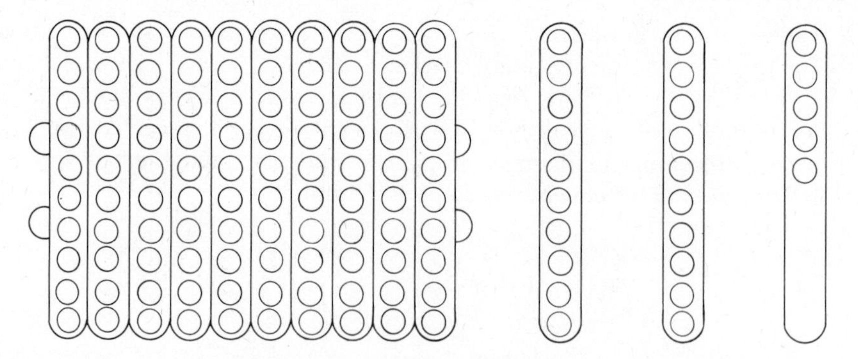

record 125 on the chalkboard as the children respond, "one hundred, two tens, five ones." Stress that the hundreds digit is placed to the left of the tens digit.

As numerals are recorded and numbers are represented, have students compare numbers: "Which is the largest number we have represented today?" "Which is the smallest?" "How can you tell?" Include examples such as 342 and 297 in which case the students must look at the hundreds place to decide 342 is greater than 297 and 643 and 687 where they must examine the tens place to determine 643 is less than 687.

When students can write numerals associated with objects, provide exercises in which they must write the numerals in expanded notation. Give the open sentences in various forms.

629 = _____ hundreds, 2 tens, 9 ones
205 = _____ hundreds, 0 tens, _____ ones
674 = _____ hundreds, _____ tens, _____ ones
_____ = 3 hundreds, 8 tens, 5 ones
258 = 200 + _____ + _____
350 = _____ + _____ + 0
758 = _____ + _____ + _____
_____ = 200 + 40 + 5

Other ideas for teaching place value in the hundreds include:

1. Let children play a trade-in game with the base-ten blocks. (Rectangular strips and squares, plastic counters, and money would work also.) Have them separate into groups of three or four and designate a "banker" in each group. Each player rolls a die and collects from the "banker" the number shown on the die. Each time a player accumulates ten units, he must exchange them at the "bank" for one long and when he has ten longs, he trades them for one flat. The first child to earn a flat base-ten block is the winner of the game.
2. Design an activity in which numbers are represented on an abacus. Demonstrate that when ten counters occur on the ones wire, they should be exchanged for one counter on the tens wire, and ten counters on the tens wire must be replaced by one counter on the hundreds wire. Have students count by tens and hundreds on the abacus, determine the number represented by abacus examples you give them, and represent on the abacus numbers you specify. For follow-up work on a picture abacus, include problems such as these:
   a) What number is represented on this abacus?

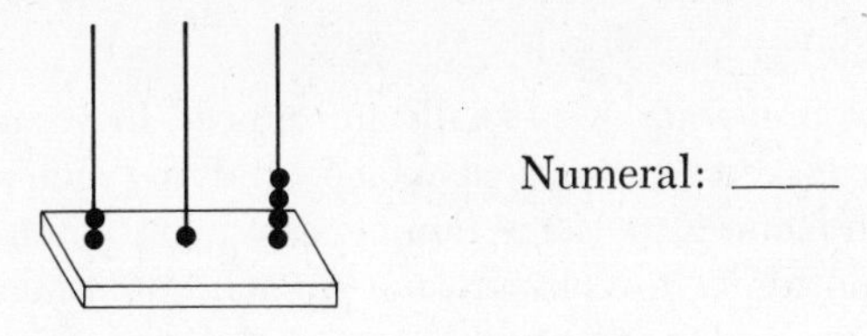

   b) Show 302 entered on the picture abacus.

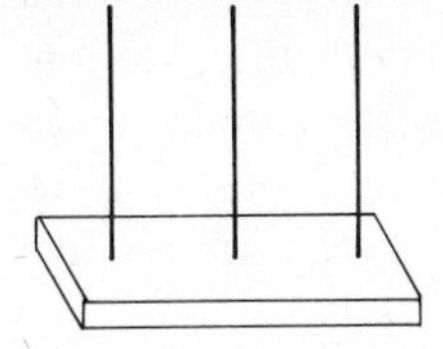

c) What is wrong with the first picture? Correct it on the picture abacus on the right.

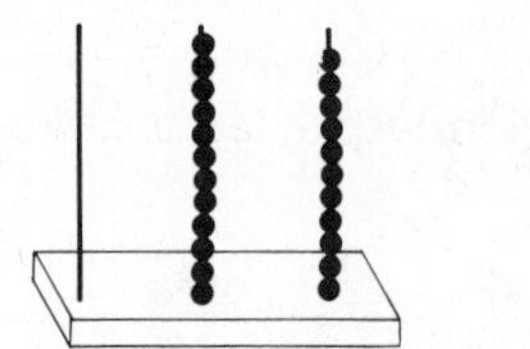
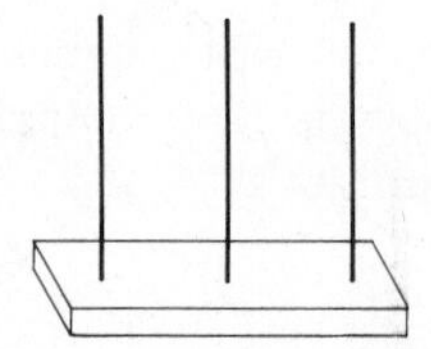

3. Give each student fourteen or fifteen of each of three colors of plastic counters. Explain that each color represents a different value, e.g., blue represents ones; red, tens; and yellow, hundreds. Tell the children ten blue counters have the same value as one red one and ten red counters are equivalent in value to one yellow one. Make a chart on the chalkboard for each child to copy on a sheet of notebook paper or distribute playing sheets.

Hundreds (yellow)	Tens (red)	Ones (blue)

Establish with the students that when representing final answers on this device there should never be more than nine counters of the same color. Describe problem situations for the students to make exchanges and to tell the number represented. For example, have them place thirteen red counters in the tens column and five blue counters in the ones column, make the proper exchange, and read the numeral from the chart. Call on students to describe their work. Provide an opportunity for them to create their own examples. The commercial *Chip-trading Activities* (see Appendix) includes many place-value activities.
4. Have students enter numerals on the pocket calculator, and ask questions on place value, face value, and total value.

Roman Numerals

The Roman system of numeration is usually introduced in second or third grade and is often tied to the task of reading a clockface. Studying such a system broadens a child's experience in naming the same numbers by many different numerals and gives students an opportunity to compare our numeration system to another one and to consequently appreciate the positive features of our own system.

Of the Roman symbols I, V, X, L, C, D, and M which are equivalent to 1, 5, 10, 50, 100, 500, and 1000, respectively, young children usually work with I, V, X, L, and C. Typical objectives include having students write Roman-numeral equivalents for our numerals and translate Roman numerals into our numerals. Guide children to learn that the Roman system (1) is repetitive as seen in the example XXX which is equivalent to 30, (2) is additive, e.g., XVI means X + V + 1, or 16,

(3) has a subtractive feature since IV means V − I, IX is X − I, XL means L − X, and XC is C − X, (4) has no zero (if none of a certain symbol is needed, that symbol is omitted), and (5) is not a place-value system, e.g., X means 10 regardless of its position in the numeral. The repetitive, additive, and subtractive features are all utilized to represent the Roman equivalent of 74: LXXIV.

When teaching Roman numerals, have students look for ways the numerals are used in their environment, e.g., in books and on clocks and buildings. Children who may benefit from a historical investigation of Roman numerals should be encouraged to do independent or small-group work. This enrichment is especially appropriate for gifted students. Also, activities with Roman numerals can easily be integrated with social studies.

Closing

Understanding the characteristics of our numeration system, the Hindu-Arabic, is a major goal in elementary school mathematics. Work with place value, grouping by tens, and expanded notation begins in first grade and these topics are extended in subsequent years and applied in the algorithms of arithmetic. Base-ten blocks, Unifix cubes, Cuisenaire rods, squares, and blocks, and an abacus are useful commercial materials for teaching place value. Other materials for teaching place value include teacher-made aids such as pocket charts and beansticks and everyday materials such as money, popsicle sticks, and straws.

Notes

1. Leonard M. Kennedy, *Experiences for Teaching Children Mathematics* (Belmont, Calif.: Wadsworth Publishing Company, Inc., 1973), p. 40.

Assignment

1. Write a guided-discovery teaching activity for introducing place value to a group of first-grade students. If possible, try it out on a child, group of children, or an entire class.
2. Examine several second- and third-grade elementary mathematics textbooks for the attention given to Roman numerals or any other historical numeration system. Do all the series

include the topic of ancient numeration systems? Did all of those which did present the topic treat it in a similar way? How many pages were devoted to the topic in each series you examined?

3. Read and summarize teaching ideas in these articles in *The Arithmetic Teacher:*
   a) "Grouping of Objects as a Major Idea at the Primary Level" by James M. Moser, May 1971, pp. 301–305.
   b) "Eight-ring Circus: A Variation in the Teaching of Counting and Place Value" by Ethel Rinker, March 1972, pp. 209–16.
4. Create a laboratory activity appropriate for first or second grade and designed for reinforcing and extending place-value concepts.
5. Create a skill lesson which makes use of the pocket calculator to reinforce the topics of place, face, and total value in a numeral.

Laboratory Session

Preparation: Study chapter 6 of this textbook.

Objectives: Each student should:

1. use base-ten blocks, an abacus, plastic chips, and straws to represent numbers in the Hindu-Arabic numeration system,
2. illustrate expanded notation for base-ten numerals, and
3. illustrate when one number is less than or greater than another.

Materials: Base-ten blocks, abacus, plastic chips, straws, plastic-bag ties

Procedure: Students should work in small groups on these problems:

1. Place fourteen straws before you. Show the grouping and exchange necessary to show $14 = 1$ ten $+ 4$ ones.
2. Use base-ten blocks to
   a) show $24 = 2$ tens $+ 4$ ones,
   b) illustrate $562 = 5 \times 100 + 6 \times 10 + 2 \times 1$,
   c) show $999 < 1000$.
3. a) With base-ten blocks, illustrate that

   $$\begin{aligned}324 &= 3 \text{ hundreds} + 2 \text{ tens} + 4 \text{ ones}\\ &= 3 \text{ hundreds} + 1 \text{ ten} + 14 \text{ ones}\\ &= 2 \text{ hundreds} + 11 \text{ tens} + 14 \text{ ones}\end{aligned}$$

   b) For what task in the early childhood curriculum do you think a problem such as (a) serves as readiness?
4. Use an abacus to illustrate that
   a) $3042 = 3 \times 1000 + 0 \times 100 + 4 \times 10 + 2 \times 1$
   b) $58,364 = 5$ ten-thousands $+ 8$ thousands $+ 2$ hundreds $+ 15$ tens $+ 14$ ones
5. When representing numbers and illustrating place value, what advantages does an abacus offer over other concrete place-value teaching aids?

6. a) Use plastic chips (one color representing ones; another color, tens; etc.) to represent 2416.
   b) Using your representation in part (a), explain the place value, face value, and total value of the digit 4 in 2416.
   c) Use plastic chips to show 2632 > 2573.

Bibliography

Davidson, Patricia S. *Chip-trading Activities.* Fort Collins, Colorado: Scott Resources, Inc., 1975.

Jungst, Dale G. *Elementary Mathematics Methods: Laboratory Manual.* Boston: Allyn and Bacon, Inc., 1975.

Kennedy, Leonard M. *Experiences for Teaching Children Mathematics.* Belmont, California: Wadsworth Publishing Company, Inc., 1973.

Reisman, Fredricka K. *Diagnostic Teaching of Elementary School Mathematics.* Chicago: Rand McNally College Publishing Company, 1977.

7

Early Work in Addition and Subtraction of Whole Numbers

Addition and subtraction are usually introduced in first grade, and children begin computing with the final algorithms for the operations in second or third grade. Much time is spent in first, second, and third grades on developing addition and subtraction concepts and skills. This chapter deals with methods and materials for introducing addition and subtraction and their properties and contains suggestions for summarizing and practicing the basic facts of the operations.

Introducing Addition

A child is ready for addition when he can perform the prenumber and number tasks discussed in chapter 5. The ability to count, read and write numerals, determine the cardinal number of a set, construct sets that represent a given cardinal number, and join sets is particularly significant. Many of the commercial, teacher-made, and ordinary materials described for teaching prenumber, number, and numeration topics are also serviceable in teaching addition and subtraction.

Some ideas for beginning work with addition follow:

1. Represent two sets and their union on the feltboard. Point to the sets as you have individual students tell the number in each of the two original sets and the set formed by joining these two sets. Place the proper felt numerals on the feltboard. Introduce the symbol " + ," explaining that it means addition, is read as "plus," and is represented by joining two sets. Several examples such as

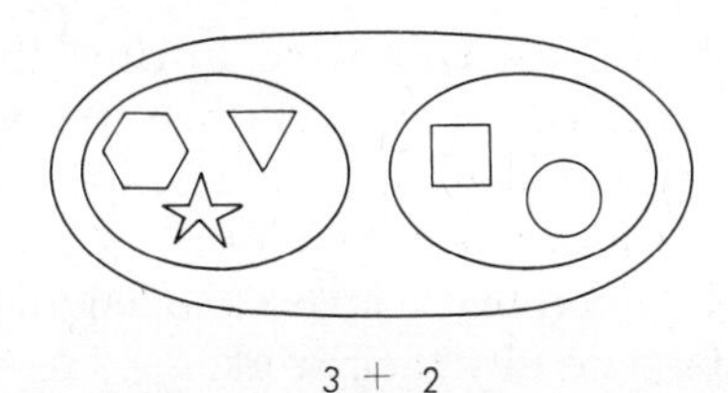

3 + 2

should be generated. Once the children can associate the set model with the correct addition expression, introduce the symbol " = ," explaining that it means "is the same as" or "is another name for." Now $3 + 2 = 5$ can be associated with the set model given above. It may help children, especially learning disabled ones, if you draw an arrow suggesting the order for reading the sentence.

$$\overrightarrow{3 + 2 = 5}$$

Write addition sentences for the other examples created in the first part of the activity.

2. Model addition sentences on an abacus. Place numeral and other symbol cards and a blank card in the chalk tray to form an open sentence, e.g., $3 + 5 = \square$. Enter three counters on the ones wire of the abacus to represent 3. Place a clothespin above the set of three before entering five more counters on the ones wire. Remove the clothespin as you tell the students the sets of three and five must be joined. Ask a student to tell the number of counters on the wire of the abacus and to place the correct numeral card over the blank card, thus producing $3 + 5 = 8$.

   Introduce the terms "addend" and "sum." In the sentence $3 + 5 = 8$, 3 and 5 are called *addends* and $3 + 5$, or 8, is the *sum* of 3 and 5.
3. Give students eight to ten cubical blocks or plastic counters, three yarn strips, and numeral and other symbol cards to make addition sentences. Have them make sets you specify and show the associated addition sentence with their cards. Also give them sentences to represent with a set model. Encourage children to create their own examples.
4. Give each child a sheet of grid paper. Instruct the students to color a certain number of squares in a row in one color and another number of squares in the same row in another color. Tell them these two sets of squares are to be joined together. Have them write the addition sentence that goes with the picture.

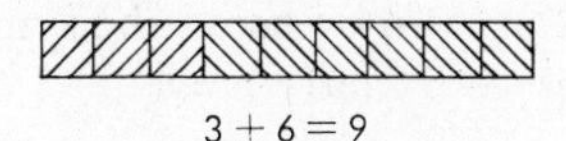

3 + 6 = 9

A similar activity can be designed using a pegboard or geoboard. Students place pegs of a certain color to represent each addend and then determine the number in the union of the two sets. On the geoboard they place rubber bands around two sets of pegs to represent the addends and another rubber band around the union of the two sets to represent the sum.

5. Give each student a set of Cuisenaire rods. Instruct the children to make a staircase having one rod of each color. Tell them the white rod represents 1 in this activity. Ask them to tell what numbers are represented by the red, green, . . . , orange rods. Instruct them to use their staircases to form a train of the 4 rod and 5 rod, find the rod that matches the length of the train, and write the addition sentence suggested by their work.

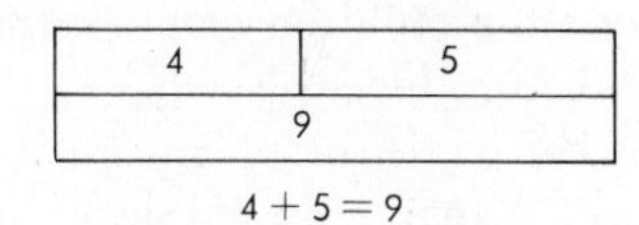

$4 + 5 = 9$

Further work on addition with Cuisenaire rods may include giving the students task cards such as the following:

> 1) Select a yellow rod and a green rod. If the white rod is "1," what whole numbers are represented by the yellow and green rods? _____
> 2) Form a train with these two rods.
> 3) Find a rod that exactly matches the length of the train. What color is it? __________ What number does it represent? _____
> 4) Write the addition sentence your work suggests: _____ + _____ = _____.
> 5) Make up a problem on addition using Cuisenaire rods for a friend to solve. Exchange problems with your friend.

Children also need experience in supplying an addend when one addend and the sum are given. Return to the Cuisenaire-rod model with the 4-rod and 5-rod train and the 9-rod equivalent. Have them remove the 5 rod. Emphasize that one rod in the train is missing as you record the associated open sentence $4 + \square = 9$. Direct the children to replace the yellow rod in the train and to tell you the numeral needed to complete the addition sentence. Next ask them to remove the 4 rod from the train and to tell what addition sentence goes with this situation. Write $\square + 5 = 9$ on the chalkboard, have the students replace the 4 rod in the train, and record the addend 4 in the frame. Point out that each of the rods in the model represents a part of an addition sentence and the missing number in the addition sentence corresponds to the rod which is missing among the three rods. Also, vary the form by using the symmetric property of equality, i.e., sometimes use $9 = \square + 5$ and $9 = 4 + \square$ instead of $\square + 5 = 9$ and $4 + \square = 9$.

6. Ask students to form a semicircle near a walk-on number line, and select students to model addition sentences on the number line. Stress the fact they must begin at 0. To solve $3 + 1 = \square$, or $\square = 3 + 1$, a child begins at 0, steps on 1, 2, and 3, pauses, and takes one more step to 4. Have a student complete the number sentence on the chalkboard. Choose several children to enact addition sentences on the number line.

7. Introduce the pictorial number line, and show the children how arrows can represent their walking on the walk-on line. When illustrating the procedure with $8 + 7 = \square$, for example,

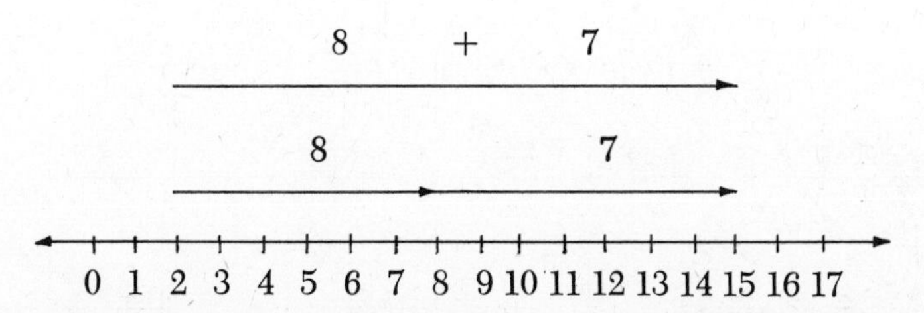

emphasize the fact there are eight steps, or units, below the "8" arrow, seven below the "7" arrow, and fifteen below the "8 + 7" arrow. When working with the number line, include examples of number-line models for which the students must write the accompanying addition sentence and open sentences which students must solve using the number line. Encourage children to create their own examples for a classmate to solve.

The number line can also be used to find the missing addend in an open addition sentence. In the example $5 + \square = 11$,

students should count the units below the dashed arrow to find the missing addend to be 6. This experience provides breadth in addition as well as readiness for subtraction.

8. When children are familiar with Cuisenaire rods, introduce them to the Papy Minicomputer. Tell them to select one rod each of the white, red, purple, and brown Cuisenaire rods. Ask the children to look at the colors of their Minicomputer boards. Note with them the colors are the same as their four Cuisenaire rods. Tell the students you want them to count from 1 through 9 on the Minicomputer using their work with Cuisenaire rods as a guide. Explain that since 1 is represented by the white rod, 1 is represented on the Minicomputer by placing a counter on the white square. Let them suggest how 2 should be entered. A counter should be placed on the red square since a red rod stands for 2. Three is represented with the rods by a train of a red and a white rod and, accordingly, it is entered on the Minicomputer by placing one counter on the red square and one counter on the white square. Continue this process until the first nine counting numbers are shown.

   Begin addition on the Minicomputer with basic facts that can be modeled on a single board. Emphasize that the basic ground rule for representing numbers on the Minicomputer is that the smallest possible number of counters must be used. Present the students with an open sentence such as $6 + 3 = \square$ and ask them to enter 6 and 3 on their Minicomputers. Guide them to exchange counters so that the fewest number of counters possible is left to indicate the final answer. The sequence of steps for solving $6 + 3 = \square$ follows: (Two different kinds of counters are used here to distinguish between the addends, but students may use the same kind of counter for both addends.)

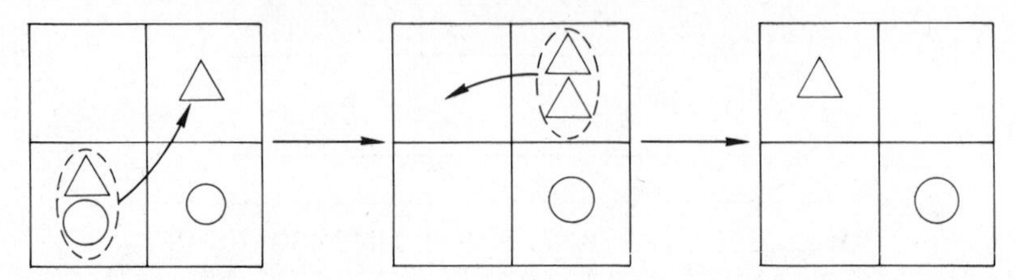

Verbalize addition facts as you guide students to exchange counters. For example, when they have entered the two addends as suggested in the first picture, say, "There are two counters on the red square. That means $2 + 2$, or 4. How should 4 be shown on the Minicomputer?" In a similar way, treat the exchange of two counters on the purple, or 4, square for one counter on the 8 square. The final picture indicates the sum of 6 and 3 is 9.

After students understand place value, have them use two Minicomputer boards. Explain that the board on the right represents ones and the other board is used to represent 10, 20, . . . , 90 in the same way 1, 2, . . . , 9 are entered on the ones board. Give them an example such as $9 + 5 = \square$ to solve.

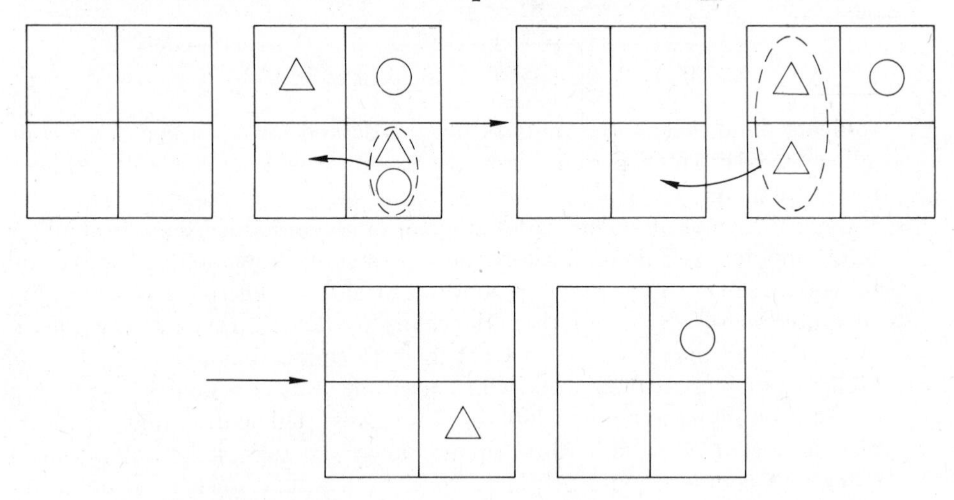

Children should read the sum as 14. Encourage them to verbalize the addition facts they use when making exchanges of counters.

9. Combine the pictorial and symbolic levels of learning by having students draw set pictures to solve addition open sentences and write addition sentences associated with set pictures. Examples may include:

a)

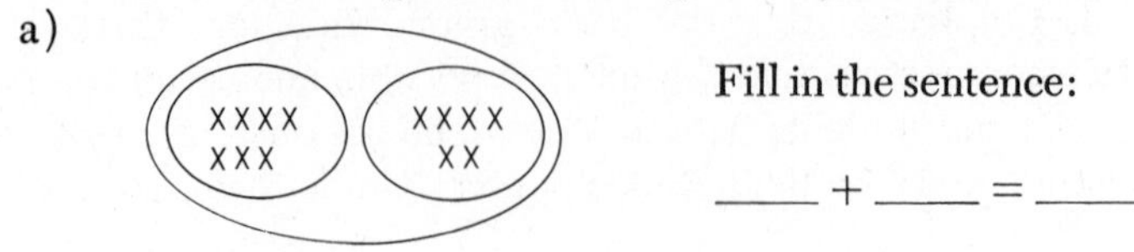

Fill in the sentence:

____ + ____ = ____

b) 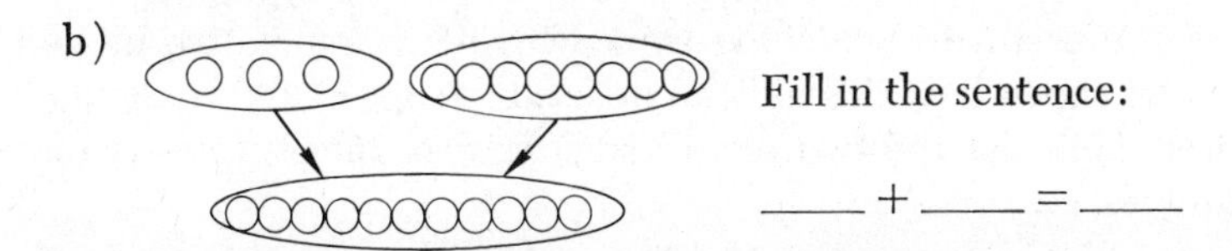

Fill in the sentence:

____ + ____ = ____

c) Draw a picture of sets to solve $8 + 9 = \square$.

10. Relate addition to everyday occurrences. As students begin an activity you might say, "Anna has two sets of pencils. There are four blue pencils in one set and seven red ones in the other. How many pencils does Anna have?" In making a jack-o'-lantern for Halloween, ask, "How many holes will we cut out for the eyes? How many will we cut out for the nose? How many will we cut out for the mouth? How many holes in all will we carve for the face?"

Properties of Addition

The properties of addition of whole numbers which are especially important at the early childhood level are the

1. *commutative property:* $a + b = b + a$, for all whole numbers *a* and *b*,
2. *associative property:* $a + (b + c) = (a + b) + c$, for all whole numbers *a*, *b*, and *c*, and
3. *identity property:* $a + 0 = a$ and $0 + a = a$, for any whole number *a*.

Guide children to explore these properties in beginning work with addition.

The Commutative Property

The commutative property should be taught while students are participating in addition activities like those described in the preceding section. A beginning activity may involve having children stand in rope loops on the floor to demonstrate that $3 + 2 = 2 + 3$, for example.

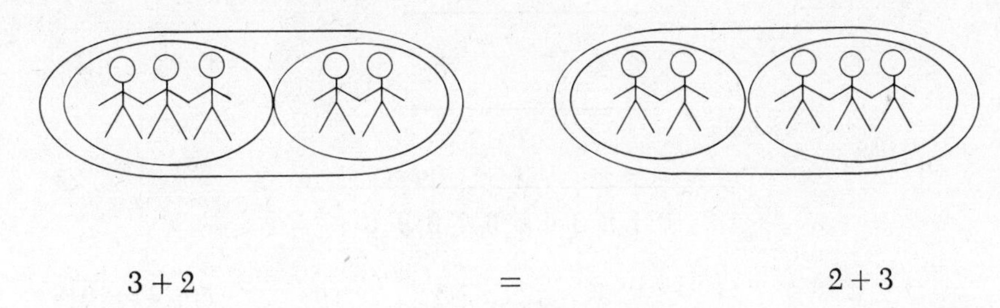

$3 + 2 \quad = \quad 2 + 3$

Have the children enact other examples such as $5 + 2$ and $2 + 5$, $6 + 0$ and $0 + 6$, and $7 + 1$ and $1 + 7$. Record their conclusions each time. Extend the pattern to situations involving larger numbers, e.g., $259 + 843$ and $843 + 259$. If they understand the pattern, they will know the two sums are equal. Since they do not yet have the skill to perform addition with large numbers, you may suggest that the students use a pocket calculator to add as a check to see if the sums are equal.

Grid paper is also useful in beginning work with the commutative property. The following picture shows how this aid can be used to suggest the commutative property. (Children can color the two sets of squares two different colors rather than shade the squares.)

$4+2=6$
$2+4=6$

Ask the children to tell how the two rows of squares are alike (the number is the same) and different (the order of the colors is reversed). After several examples are generated, ask the children if they see a pattern among all the examples they have worked. Tell them that these are examples of the order (commutative) property of addition. Encourage them to state the pattern in words. If they have difficulty, verbalize for them: "The order of the addends does not affect the sum."

Cuisenaire rods afford further concrete experience. Have students use the rods to find $8+4=\square$ and $4+8=\square$, for example. Emphasize that, although the trains of rods contain rods in reversed order, the same replacement rod is needed in each case.

Ordinary materials can also be used to suggest the commutative property. Place five clothespins on one index card and six clothespins on another. Hold the two cards with attached clothespins in front of the class, and tell the students the sets of clothespins are to be combined. Instruct them to write the addition sentence represented. Then exchange positions of the two cards and have students write the new sentence. They should note the sums are the same. Next let several students take turns presenting examples with the cards and clothespins while other members of the class continue to write the pairs of addition sentences.

As children work with the number line, present them with problems such as $4+5=\square$ and $5+4=\square$ to solve on the same number line. Instruct them to draw an arrow diagram of one sentence above the number line and a diagram of the other below the number line. Ask them to solve each open sentence and compare the sums.

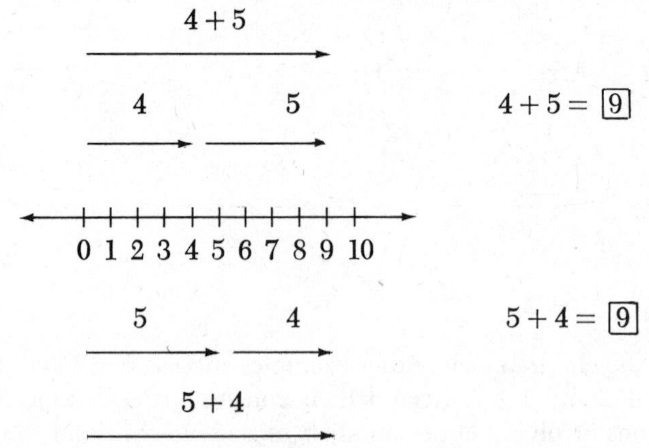

A worksheet such as the following combines the pictorial and symbolic levels of learning.

Draw a set of 7 objects.	Draw a set of 4 objects.

Join the sets above by drawing a ring around them.

Complete the open sentence that tells what you did.

$$7 + 4 = \square$$

Draw a set of 4 objects.	Draw a set of 7 objects.

Join the two sets above by drawing a ring around them.

Complete the open sentence that tells what you did.

$$4 + 7 = \square$$

Compare $7 + 4$ and $4 + 7$. Does $7 + 4 = 4 + 7$? ____ This is an example of the commutative property of addition.

Make up another example of the commutative property of addition. Let a friend check your work.

Notice that commutativity for addition holds because set union is commutative: $A \cup B = B \cup A$. Learning the commutative property allows children to condense the number of individual basic addition facts they must memorize. For example, if a child knows $2 + 3 = 5$, $3 + 6 = 9$, and $7 + 8 = 15$ and understands that addition is commutative, he will also know that $3 + 2 = 5$, $6 + 3 = 9$, and $8 + 7 = 15$ without having to treat these as new facts to be learned.

The Associative Property

When studying this property, a child begins adding three addends and is introduced to parentheses in numerical expressions. Teachers should stress that the numbers represented by the numerals in the parentheses are to be added first. In a beginning teaching activity, you might distribute grid paper to the children and guide them to color and group three sets of squares and record the sentences represented. A student's work may look like this:

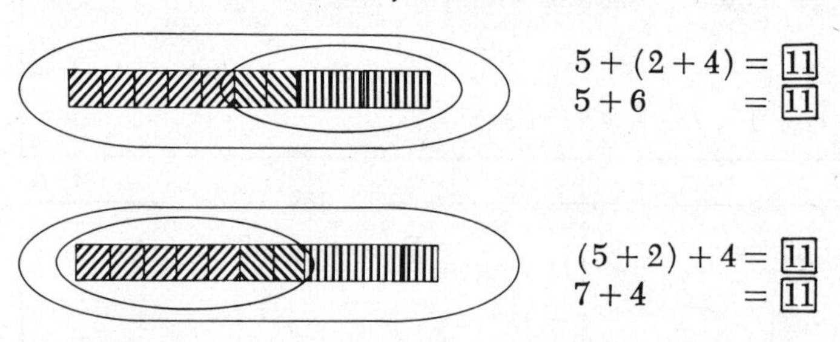

Ask the children to look at their two pictures and corresponding addition sentences and to tell how they are alike (the number, or sum, is the same) and how they are different (the grouping has changed). Have students create other examples to explore. Stress that three numbers are to be added but only two can be added at a time and that the same sum results when the first two are grouped together as when the second two are grouped together.

To provide further work in recording addition expressions and grouping addends, place three paper cups in a row in front of the students. Put straws in each cup, cup your hands (to resemble parentheses) around the last two cups and then around the first two cups, and instruct the children to record the three addends, place parentheses around the two numbers which are to be added first as suggested by each grouping of the cups, and perform the additions. The concrete model and a child's accompanying work may look like this:

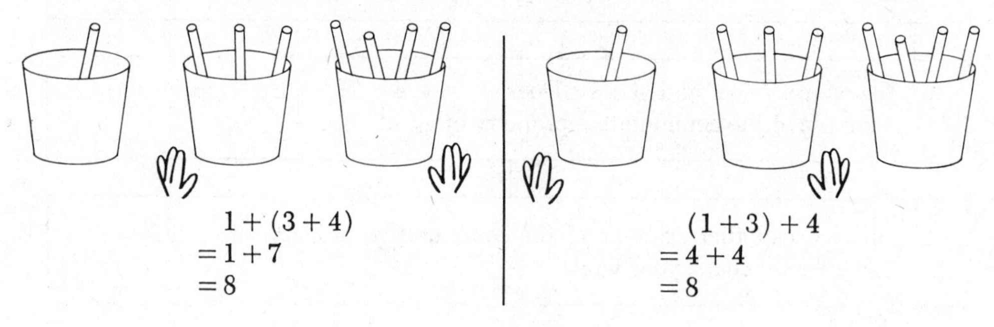

Let students create other examples to work. Summarize with "It does not matter which numbers are grouped together when we add—the sum is always the same." Tell the students this pattern is called the grouping, or associative, property of addition.

The associative property should be a part of children's work with Cuisenaire rods. Present the students with a problem such as $4 + 2 + 3 = \square$. Let them use the rods to find the sum. Some may add 4 and 2 first and show their work with the rods this way:

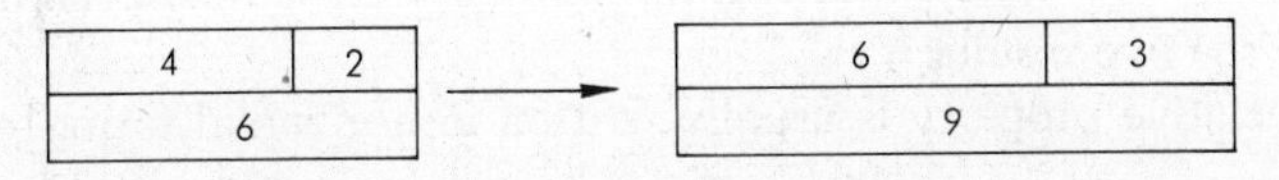

Others may add 2 and 3 first.

2 3
5
4 5
9

Choose students to explain their work to the class, and record their suggestions:

$$4 + 2 + 3 = (4 + 2) + 3 = 6 + 3 = 9$$

$$4 + 2 + 3 = 4 + (2 + 3) = 4 + 5 = 9$$

Pictorial and symbolic exercises are important also. Prepare worksheets with problems such as these:

1. Look at the picture and complete the addition sentence represented by the picture. Show the grouping with parentheses.

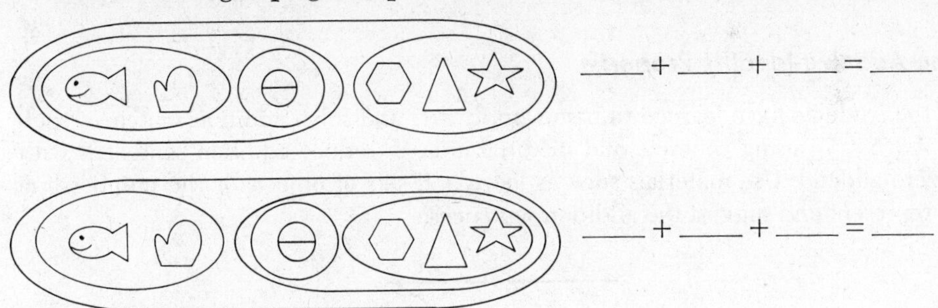

____ + ____ + ____ = ____

____ + ____ + ____ = ____

2. Are the sums in problem 1 the same? ____
3. How are parts (a) and (b) of problem 1 different?

After children learn place value, present them with examples that illustrate the use of the associative property in simplifying computations.

$(4 + 9) + 1 = 4 + (\_\_\_\_ + 1)$
$= 4 + \_\_\_\_$
$= \_\_\_\_$

$13 + 6 = (10 + \_\_\_\_) + 6$
$= 10 + (\_\_\_\_ + 6)$
$= 10 + \_\_\_\_$
$= \_\_\_\_$

Children often enjoy enacting the associative property with numeral and symbol cards. Line seven students up facing the other members of the class and give each of the seven a numeral, operation symbol, or grouping symbol in an expression such as $5 + (2 + 6)$. Have the two students holding the parentheses cards step forward. Call on a student to tell which symbols are dealt with next. (The students holding the 2, +, and 6 cards should step forward.) Ask a child to select the correct numeral card representing the sum of 2 and 6 and to stand in front of the "2 + 6" students. Then instruct the two children holding the "5" and " + " symbols to step forward in line with the child holding the "8." Choose another student to find the correct numeral card for the sum of 5 and 8 and to stand in front of the "5 + 8" children. Next, have students evaluate the sum $(5 + 2) + 6$ by a similar procedure and compare the two results.

The associative property is usually written in horizontal form, but consider a vertical addition problem such as $\begin{array}{r} 6 \\ +7 \\ 3 \\ \hline \end{array}$. (Drawing an arrow pointing downward beside the vertical addition may help a learning disabled child as well as other children read such notation.) If a student adds downward, he thinks, "$6 + 7 = 13$ and $13 + 3 = 16$, so $(6 + 7) + 3 = 16$." To check the addition, he can add upward, thus changing the grouping and the order of the addends. This check is justified by the associative and the commutative properties. Help students become aware that sometimes it is more advantageous to find the sum with one grouping than with the other. For instance,

$$\begin{array}{r} 9 \\ +\ \left.\begin{array}{r}5 \\ 5\end{array}\right\}10 \\ \hline 19 \end{array} \quad \text{suggests an easier way to add than} \quad \begin{array}{r} \left.\begin{array}{r}9 \\ +\ 5\end{array}\right\}14 \\ 5 \\ \hline 19 \end{array}$$

The grouping in the first example promotes quick mental computation.

The Additive-Identity Property

If the students have learned to name empty sets and solve addition sentences such as $4 + 5 = \square$ using concrete and pictorial aids, introduce addition sentences with a zero addend. Use materials such as loops and sets of objects or the number line to represent and suggest the addition sentences.

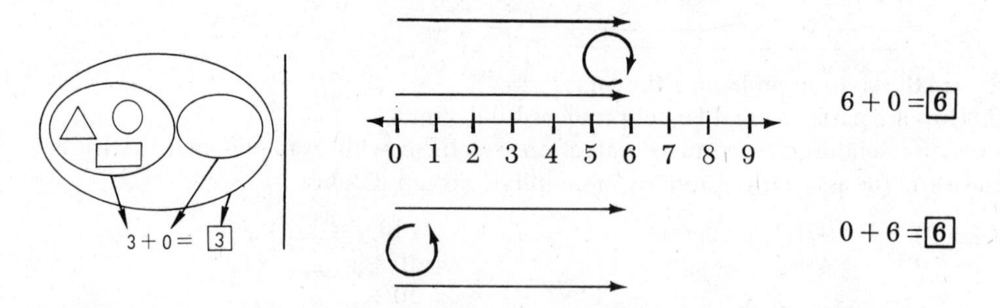

When you are convinced the children are ready to examine the symbolic pattern involved when zero is an addend, summarize the basic facts as follows:

$$\begin{array}{lcl} 0+0=0 & & \\ 1+0=1 & \text{and} & 0+1=1 \\ 2+0=2 & \text{and} & 0+2=2 \\ \cdot & & \cdot \\ \cdot & & \cdot \\ \cdot & & \cdot \\ 9+0=9 & \text{and} & 0+9=9 \end{array}$$

Ask the students to use the pattern to predict the solution to equations such as $500 + 0 = \square$ and $0 + 2000 = \square$. Examples involving these larger numbers allow children to make generalizations for finding sums they do not yet have the skill to compute.

Introducing Subtraction

Two methods for introducing a student to subtraction is the removal-of-a-subset model and the missing-addend approach.* The following example suggests how set separation is involved in the removal-of-a-subset method.

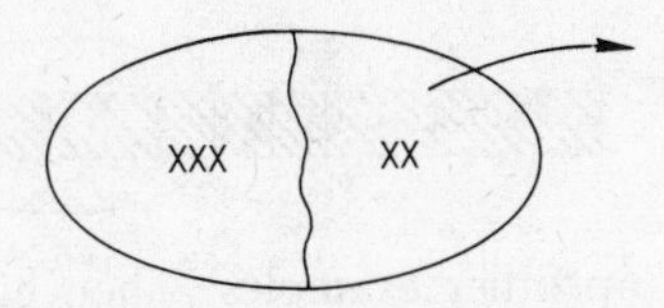

In the example, two elements are removed from a set of five, leaving three elements. The associated mathematical sentence is $5 - 2 = 3$. After a student has learned to subtract using the removal-of-a-subset model, introduce the missing-addend approach which helps him relate the new operation of subtraction to previously learned concepts and skills in addition. He should already be able to solve addition sentences with an addend missing before he begins to use this method of subtraction. The example

$$7 - 2 = \square \longrightarrow \square + 2 = 7 \longrightarrow \boxed{5} + 2 = 7 \longrightarrow 7 - 2 = \boxed{5}$$

shows how open subtraction sentences can be solved by relating subtraction to addition.

Both interpretations of subtraction are utilized in these teaching suggestions:

1. Place several artificial flowers in a vase and ask the students to tell the number of flowers in the vase. Select a child to remove some of the flowers and to tell the number he takes away. Choose another student to state the number of flowers left in the vase. Repeat this procedure in two or three examples. Then place numeral and other symbol cards in the chalk tray to construct a subtraction sentence represented by one of the situations the

* Even though the terminology "removal-of-a-subet" is commonly used, it is a misnomer in that subsets are not actually removed but elements are.

students have enacted with the flowers. In the case of $\boxed{8}\,\boxed{-}\,\boxed{5}\,\boxed{=}\,\boxed{3}$, for example, tell the students this is a subtraction sentence and is read as "eight minus five equals three, or eight take away five is three." Stress that three objects are left when five objects have been taken away from a set of eight. Have students read the sentence as you point to it. Construct subtraction sentences for the other examples the students have worked with the flowers.

2. Prepare a functional bulletin board of a zebra with several stripes. Have students determine the number of stripes, remove some of the stripes (and pin them elsewhere on the bulletin board), tell the number removed and the number left, and pin numerals on the bulletin board to make the associated subtraction sentence. Vary the number of stripes so that several different subtraction sentences can be generated.
3. Give each child a sheet of grid paper. Instruct the students to shade or color a certain number of squares, e.g., twelve, and then draw a ring around some of the twelve squares. Tell them the squares in the loop are to be taken away. Instruct each student to determine the number left in the set and to write below the picture the subtraction sentence represented by the take-away story. The case for $12 - 5 = \boxed{7}$ should look like this:

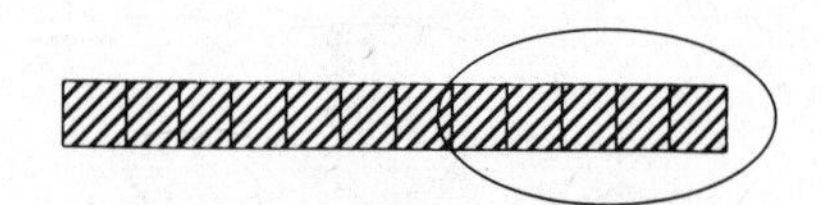

Let children make up other examples. Then give them open subtraction sentences to solve using grid paper. The geoboard or pegboard could be used in a similar activity.

4. Guide the students in solving subtraction sentences with Cuisenaire rods. Have the students build a staircase with the rods and tell them the rods in the staircase represent 1, 2, . . ., 10 in this activity. Help them solve $6 - 2 = \square$, for example, by instructing them to place the 2 rod below the 6 rod so that one end of each rod lines up and then to find the missing rod which completes the train with the 2 rod.

6

4 2

Emphasize that the missing rod tells the answer in the subtraction sentence. Now ask the children to remove the 4 rod and write a missing-addend sentence by placing a frame $\square$ in the same place in the sentence as the missing rod occurs in the train of rods. Record $\boxed{4} + 2 = 6$ below $6 - 2 = \boxed{4}$ on the chalkboard, stressing that these two sentences are equivalent. Use these two sentences to introduce the vocabulary in a subtraction sentence: in $6 - 2 = \boxed{4}$, 6 is called the sum; 2, an addend; and 4, the missing addend. Point out these same terms in the related addition sentence. It is appropriate to later use terminology for subtraction independent of addition, e.g., in $6 - 2 = 4$, 6 is the minuend; 2, the subtrahend; and 4, the difference. Give other Cuisenaire-rod models for which the students write the subtraction and addition related sentences. Stress the pattern

$$\text{addend} + \text{addend} = \text{sum}$$
means
$$\text{sum} - \text{addend} = \text{addend}$$

Let students use the subtraction-addition pattern to suggest steps in solving subtraction sentences with the variable in three different places. Suggested board work includes:

$12 - 3 = \square$	$9 - \square = 3$	$\square - 5 = 8$
$\square + 3 = 12$	$3 + \square = 9$	$8 + 5 = \square$
$\boxed{9} + 3 = 12$	$3 + \boxed{6} = 9$	$8 + 5 = \boxed{13}$
$12 - 3 = \boxed{9}$	$9 - \boxed{6} = 3$	$\boxed{13} - 5 = 8$

5. On the feltboard, separate a set into two subsets. Point to one of the subsets and ask a student to construct on the feltboard the subtraction sentence that corresponds to the removal of those elements. Do the same for the other subset. Now point to both subsets and say, "If these subsets are joined, what two addition sentences are represented?" Select students to construct the addition sentences on the feltboard. The following is an example of a feltboard model and associated sentences:

Tell the students these four equations are an addition and subtration fact team and all four can be written from the same set picture.

6. Arrange students near the walk-on number line. Choose a student to begin at zero, take a certain number of steps, turn around, and go back a specified number of steps as another student records the subtraction sentence on the chalkboard. Then draw an arrow diagram on a pictorial number line to model a sentence such as $8 - 6 = 2$.

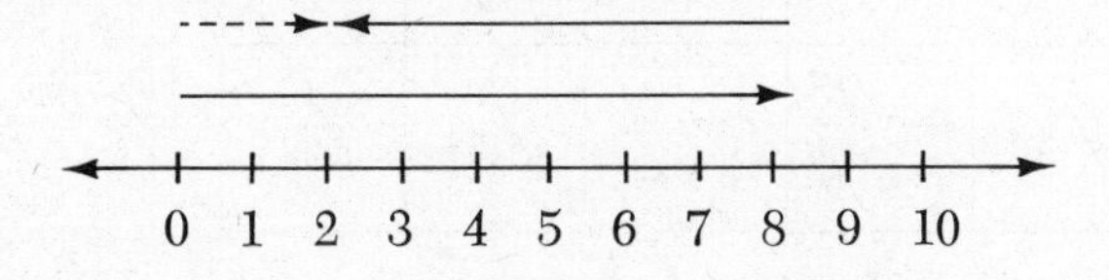

Relate the picture to their walks on the number line. Emphasize that the arrow pointing to the left has six units below it to depict the removal of six of the eight units.

Another representation of $8 - 6 = 2$ on the number line involves the missing-addend interpretation of subtraction. Draw the 8 and 6 arrows and then indicate the missing arrow with a dashed line.

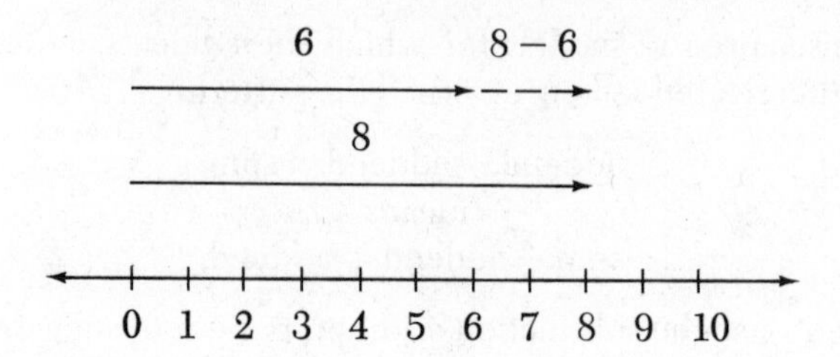

Point out that the answer in the subtraction sentence corresponds to the number of units, or steps, below the dashed arrow.

7. For beginning work in subtraction on the minicomputer, each child needs a Minicomputer board and two different kinds of counters. You may need to review the students on representing numbers on the Minicomputer. Demonstrate the procedure with $6 - 3 = \square$, for example. Children may already know the answer, but simple problems should be modeled first so the procedure can be learned before more difficult subtractions are performed on the device. Represent 6 and 3 with distinguishable counters. The steps in solving $6 - 3 = \square$ follow:

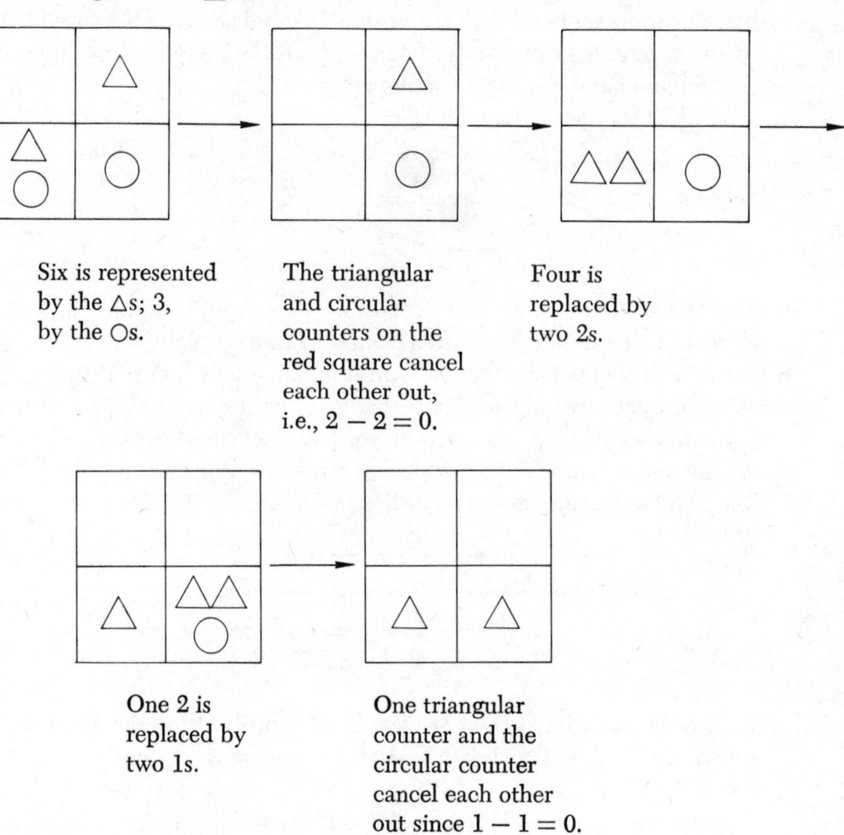

Six is represented by the △s; 3, by the ○s.

The triangular and circular counters on the red square cancel each other out, i.e., $2 - 2 = 0$.

Four is replaced by two 2s.

One 2 is replaced by two 1s.

One triangular counter and the circular counter cancel each other out since $1 - 1 = 0$.

The difference 3 is read from the Minicomputer.

The exchange of counters in subtraction proceeds from squares of greatest value to those of least value, the same pattern applied in the regrouping step in the algorithm for subtracting numbers named by multi-digit numerals. Addition on the Minicomputer similarly transfers to the addition algorithm in that the exchange of counters begins with those on the squares of least value in a general right-to-left action.

Properties of Subtraction

The properties of subtraction are examples of Gagné's rule, or principle, learning. Knowing generalizations in subtraction should help teachers of grades one through three guide the thinking of children as they explore subtraction. Be sure to have students compare the behavior of the subtraction operation to that of the operation of addition they have previously explored.

Some properties and ideas for teaching them follow. Learning every one of these properties may not be an appropriate goal for all students, but precocious learners should explore most, if not all, of them.

1. Subtraction is not commutative. The difference $4 - 3$, for example, means the number left when three elements are removed from a set of four; thus $4 - 3 = 1$. On the other hand, four elements cannot be removed from a set of three, and $3 - 4$ is not a whole number. Thus $4 - 3 \neq 3 - 4$. Some students will be curious as to how to find a difference such as $3 - 4$. Even though negative numbers are not usually introduced before third grade, satisfy younger students' curiosity by dealing briefly with them. Ask them to imagine they are taking a trip on an elevator with the ground level labeled as 0; the floors above ground floor, as 1, 2, 3, and so on; and the floors below ground level, as -1, -2, -3, etc. Tell them that the " $-$ " symbol is read as "negative" and that negative numbers tell positions on the other side of zero. This diagram represents a ride on the elevator from ground floor to third and down four floors:

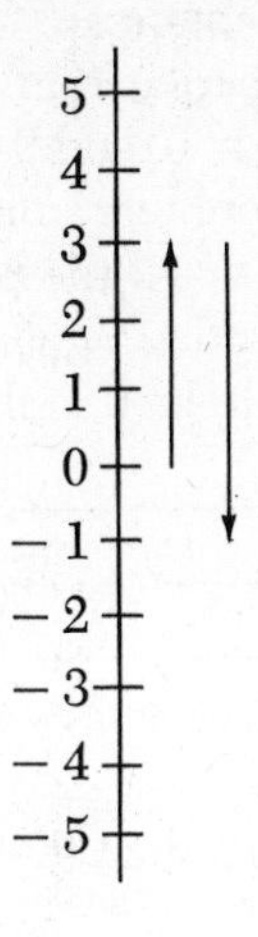

The picture shows $3 - 4 = -1$. The number line and a thermometer also provide good models for negative numbers.

Although, in general, changing the order of minuend and subtrahend affects the difference, challenge students to think of an exception to the rule. When $a = b$, $a - b$ does equal $b - a$, for whole numbers a and b.

2. Guide students to discover that subtraction is not associative by having them compare examples such as $7 - (5 - 1) = \square$ and $(7 - 5) - 1 = \square$. Suggested board work includes:

$$7 - (5 - 1) = \square \qquad (7 - 5) - 1 = \square$$
$$7 - 4 = \boxed{3} \qquad 2 - 1 = 1 \ \square$$
$$7 - (5 - 1) = \boxed{3} \qquad (7 - 5) - 1 = \boxed{1}$$
$$7 - (5 - 1) \neq (7 - 5) - 1$$

Children should compute the differences, referring to concrete or pictorial materials if necessary.

3. The zero property (one-sided) of subtraction is an important one for all children to explore. The number line affords an especially good model for this property. The subtraction sentence $5 - 0 = 5$ is shown on this arrow diagram:

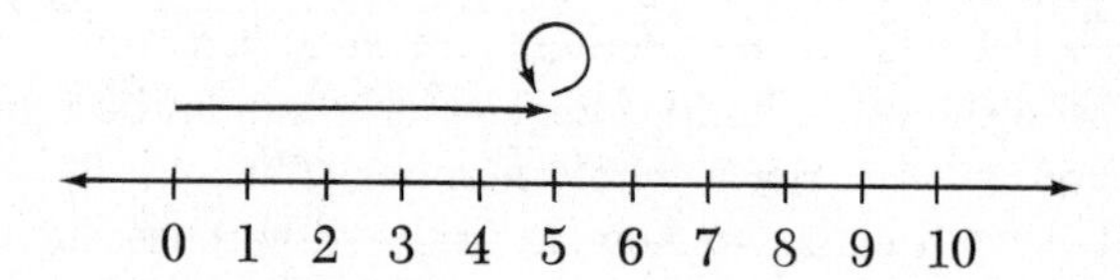

Subtraction of zero can also be associated with removing no objects from a set of objects, e.g., cut-outs on a feltboard or pegs on a pegboard. Provide enough examples to help students generalize that any number minus zero is the original number. Note that this property is one-sided in that $a - 0 = a$ for all whole numbers a, but $0 - a$ is not necessarily even a whole number. This one-sidedness is not characteristic of addition since $a + 0 = a$ and $0 + a = a$ for every whole number a.

4. All children also need experience in the same-number property of subtraction, e.g., $2 - 2 = 0$, $6 - 6 = 0$, and $9 - 9 = 0$. Students may learn this property by removing all the counters from a given set of counters, all the objects from a set of felt objects, or all the pegs from a set of pegs on a pegboard. The number line also provides a good model for this property. The solution to $6 - 6 = \square$ can be seen to be 0 in the following arrow diagram:

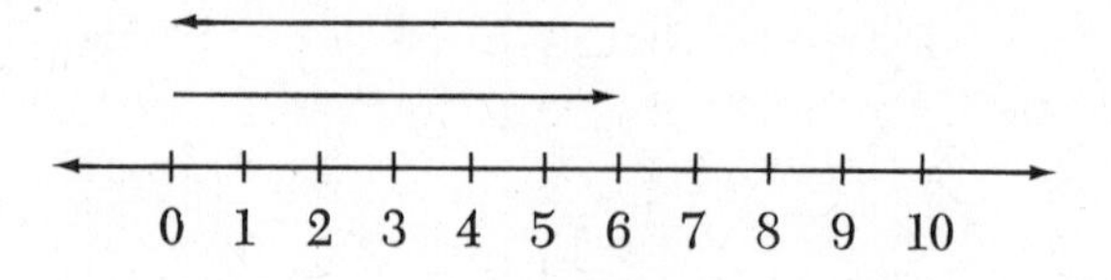

Guide the children to generalize that any number minus itself is zero. Remember to extend the pattern to include examples with larger numbers, e.g., $2348 - 2348 = \square$.

5. Students should examine the rearrangement property[1] of addition and subtraction after they have gained symbolic experience in addition and subtraction. Use a feltboard model to compare $(5+4) - (3+1) = \square$ and $(5-3) + (4-1) = \square$. For $(5+4) - (3+1) = \square$, join a set of five and a set of four and then remove the elements of the set formed by joining a subset of three of the five-element set with a subset of one member of the four-element set. The difference is represented by the number of elements left in the original two sets. The solution is seen to be 5.

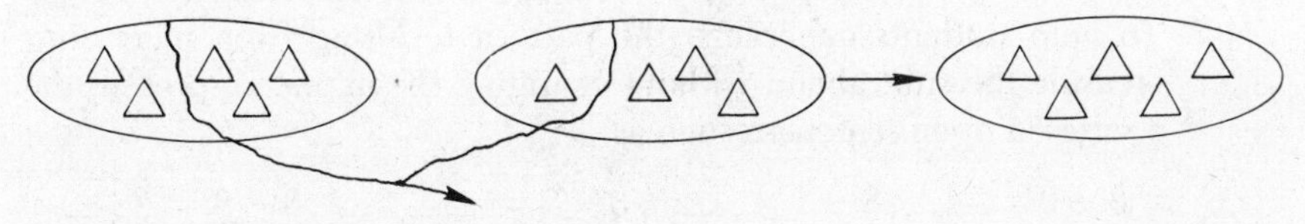

To solve $(5 - 3) + (4 - 1) = \square$, remove three elements from a set of five and one element from a four-element set which is disjoint with the five-member set; join the subsets which are left. The work on the feltboard looks like this:

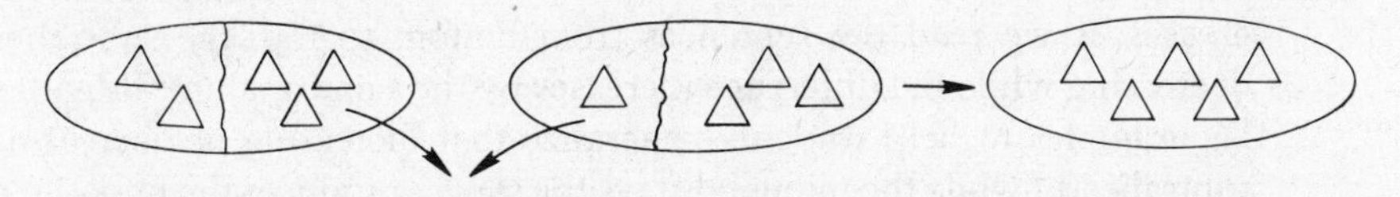

The solution again is 5. Write the symbolic sentences on the chalkboard, pointing out that the same numbers occur in the equations but they are rearranged from two additions and one subtraction to two subtractions and one addition. Give other examples of this property for students to examine, each time judiciously constructing the examples to avoid negative numbers.

The rearrangement property provides readiness for the vertical subtraction algorithm. This topic is explored in chapter 9.

6. The pattern resulting from increasing the minuend and holding the subtrahend constant can be investigated on the number line or with cubical blocks or plastic counters. Ask students to make, for example, a set of seven blocks and a set of four directly below it. Call on a student to tell how many more blocks there are in the seven-element set and to write the associated subtraction sentence on the chalkboard. Instruct children to add a block to the set of seven but not to the set of four and to determine the new difference. Select a child to record the associated subtraction sentence on the chalkboard. Have them continue adding blocks to the top set and writing the equations. Their work should look like this:

$$\begin{array}{r} 7 \\ -4 \\ \hline 3 \end{array} \qquad \begin{array}{r} 8 \\ -4 \\ \hline 4 \end{array} \qquad \begin{array}{r} 9 \\ -4 \\ \hline 5 \end{array} \qquad \begin{array}{r} 10 \\ -4 \\ \hline 6 \end{array}$$

Question the children on how increasing the minuend affects the difference. Some students should suggest that increasing the minuend produces the same effect in the difference. Extend the pattern to an example such as $(7 + 10) - 4 = 3 + \square$. Guide students to generalize that decreasing the minuend also produces the same effect in the difference by having them remove some blocks from the top set in one of preceding examples and compare the difference represented before and after the blocks were taken away.

7. To help students generalize the pattern resulting from increasing or decreasing the subtrahend without changing the minuend, present them with a series of open sentences such as

$$8 - 0 = \_\_\_\_, 8 - 1 = \_\_\_\_, 8 - 2 = \_\_\_\_, \ldots, 8 - 8 = \_\_\_\_.$$

Have students draw pictures representing the minuend and subtrahend to help them visualize the effect on the difference of increasing or decreasing the subtrahend, and then let them fill in the differences. Ask questions about how the sentences are different and how they are alike. As we read downward, the difference decreases by the same amount the subtrahend increases; if we read the sentences from bottom to top, the subtrahend is decreasing while the difference increases by the same amount. Provide similar examples to help students generalize that increasing or decreasing the subtrahend (while the minuend stays the same) produces the opposite effect on the difference.

8. An important subtraction property which is basic to later work with the equal-additions subtraction algorithm involves increasing or decreasing the minuend and subtrahend by the same amount. Guide students in exploring this property by having them make a set of six plastic counters and a set of four, tell how many more in the set of six, add a counter to each set, tell the difference in the numbers represented, and continue for several more times adding the same number of counters to each set and determining the difference. Write the associated subtraction sentences on the chalk board. A sample of the work follows:

OOOOOO OOOOOOO OOOOOOOO OOOOOOOOO

OOOO OOOOO OOOOOO OOOOOOO

$$\begin{array}{r} 6 \\ -4 \\ \hline 2 \end{array} \qquad \begin{array}{r} 7 \\ -5 \\ \hline 2 \end{array} \qquad \begin{array}{r} 8 \\ -6 \\ \hline 2 \end{array} \qquad \begin{array}{r} 9 \\ -7 \\ \hline 2 \end{array}$$

Let students describe the pattern and make the generalization. Emphasize that the same number has been added to minuend and subtrahend and the difference has remained unchanged. Now direct them to remove the same number of counters from each set in one pair of sets and compare the difference in number of elements in the two sets before and after the elements were removed. Children should discover that the difference also stays the same when the same number is subtracted from minuend and subtrahend. Extend the subtraction pattern to examples such as "If $9 - 2 = 7$, then $(9 + 50) - (2 + 50) = \square$" and "If $14 - 5 = 9$, then $21 - \square = 9$."

Summarizing the Basic Facts of Addition and Subtraction

The basic facts of addition are $0 + 0 = 0$ through $9 + 9 = 18$, and the basic subtraction facts are those which are the related sentences of the basic addition facts. For example, $0 - 0 = 0$, $4 - 3 = 1$, $12 - 9 = 3$, and $18 - 9 = 9$ are basic subtraction facts since $0 + 0 = 0$, $1 + 3 = 4$, $3 + 9 = 12$, and $9 + 9 = 18$ are basic facts of addition. These facts are important tools throughout one's life.

When students have conceptualized the operations by manipulating materials, drawing pictures, and writing associated addition and subtraction sentences, begin to summarize the basic facts at the symbolic level. Present the facts in a systematic manner. Here are some suggestions:

1. Present families of addition facts such as all facts in which 6 is the sum, e.g., $6 + 0 = 6$, $0 + 6 = 6$, $5 + 1 = 6$, $1 + 5 = 6$, $4 + 2 = 6$, $2 + 4 = 6$, $3 + 3 = 6$.
2. Generate sentences in which the first or second addend remains constant, e.g., $2 + 0 = 2$, $2 + 1 = 3$, $2 + 2 = 4$, . . ., $2 + 9 = 11$ and $0 + 7 = 7$, $1 + 7 = 8$, $2 + 7 = 9$, . . ., $9 + 7 = 16$. Note with the students that when an addend increases the sum increases by the same amount.
3. Have students find the sums in a pattern such as this: $4 + 0 = 4$, $5 + 1 = 6$, $6 + 2 = 8$, $7 + 3 = 10$, $8 + 4 = 12$, $9 + 5 = 14$. Have students tell how the related sentences change from one to the next.
4. Present the "doubles" facts ($0 + 0 = 0$, $1 + 1 = 2$, $2 + 2 = 4$, . . ., $9 + 9 = 18$) and then the "doubles plus one" facts, helping the children to understand that the "doubles plus one" facts can be found by adding one to the more-easily-remembered "doubles" facts. For example, $7 + 8$ is 15 because $7 + 7$ is 14 and 8 is one more than 7. Sums in which the addends differ by 2 can also be related to the "doubles" facts. The 6 and 8, for example, in $6 + 8$ can be thought of as "sharing" a one, and thus $6 + 8 = 7 + 7 = 14$. An alternative to the sharing strategy for computing sums in which the addends differ by 2 is a "doubles plus two" thinking process. To compute $6 + 8$ by this method, a child thinks, "$6 + 8 = 6 + (6 + 2)$ which is the same as $(6 + 6) + 2$, or 14." Similar strategies can be developed for other basic facts. Organizing the facts in this way helps students form strategies for remembering the facts. Such strategies especially benefit learning disabled children who have difficulty memorizing the basic facts.

A meaningful exercise for students is to complete a composite table of basic addition facts after you enter the addends on the side and top of the table (first addend on the side and second addend on the top).

+	0	1	2	3	4	5	6	7	8	9
0										
1										
2										
3										
4										
5										
6										
7										
8										
9										

It is customary to have students fill in the table by subsets. For instance, you might have them first fill in the sums less than or equal to five, and then design practice activities dealing with these facts. Next, have the students extend the table through all facts in which 10 is the sum. Practice activities should occur again for this group of facts. For the remaining facts in the table, guide students to use their knowledge of the facts with sums less than or equal to ten, place value, and the commutative and associative properties of addition. For instance, provide series of open sentences to help the children relate the more difficult facts to the easier ones.

$$\begin{aligned} 5+8 &= 5+(\_\_+3) \\ &= (5+5)+\_\_ \\ &= \_\_+3 \\ &= \_\_ \end{aligned}$$

$$\begin{aligned} 8+9 &= (\_\_+3)+(\_\_+4) \\ &= (5+5)+(\_\_+4) \\ &= \_\_+7 \\ &= \_\_ \end{aligned}$$

Students may also use the "doubles plus one" strategy to solve an example such as $8+9=\square$. When a student finds a sum, he should enter it on his composite table. By determining the more difficult sums in this manner, children are applying previously learned concepts and skills to a new learning task—a problem-solving technique that should be encouraged in all learning situations. The completed composite table looks like this:

+	0	1	2	3	4	5	6	7	8	9
0	0	1	2	3	4	5	6	7	8	9
1	1	2	3	4	5	6	7	8	9	10
2	2	3	4	5	6	7	8	9	10	11
3	3	4	5	6	7	8	9	10	11	12
4	4	5	6	7	8	9	10	11	12	13
5	5	6	7	8	9	10	11	12	13	14
6	6	7	8	9	10	11	12	13	14	15
7	7	8	9	10	11	12	13	14	15	16
8	8	9	10	11	12	13	14	15	16	17
9	9	10	11	12	13	14	15	16	17	18

The teacher should stress certain number properties so that some of the one-hundred basic facts can be related. If a student has learned the additive-identity property, he should be able to determine all the sums in the first row and first column of the composite table. When he learns the commutative property and any fact in which the sum is entered above the diagonal from the upper left to lower right corner of the table, he will automatically know a corresponding addition fact in which the sum is entered below the diagonal. The following statement by Bruner indicates the importance of teaching number properties to promote the categorization of certain facts:

> Teaching specific topics or skills without making clear their context in the broader fundamental structure of a field of knowledge is uneconomical in several deep senses. In the first place, such teaching makes it exceedingly difficult for the student to generalize from what he has learned to what he will encounter later. In the second place, learning that has fallen short of a grasp of general principles has little reward in terms of intellectual excitement. The best way to create interest in a subject is to render it worth knowing, which means to make the knowledge gained usable in one's thinking beyond the situation in which the learning has occurred. Third, knowledge one has acquired without sufficient structure to tie it together is knowledge that is likely to be forgotten. An unconnected set of facts in terms of principles and ideas from which they may be inferred is the only known way of reducing the quick rate of loss of human memory.[2]

Basic facts of subtraction can be read from the table also. It is important to emphasize the inverse relationship of addition and subtraction, and the mere reading of the table reinforces that relationship. For example, to solve $12 - 7 = \square$, 12 is located below the 7 and then the difference (or missing addend) 5 is seen at the far left.

When the entire table is completed, use it to reinforce certain properties of addition and subtraction and to allow students to discover other patterns. These are important considerations when guiding students to use the table and to discover relationships in it:

1. Select children to read addition facts on the table. For a learning disabled child, you might copy the table on an overhead-projector transparency and provide visual cues by placing arrows by each addend and the "+" sign. For example, if you asked him to complete the sentence $8 + 3 = \square$, you could mark the table this way:

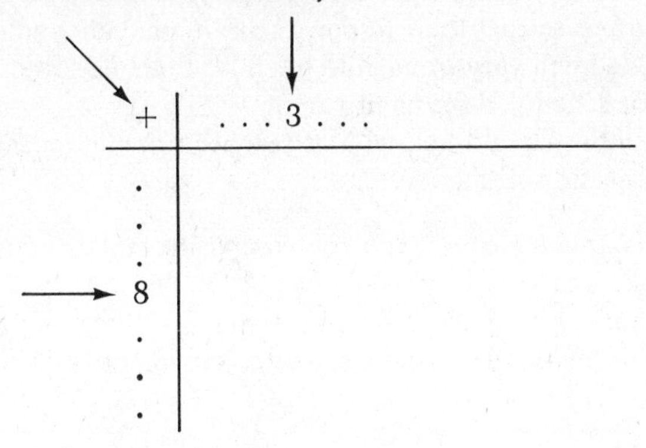

2. Have various students read subtraction facts on the table.
3. Ask students to use the table to find all the addition facts that have a sum of 10. Ask the children, "How many facts on the table have 10 as a sum? Can you describe a pattern for locating all the sums of 10?" Then have them write each of the corresponding facts of subtraction. Students should also use the table to list all the subtraction facts having a given difference. The ten facts $6 - 0 = 6$, $7 - 1 = 6$, $8 - 2 = 6$, . . ., $15 - 9 = 6$, for example, all have the difference 6.
4. Have students examine the sums in the first row and first column and state the property of addition (additive-identity property) represented by these examples.
5. Ask the children, "Why is the first row the same as the first column, the second row the same as the second column, . . ., the tenth row the same as the tenth column? What property is illustrated by these examples?" Students can also investigate commutativity of addition by drawing a diagonal from the upper left-hand sum in the table to the sum in the lower right-hand corner and observing the symmetry on each side of the diagonal—corresponding entries are the same. You may use this opportunity to bring into the discussion even and odd numbers if these concepts have been introduced earlier. When the students draw the diagonal, have them identify what kind of numbers are represented along the diagonal (even numbers) and which are represented directly above and below the diagonal (odd numbers).
6. Ask students to find all subtraction sentences on the table which have the minuend and subtrahend the same and to verbalize the pattern.
7. As you point to the 0 addend (subtrahend) at the top of the table, ask students to tell what the difference is each time the subtrahend is 0. They should generalize the zero property of subtraction.
8. Students can also use the table to investigate patterns in addition of even and odd numbers. (See chapter 14.)

Practice in the Basic Facts

After students have formed the concepts of addition and subtraction and have gained experience in relating the basic facts, they need opportunity to practice the facts in a variety of settings so that their responses are immediate and correct. The amount of practice needed will vary from child to child. Practice sessions should be kept brief and should be scheduled frequently.

The following activities should help you as teacher provide meaningful practice exercises on addition and subtraction.

1. After the students have completed the composite table discussed earlier, have them make a tape recording of some of the facts. A child records $5 + 2 = 7$, for example, by saying "Five plus two equals (pause) seven." At a later time he listens to the tape and writes down the sum or difference

during the pause on the tape. It is better that he make several different recordings rather than one extremely long one.

2. Have children form pairs to practice the basic facts with flash cards. For every fact missed, ask the student to write it correctly five times.
3. Involve the students in an individual activity with flash cards. Have them make a grid of several sums or differences, e.g.,

12	5	1
8	0	4
15	7	9

Give them a set of flash cards on which addition or subtraction expressions are written. Ask them to place the cards having the sums or differences indicated on the grid on the proper numeral.[3]

4. Play "Have You Seen My Geese?" with the students. Have six or seven arrange themselves in a circle at the front of the room. Give each student a large numeral card to hold. Explain the rules of the game. You will be the farmer and will ask someone in the circle, e.g., Jon, "Have you seen my geese?" Jon will then say, "No. How many have you?" Then you say, for example, "Three more than you have." Jon then adds three to his number, looks around the circle to find the child holding the numeral representing the sum of his number and three, and chases that student around the outside of the circle. If he catches the child before that child runs around the circle and back to his original place, that student must leave the circle and select another child in the class to take his place. If Jon cannot catch the child, he chooses another student to replace him. You then ask another child, "Have you seen my geese?" and the game proceeds as before. When working with subtraction facts, change your reply to "No. How many have you?" to a statement such as "Five less than you have."[4]
5. Give students handouts of puzzles in which they compute the sums horizontally and vertically.[5] Examples are

4	2	
6	4	

and

3		
1	5	
	9	13

6. Have students fill in incomplete grids in which the addends are in random order or missing.

+	7	1	6	5
4				
3				
2				

+	6	3
9		
		8
7		

7. Provide practice in solving open sentences such as $\square + 3 = 10$, $2 + \square = 11$, $6 - \square = 2$, and $\square - 5 = 9$. Construct each open sentence by placing a blank card and numeral and symbol cards in the chalk tray. Select different students to place the correct numeral card over the blank card to complete the sentence. In some situations, you may need to use pictorial or concrete aids to model the sentence. For instance, to help students visualize the process involved in solving $\square + 3 = 10$, you might use this model:

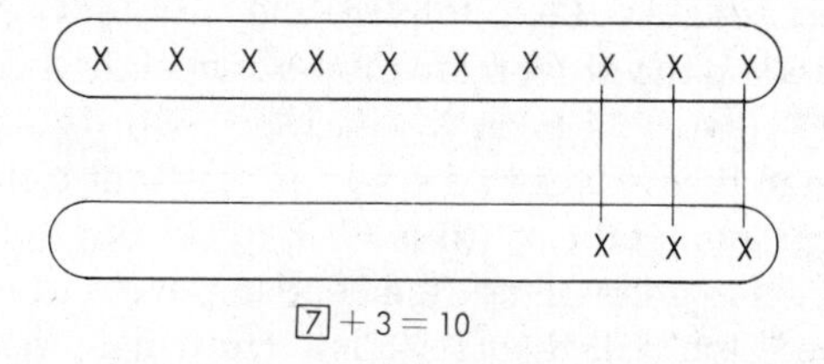

8. Make a bulletin board of two basketball goals with a numeral on each net and two benches with players who have addition and subtraction expressions on their uniforms.

Students are to choose the five players on the left bench who have the sum or difference 6 to go under the "6" net to form one team and the players on the right bench who have a sum or difference 7 to make the other team. Children pin the proper players under each goal. Change the numerals on the goals and the players' sums and differences frequently.

9. Play the game "Let's Play Teacher." Designate one student as the "teacher." He holds up flash cards of basic addition and subtraction facts and calls on a child to read the front of the card. (The flash cards have the answers on the back.) Each student records the completed addition or subtraction sentences on his paper. After ten flash cards have been shown, the "teacher" calls out the answers on the cards he has used and the other children check their work. Select another student to be the "teacher" and continue for several more turns.
10. Prepare a large grid of numerals on which students take turns tossing two bean bags. They should tell the sum of the two numbers represented by the numerals on which the two bean bags fall and write the addition sentence on paper. They can also subtract the smaller from the larger and record the subtraction sentence. A similar practice session can be planned in which students spin a spinner.
11. Write several addition and subtraction problems in vertical form at the bottom of the chalkboard. Select children to write the sums and differences on the chalkboard. If a student gets all the answers correct, place a "happy face" in the chalk tray; if he misses any, place a "sad face" in the chalk tray and have him write each fact he missed at least five times.
12. To give practice in some of the properties of addition and subtraction, play the game "Hint." Tell the students you are going to write a hint on the chalkboard and if someone knows the answer, he should raise his hand. Some hints are:

    If $12 - 5 = 7$, then what is $13 - 5$?
    If $15 - 8 = 7$, then what is $16 - 9$?
    If $14 - 7 = 7$, then what is $14 - 8$?
    If $2 + 8 = 10$, then what is $3 + 9$?
    If $7 + 9 = 16$, then what is $8 + 9$?

    Each time ask children to explain how the hint helped them answer the question.[6] For example, a student may answer the second question as "7, because 1 has been added to both 15 and 8 so the answer stays the same." This is an example of one of the subtraction properties considered earlier in this chapter.

Closing

Young children should begin their study of addition by joining sets of concrete objects. Iconic and symbolic experience should follow. Subtraction, introduced after a student has gained some skill in addition, can be taught by a set-separation model and by relating subtraction to addition in the missing-addend approach. Students should be guided to discover properties of addition and subtraction as they explore the operations. Especially important are the commutative, associative, and identity properties of addition since these properties are used throughout a student's study of mathematics. Meaningful practice opportunities should ensue after a child understands the basic facts. Mastery of the basic facts of addition and subtraction allows students to solve more advanced problems such as $\begin{array}{r}85\\ \underline{+\,24}\end{array}$ and $\begin{array}{r}67\\ \underline{-\,25}\end{array}$ (see chapter 9) and to apply addition and subtraction in their everyday lives through the solving of story problems (see chapter 15).

Notes

2. Jack E. Forbes and Robert E. Eicholz, *Mathematics for Elementary Teachers* (Menlo Park, California: Addison-Wesley Publishing Company, Inc., 1971), p. 121.
3. Jerome S. Bruner, *The Process of Education* (New York: Vintage Books, 1960), pp. 31–32.
4. Enoch Dumas, *Math Activities for Child Involvement* (Boston: Allyn and Bacon, Inc., 1971), p. 89.
5. Mary E. Platts, *Plus: A Handbook of Classroom Ideas to Motivate the Teaching of Primary Mathematics* (Stevensville, Michigan: Educational Service, Inc., 1975), p. 104.
6. Foster E. Grossnickle and John Reckzeh, *Discovering Meanings in Elementary School Mathematics* (New York: Holt, Rinehart and Winston, Inc., 1973), p. 153.
7. Platts, *Plus*, p. 105.

Assignment

1. Write a brief activity to help children
   a) form the concept of addition,
   b) form the concept of subtraction,
   c) develop skill in the basic facts of addition,
   d) develop skill in the basic facts of subtraction.
2. Read the article, A "two-dimensional abacus—The Papy Minicomputer" by Jean Van Arsdel and Joanne Lasky in the October, 1972, issue of *The Arithmetic Teacher*. Make two Minicomputer boards and practice adding and subtracting with them. Do you think this aid promotes concepts, skills, or both concepts and skills in addition and subtraction? Explain.
3. Begin a collection of games and puzzles that can be used to help students master the basic facts of addition and subtraction.
4. Read the article "Using Functional Bulletin Boards in Elementary Mathematics" by William E. Schall in the October, 1972, issue of *The Arithmetic Teacher*, and generate at least five ideas for functional bulletin boards related to addition and subtraction of whole numbers.

5. Study Cawley's Interactive Model, the "Do-See-Say" method, on page 258 of *Clinical Teaching: Methods of Instruction for the Retarded* by Robert M. Smith (published by McGraw-Hill Book Company in 1974), and write a sequence for teaching a child addition using the procedure outlined.

Laboratory Session

Preparation: Study chapter 7 of this textbook.

Objectives: Each student should

1. use plastic counters, a pegboard, Cuisenaire rods, number line, and Minicomputer to solve open sentences involving the basic facts of addition and subtraction of whole numbers,
2. illustrate the commutative, associative, and identity properties of addition,
3. provide a set model of the rearrangement property of addition and subtraction, and
4. provide a set model of the property of whole numbers: if $a - b = c$, then $(a + n) - (b + n) = c$ and $(a - n) - (b - n) = c$, for whole numbers a, b, c, and n where $a \geqq b$ and $b \geqq n$.

Materials: plastic counters, yarn strips, Cuisenaire rods, pegboard, Minicomputer boards.

Procedure: Students should form small groups and work the following problems:

1. Use plastic counters and yarn loops to
   a) solve $2 + 5 = \square$,
   b) solve $6 - 4 = \square$,
   c) solve $3 + \square = 11$,
   d) solve $\square - 3 = 5$,
   e) show $2 + 5 = 5 + 2$,
   f) demonstrate $3 + (1 + 7) = (3 + 1) + 7$,
   g) show $6 + 0 = 6$.
2. With Cuisenaire rods
   a) solve $\square = 5 + 8$,
   b) solve $12 - 3 = \square$,
   c) solve $\square + 7 = 13$,
   d) solve $8 - \square = 6$,
   e) show all the basic facts with a sum of 8,
   f) demonstrate $6 + 3 = 3 + 6$. This example illustrates what property?
   g) show $2 + (1 + 8) = (2 + 1) + 8$. This example illustrates what property?
3. Use a pegboard to
   a) solve $7 + 4 = \square$,
   b) solve $9 - 5 = \square$,
   c) solve $\square + 4 = 10$,
   d) solve $13 - \square = 8$,
   e) show $5 + 1 = 1 + 5$,
   f) demonstrate $1 + (2 + 3) = (1 + 2) + 3$.
4. Draw an arrow diagram on a number line to
   a) solve $5 + 4 = \square$,
   b) solve $11 - 6 = \square$,
   c) solve $7 - 0 = \square$,
   d) demonstrate $8 + 0 = 8$ and $0 + 8 = 8$. This example illustrates what property?
   e) show $2 + 9 = 9 + 2$,
   f) show $6 + 0 = 0 + 6$. What property is illustrated here?

5. a) With plastic counters and loops, illustrate the rearrangement property of addition and subtraction with $(6+7)-(2+4)=(6-2)+(7-4)$. Explain why this example is not easily modeled on a number line.
   b) Use the plastic counters to demonstrate the property of subtraction in which the same number is added to or subtracted from both minuend and subtrahend. Model this series of open sentences:
   $8-2=\square$, $9-3=\square$, $10-4=\square$, $11-5=\square$, $12-6=\square$.
6. Perform the operations on the Papy Minicomputer.
   a) $6+3=\square$, b) $7+6=\square$,
   c) $8-5=\square$, d) $17-9=\square$.

Bibliography

Arsdel, Jean Van, and Lasky, Joanne. "A Two-dimensional Abacus—the Papy Minicomputer." *The Arithmetic Teacher* 19 (October 1972): 445–51.

Banks, J. Houston. *Elementary-School Mathematics: A Modern Approach for Teachers.* Boston: Allyn and Bacon, Inc., 1966.

Bruner, Jerome S. *The Process of Education.* New York: Vintage Books, 1960.

Dumas, Enoch. *Math Activities for Child Involvement.* Boston: Allyn and Bacon, Inc., 1971.

Forbes, Jack E., and Eicholz, Robert E. *Mathematics for Elementary Teachers.* Menlo Park, California: Addison-Wesley Publishing Company, Inc., 1971.

Grossnickle, Foster E., and Reckzeh, John. *Discovering Meanings in Elementary School Mathematics.* New York: Holt, Rinehart and Winston, Inc., 1973.

Jungst, Dale G. *Elementary Mathematics Methods: Laboratory Manual.* Boston: Allyn and Bacon, Inc., 1975.

Platts, Mary E. *Plus: A Handbook of Classroom Ideas to Motivate the Teaching of Primary Mathematics.* Stevensville, Michigan: Educational Service, Inc., 1975.

Early Work in Multiplication and Division of Whole Numbers

In most schools multiplication is introduced in second grade and division, in third grade. This situation may not always be the case, however, since some research indicates that children may be able to work with division as a sharing process prior to understanding multiplication.[1] In this chapter various approaches and materials to help a student conceptualize multiplication and division, properties of these operations, and ways to summarize and provide practice in the basic facts are discussed. These operations are extended in chapter 10 and applied in story situations in chapter 15.

The prenumber and number concepts and skills discussed in chapter 5 and the processes involved in addition and subtraction described in chapter 7 serve as readiness for multiplication and division. Especially important is experience with equivalent sets and in skip counting (counting by twos, threes, and so on) forward and backward. The number pattern 0, 3, 6, 9, . . ., 27, for example, generates all the products in the basic facts in which three is a factor, and the numbers in the pattern in reverse order 27, 24, 21, 18, . . . , 0 are important dividends when dividing by three. These symbolic patterns can be illustrated with sets of three or on the number line, e.g.,

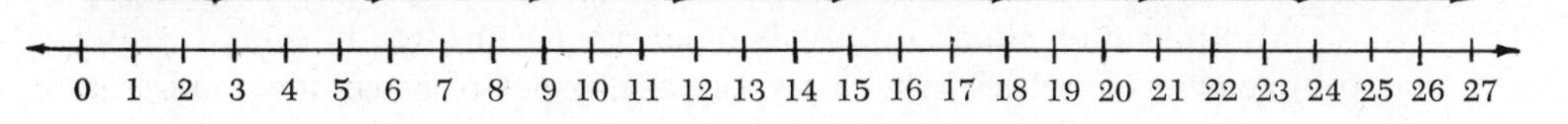

Counting by fives is useful in work with time and money.

Introducing Multiplication

Approaches

Three interpretations of multiplication should help a child conceptualize the operation.

1. *sets of equivalent, disjoint sets*
   When equivalent, disjoint sets are joined, the number in the union can be determined by counting, by addition of equal addends, or by multiplication. This situation is illustrated with $4 \times 3 = 12$.

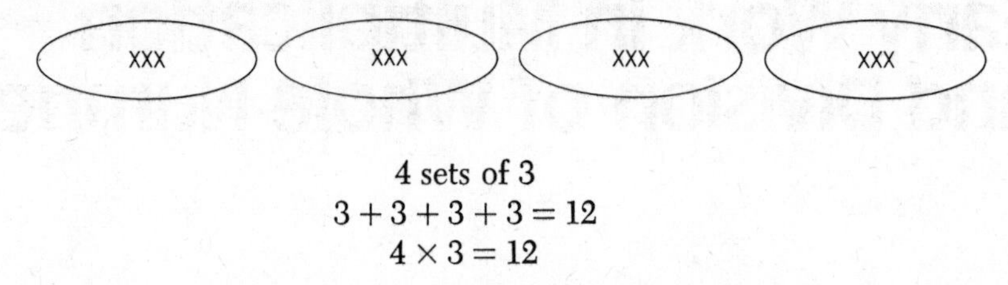

$$\begin{gathered} \text{4 sets of 3} \\ 3 + 3 + 3 + 3 = 12 \\ 4 \times 3 = 12 \end{gathered}$$

   In the sentence $4 \times 3 = 12$, 4 and 3 are called factors, with 4 telling the number of sets, or the number of addends, and 3, the number in each set, or the name of each addend; 12 is the product and denotes the number in the combined sets, or the sum in the addition sentence. Later, children should extend their vocabulary to multiplier and multiplicand, the multiplier being the first factor and the multiplicand, the second factor.* The horizontal form is used initially, but students should soon see the vertical computational form: $\begin{array}{r} 3 \\ \times\ 4 \\ \hline 12 \end{array}$

2. *arrays*
   An array is a rectangular arrangement of items in rows and columns. The following 5 by 2 array shows $5 \times 2 = 10$.

$$\begin{array}{cc} \cdot & \cdot \\ \cdot & \cdot \\ \cdot & \cdot \\ \cdot & \cdot \\ \cdot & \cdot \end{array}$$

   There are five rows, two columns, and ten dots in the array. You should consistently make arrays such that the number of rows agrees with the first factor in the associated multiplication sentence and the number of columns is the same as the second factor.

   Multiplication situations involving arrays occur often in everyday life: arrangement of bottles in a carton or case, panes of a window, and seats

---

* Occasionally you might see $a \times b$ interpreted as a multiplied by b, in which case b is the multiplier and a, the multiplicand.

in a theater. Story problems that come later in a child's school experience with multiplication should include this interpretation of multiplication. The array approach also helps students discover many of the properties of multiplication and provides readiness for the formula for area of rectangular regions.

3. *Cartesian product of two sets*
Readiness for this approach may be provided by preliminary work with pairing the elements of two sets. Students can be asked to pair the cups in a set of cups, for example, with the saucers in a set of saucers in all possible ways by physically placing each cup in each saucer. After children demonstrate the pairings, you may show the possible pairs with ribbon or yarn strips.

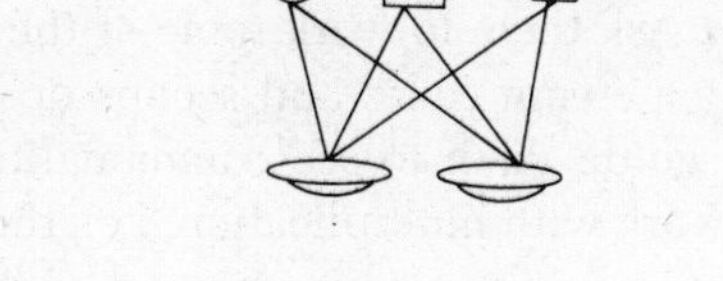

In the example, there are three in one set, two in another, and six ways to pair each number of the first set with each member of the second set. This model represents $3 \times 2 = 6$.

In one study Cartesian-product problems were found to be more difficult for second-grade students to represent and solve than equal-addends problems. Furthermore, results of the study indicate that multiplication problems using the Cartesian-product approach can be solved more often by high achievers than low achievers and by students with above-average intelligence.[2] These results suggest that teachers should not expect all students to master the Cartesian-product approach to multiplication. All children do, however, need instruction in simple situations involving Cartesian product since this conceptualization is useful in solving certain real-life problems.

Teaching Suggestions

The following are some ways to help children understand the set-of-equivalent-disjoint-sets, array, and Cartesian-product interpretations of multiplication.

1. On the feltboard, place a certain number of triangular cut-outs, e.g., 4, to represent ice cream cones and two cut-outs resembling scoops of ice cream on each "cone."

Tell the children there are four "cones" with two "scoops" in each, or four sets of two. Have students demonstrate other examples such as five sets of three and three sets of four. Now return to the original example. Ask the students how they would determine the number of "scoops" in all. Some may suggest, "Count the scoops," "Count by twos," or "Add two four times." Assure the children all of these are correct. Place $2 + 2 + 2 + 2 = 8$ on the feltboard. Tell the students there is another way to find the number in four sets of two as you introduce the sentence $4 \times 2 = 8$ on the feltboard. Tell them this is a multiplication sentence, "×" is a multiplication symbol, and the sentence is read "four times two equals eight." Construct repeated addition and multiplication sentences for the other sets the students have made, each time emphasizing the number of equivalent sets, the number in each set, and the number in the union of the sets.

2. Give children dried beans and egg cartons or plastic counters and yarn strips to use in an activity on the set-of-equivalent-disjoint-sets interpretation of multiplication. Ask them to work some of the same examples given earlier with the felt ice cream cones and scoops of ice cream. Through a questioning process, guide them to see commonalities in the two experiences as with future work with multiplication; i.e., they should understand, for example, that 4×2 can always be renamed as 8 in ordinary multiplication.
3. Have students form trains of equivalent Cuisenaire rods and verbalize the multiplication sentences represented as you (or a student) record(s) them on the chalkboard. The model for $4 \times 3 = 12$ looks like this:

<table>
<tr><td>3</td><td>3</td><td>3</td><td>3</td></tr>
<tr><td colspan="3">10</td><td>2</td></tr>
</table>

4. Have students represent sets of equivalent, disjoint sets on the walk-on number line; then show them how to multiply on the pictorial number line. The product 5×3 can be seen to be 15 when five sets of three are depicted this way:

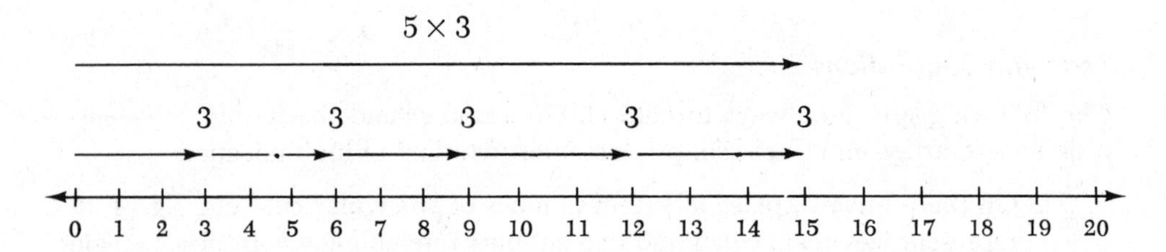

After students have worked two or three examples such as this one, review some addition situations with which multiplication cannot be associated. For instance, present an arrow diagram of the union of a set of four and a set of seven and have the students state the addition sentence represented. Then give them a worksheet or task cards on which there are both multiplication and addition problems. Some sample problems follow:

a) 1) Draw an arrow diagram to show 5 sets of 4.

2) Write the multiplication sentence that goes with part (1).
$\_\_\_\_ \times \_\_\_\_ = \_\_\_\_$

3) Write the addition sentence represented in (1).
$\_\_\_\_ + \_\_\_\_ + \_\_\_\_ + \_\_\_\_ + \_\_\_\_ = \_\_\_\_$

b) Write the multiplication sentence represented in each diagram.

1)
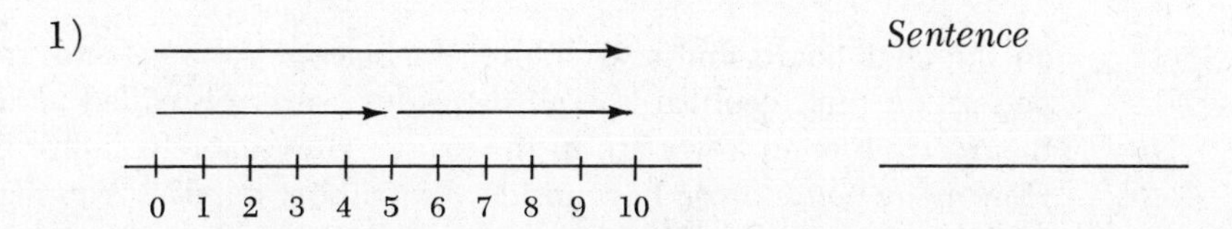

2)
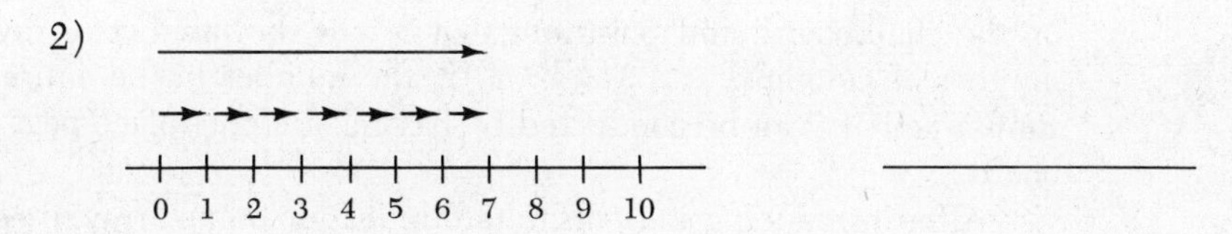

3)
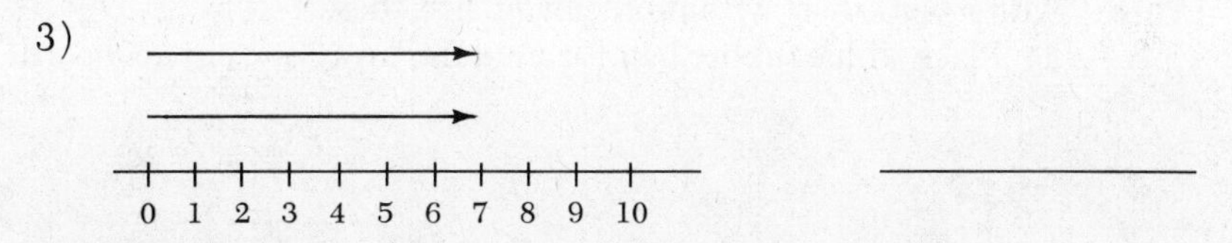

c) Solve each open sentence by drawing an arrow diagram on a number line.
1) $4 + 4 + 4 = \square$
2) $7 \times 3 = \square$
3) $7 + 3 = \square$

5. Model multiplication on the Minicomputer. For the example 5×3, place five counters on each square needed to represent three. Make conversions beginning with the counters on the square of least value. The steps for finding $5 \times 3 = 15$ appear this way:

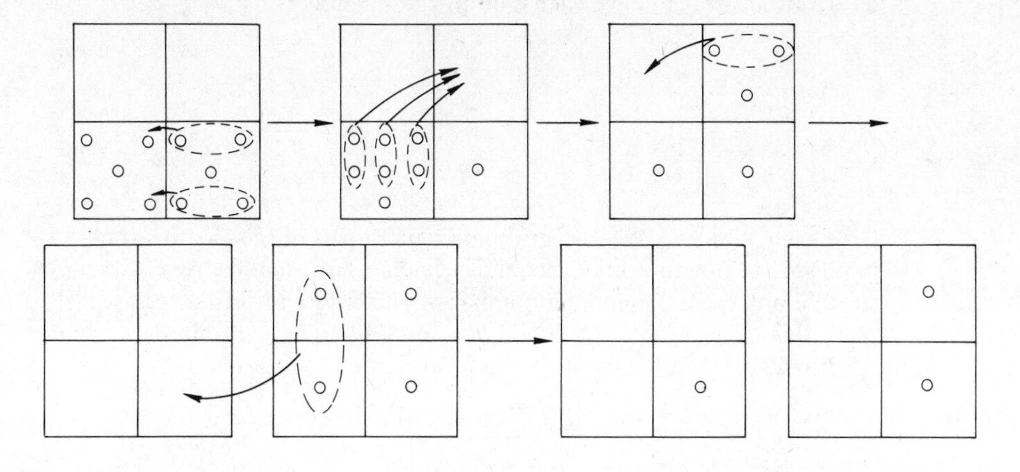

6. The array approach is utilized when modeling multiplication on the geoboard and pegboard. Make a pattern such as

   · · · ·
   · · · ·
   · · · ·

   on the chalkboard and ask children to place a rubber band around this pattern on their geoboards. Tell them this pattern is called an array. Ask them, "How many rows are in the array? How many columns are there? How many pegs do you have inside your rubber band?" (You may have to explain what rows and columns are also.) Write the sentence $3 \times 4 = 12$ on the chalkboard, and point out that 3 tells the number of rows; 4, the number of columns; and 3×4, or 12, the number in the entire array. A similar activity can be conducted by having students place pegs in a pegboard.

   After introducing students to multiplication on an array, present them with a worksheet with problems such as these:

   a) Place a blue rubber band around this array on your geoboard.

      · · · · ·
      · · · · ·
      · · · · ·

      There are ___ rows and ___ columns. Altogether there are ___ pegs inside the rubber band.

   b) On your geoboard place a red rubber band around an array with 4 rows and 3 columns. Draw the array of dots on your paper with a red felt-tip pen. There are ___ dots in your picture. Your picture shows that ____ $\times$ ____ $=$ ____.

   c) Use a green rubber band to show a 5×7 array on your geoboard. Then draw the array using a green felt-tip pen. The 5×7 array has ___ green dots. Write the multiplication sentence shown by your array.

   d) Draw an array to solve each multiplication sentence.

      Array

      a) $3 \times 8 = \square$
      b) $4 \times 5 = \square$
      c) $1 \times 9 = \square$
      d) $9 \times 1 = \square$

7. Give each student a sheet of grid paper and a pair of scissors. Instruct the pupils to cut out an array (rectangle) having, for example, two rows and six columns and to count the number of small squares in the array. Tell them this array of 2 rows, 6 columns, and 12 squares in all shows that $2 \times 6 = 12$.

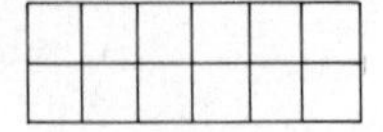

After several such arrays are cut out and associated with multiplication sentences, let each child color and assemble them to make an art design.

8. Give instruction in the use of the Cartesian product of sets to solve multiplication equations. Have a group of boys and a group of girls stand in front of the classroom. Tell the students in the class you want them to tell all the ways each boy can be paired with each girl. As children suggest pairs, have the boy and girl in the pair stand together. As pairs are generated, indicate them on the chalkboard by drawing line segments between members of the two sets and by writing ordered pairs of names, e.g.,

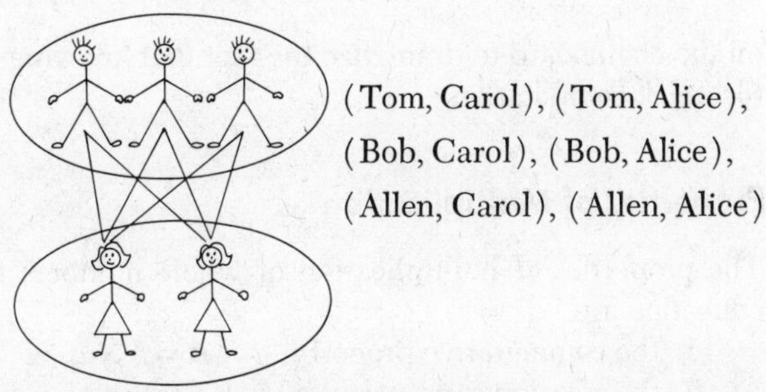

(Tom, Carol), (Tom, Alice),
(Bob, Carol), (Bob, Alice),
(Allen, Carol), (Allen, Alice)

Tell them the number of pairs can be found by multiplying 3 and 2. Emphasize that there are three in one set, two in another set, and six ways each member of the first set can be paired with each member of the second set and that this situation represents $3 \times 2 = 6$.

9. Have students pair each member of a set of riders with each element of a set of toy horses. Choose a student to place one rider on each of the horses, another student to place another rider on each of the horses, and so on. Ask the children to state the associated multiplication sentence for each example constructed. Also, present the students with multiplication sentences to be solved by pairing the horses and riders.

10. Ask children to tell their favorite cakes and frostings. Itemize them on the chalkboard and list the pairs of cakes and frostings as students suggest them. The case for pairing chocolate, yellow, and white cakes with chocolate and butter frostings can be presented this way:

	C	B
C	(C, C)	(C, B)
Y	(Y, C)	(Y, B)
W	(W, C)	(W, B)

You can also draw a set of three and a set of two and then indicate the pairs with line segments as was done previously, but the arrangement into rows and columns bridges the gap between the Cartesian-product and array models of multiplication. In the example, the pairs are listed in a 3×2 array.

Other Cartesian-product situations include pairing shirts with slacks, colors of cars with colors of vinyl tops, kinds of sandwiches with beverages, and flavors of ice cream with kinds of cones.

In order to conceptualize multiplication, children should represent examples using diverse materials and approaches. The teacher, however, should be careful to integrate these experiences through appropriate questioning. Teachers should also help students become aware that multiplication in many cases is a shortcut to addition just as addition is for counting. For instance, ask the children if they would rather add or multiply when finding the number of members in nine sets of eight. Write

$$8+8+8+8+8+8+8+8+8=\square$$
$$\text{and } 9\times 8=\square$$

on the chalkboard to dramatize the fact that knowing the multiplication fact allows for quick computation.

Properties of Multiplication

The properties of multiplication of whole numbers important in early childhood education are

1. the commutative property: $a \times b = b \times a$,
2. the associative property: $a \times (b \times c) = (a \times b) \times c$,
3. the identity property: $a \times 1 = a$ and $1 \times a = a$,
4. the zero property: $a \times 0 = 0$, and
5. the distributive properties of multiplication with respect to addition and to subtraction:
   $a \times (b+c) = (a \times b) + (a \times c)$;
   $(a+b) \times c = (a \times c) + (b \times c)$;
   $a \times (b-c) = (a \times b) - (a \times c)$, $b \geqq c$; and
   $(a-b) \times c = (a \times c) - (b \times c)$, $a \geqq b$.
   All variables in the equations represent whole numbers. These properties should be a part of explorations in multiplication such as those just described. You should construct examples and ask questions so that children may discover patterns in multiplication and make generalizations. Some students will discover these properties with little or no guidance in which case they should be allowed to verbalize their discoveries and share them with their peers to the benefit of all class members.

The Commutative Property

The commutative property is usually introduced in second grade and at that time is often referred to as the order property of multiplication. When beginning instruction on the property, you may use everyday materials that suggest arrays, e.g., an empty drink-bottle carton, a partitioned candy box, or plastic holders for frozen fruit juice. Present an array and have students write the multiplication sentence associated with it. Then turn the material ninety degrees and ask them to write the sentence represented by the array in that position. Have children compare the products as you write the sentences on the chalkboard. Stress that the order of the factors is different but the product is the same. Extend the pattern to examples such as 248×659 and 659×248. After predicting whether or not the products would be equal, students should check their predictions by performing the computations on a hand-held calculator. Tell the children the examples they have considered are examples of the order property of multiplication. Encourage them to state it in words.

Structured materials such as Cuisenaire rods, the number line, and arrays on grid paper, a geoboard, or a pegboard afford many opportunities for discovering the commutative property. The following pictures suggest array, Cuisenaire rod, and number-line models illustrating $3 \times 4 = 4 \times 3$:

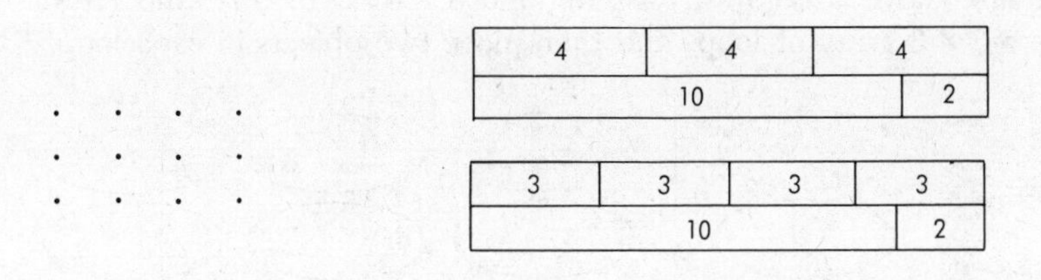

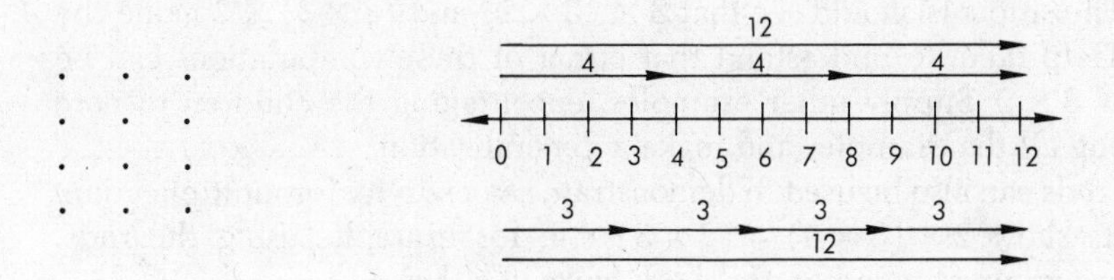

Instead of making two different arrays on grid paper, a geoboard, or a pegboard to demonstrate that the order of the factors does not affect the product, students may rotate the material ninety degrees after they have constructed one array.

Knowing the commutative property allows children to condense the number of basic multiplication facts they must memorize. For instance, if they know $7 \times 0 = 0$, $7 \times 1 = 7$, $7 \times 2 = 14, \ldots, 7 \times 9 = 63$ and the commutative property, they will automatically be able to find $0 \times 7 = 0$, $1 \times 7 = 7$, $2 \times 7 = 14, \ldots, 9 \times 7 = 63$.

The Associative Property

One way to teach this property is to combine the set-of-sets and array approaches. Write the sentence $3 \times (3 \times 2) = \square$ on the chalkboard. By appealing to their previous experience with parentheses in associativity for addition, establish with the students that 3×2 must be determined first. Have a child show three sets of two on the feltboard, and record $3 \times 6 = \square$ on the chalkboard. Choose another student to put two more sets of six on the feltboard to find the solution to $3 \times 6 = \square$. The feltboard work has progressed from

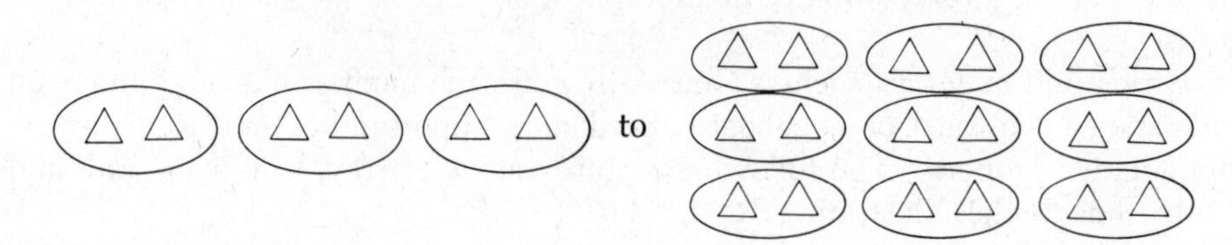

Point out that there are three rows with 3×2, or 6, in each row. Complete the sentences on the chalkboard: $3 \times (3 \times 2) =$ 18 and $3 \times 6 =$ 18 . To find $(3 \times 3) \times 2 = \square$, make a 3×3 array of loops and then place two objects in each loop.

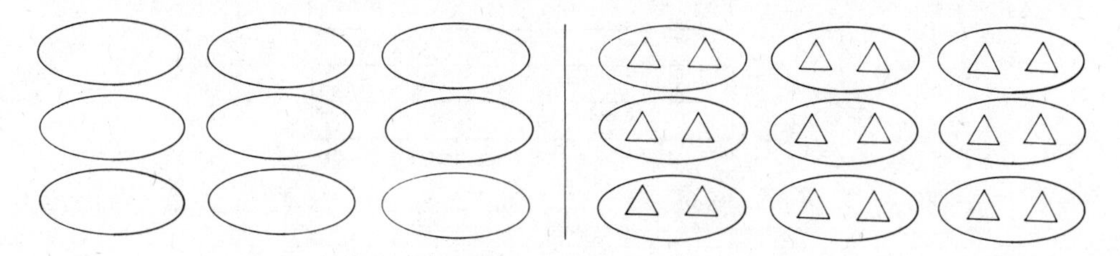

Complete the accompanying sentences on the chalkboard: $(3 \times 3) \times 2 =$ 18 and $9 \times 2 =$ 18 . The students should see that $3 \times (3 \times 2)$ and $(3 \times 3) \times 2$ name the same number. Help them to understand that either of these computations can be done to find $3 \times 3 \times 2$. Supply other examples, encouraging the children to note similarities among all the examples and make a generalization.

Cuisenaire rods can also be used to demonstrate associativity for multiplication. Let the students show $2 \times (3 \times 4) = (2 \times 3) \times 4$, for example, using the rods. When 3 and 4 are grouped together, the work looks like this:

4	4	4
10		2

→

10	2	10	2
10	10		4

The product $(2 \times 3) \times 4$ is found to be 24 this way:

3	3
6	

→

4	4	4	4	4	4
10		10		4	

Verbalize that the product is the same regardless of which factors are grouped together.

Another suggestion is to engage children in symbolic work dealing with the associative property. Have them write three numerals, e.g., 4, 1, and 2, on their paper, cut out the numerals, and put them in a row. Then instruct them to choose two of the numbers, multiply them, write down the product, multiply the product with the remaining number, and record that product. Select children to describe their work as you record the multiplication sentences on the overhead projector or chalkboard. Point out various groupings of the factors, stressing that all the final products of the three factors are the same.

The associative property is useful in computing some products mentally. For example, in computing $28 \times 2 \times 5$, the product is quickly seen to be 280 if $28 \times 2 \times 5$ is viewed as $28 \times (2 \times 5)$. On the other hand, the grouping $(28 \times 2) \times 5$ would probably necessitate pencil and paper or a calculator. The associative property also helps students relate basic multiplication facts, e.g.,

$$\begin{aligned} 6 \times 3 &= (2 \times 3) \times 3 \\ &= 2 \times (3 \times 3) \\ &= 2 \times 9 \\ &= 18. \end{aligned}$$

The Identity Property of Multiplication

Teachers should help all children to generalize that the product of any number and one is the number itself. This property can be illustrated with arrays, the set-of-equivalent-disjoint-sets approach, and the Cartesian-product interpretation of multiplication.

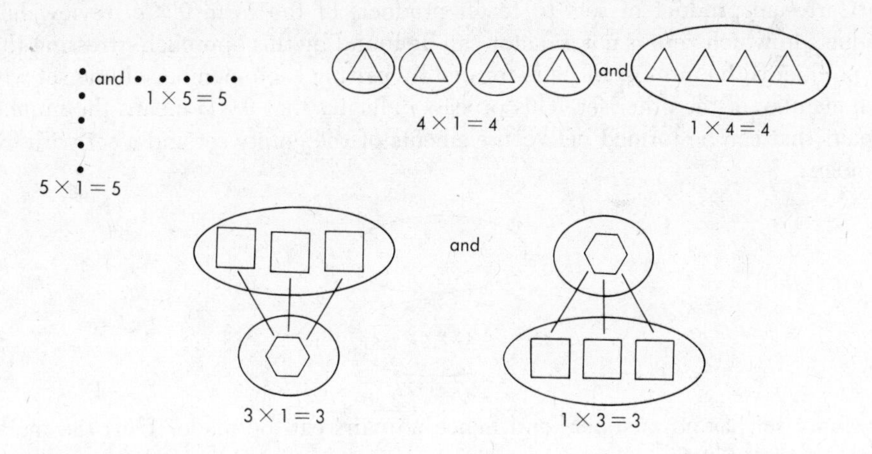

The number line also provides a good model for the identity property. The next picture shows $8 \times 1 = 8$ and $1 \times 8 = 8$.

Teaching the identity property is important in that it enables students to categorize all products having one as a factor. For instance, if they have learned

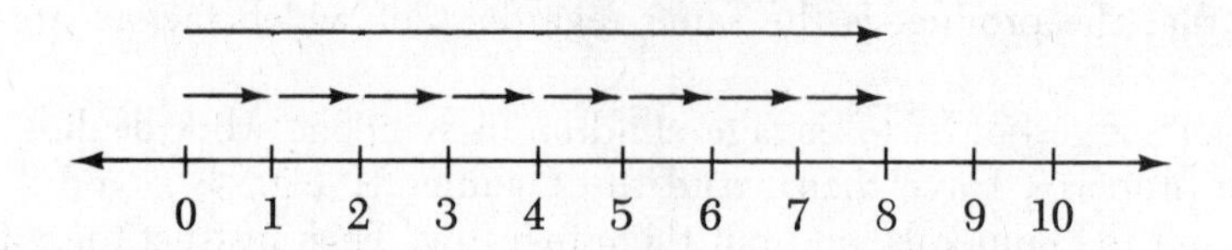

the identity property, they will know the seventeen basic facts $1 \times 1 = 1$, $2 \times 1 = 2$, $1 \times 2 = 2$, $3 \times 1 = 3$, $1 \times 3 = 3, \ldots, 9 \times 1 = 9$, $1 \times 9 = 9$.

The Property of Zero in Multiplication

To teach the property $a \times 0 = 0$ for any whole number a, you may use the set-of-equivalent-disjoint-sets approach to multiplication. To find $5 \times 0 = \square$, for instance, review examples such as $5 \times 2 = \square$ and $5 \times 1 = \square$ using the set-of-sets interpretation. As you show five loops, explain that 5×0 means the number of members in five empty sets.

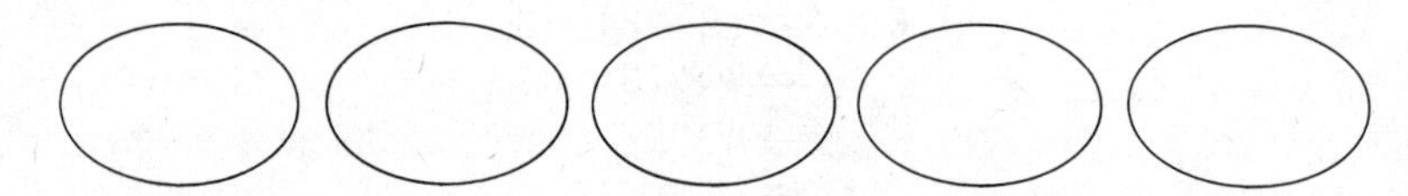

Students should then conclude $5 \times 0 = 0$. Present them with other examples in which 0 is the second factor. To help children find products of the form $0 \times a$, where a is a whole number, two strategies can be utilized. You can teach the $a \times 0$ facts and then appeal to a student's understanding of the commutative property to rename products of the form $0 \times a$ as $a \times 0$, or you may use the Cartesian-product approach. To help students solve $0 \times 5 = \square$ by the former strategy, guide them to rename 0×5 as 5×0 and then use $5 \times 0 = 0$ to conclude $0 \times 5 = 0$. When using the Cartesian product of sets to teach products of the form $0 \times a$, review how products in which zero is not a factor can be found by this approach, stressing that the product is the number of pairs formed by pairing each member of one set with each member of the other set. This process indicates that 0×5 means the number of pairs that can be formed between elements of the empty set and a set with five members.

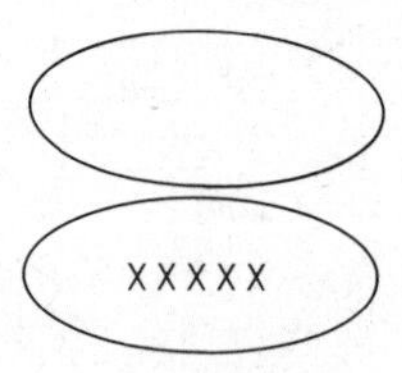

An empty set has no members and hence no pairs can be made. Thus the model shows that $0 \times 5 = 0$.

This property, like the commutative and identity properties, allows a student to place the basic facts of multiplication in categories. The facts $0 \times 0 = 0$, $1 \times 0 = 0$, $0 \times 1 = 0, \ldots, 9 \times 0 = 0$, $0 \times 9 = 0$ all fall under the heading of the zero property of multiplication.

The Distributive Property

Any distributive law combines two operations. In this case, multiplication and addition or multiplication and subtraction are involved. Actually there are two distributive properties of multiplication with respect to addition. The equation $2 \times (6+3) = (2 \times 6) + (2 \times 3)$ is an example of the left-hand distributive property of multiplication over addition since the multiplication of 2 in $2 \times (6+3)$ is to the left of $6 + 3$; analogously, $(2+6) \times 3 = (2 \times 3) + (6 \times 3)$ exemplifies right-hand distributivity of multiplication over addition. Examples of the left-hand and right-hand distributive properties with respect to subtraction are $5 \times (4-2) = (5 \times 4) - (5 \times 2)$ and $(7-3) \times 2 = (7 \times 2) - (3 \times 2)$, respectively. Only superior learners in the early childhood years, however, should be expected to use terminology reflecting an understanding of right-hand and left-hand distributive situations. When distributivity is introduced in second or third grade, the terms *common-factor property* or *multiplication-addition* and *multiplication-subtraction principle* are often used to name this property.

Arrays on grid paper, a geoboard, a pegboard, or feltboard are especially functional in helping students discover how expressions involving multiplication and addition or multiplication and subtraction can be renamed. To help toward this end, have students cut out from grid paper a 7×5 array, for example, and fold it either horizontally or vertically along the sides of the squares in the grid. Call on students to tell the sizes of the arrays they have made. Emphasize that the two arrays each child has formed can be joined to make the 7×5 array. These are two examples of possible work of students and the associated mathematical sentences.

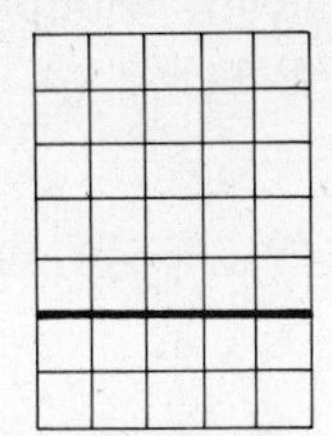

$7 \times 5 = (5 \times 5) + (2 \times 5)$
$(5+2) \times 5 = (5 \times 5) + (2 \times 5)$

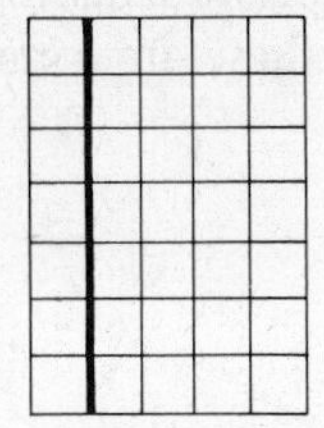

$7 \times 5 = (7 \times 1) + (7 \times 4)$
$7 \times (1+4) = (7 \times 1) + (7 \times 4)$

Explain that 7×5 is associated with the entire array whereas $(5 \times 5) + (2 \times 5)$ and $(7 \times 1) + (7 \times 4)$ indicate two smaller arrays are joined to make the 7×5 array. Note that a vertical fold of the paper corresponds to sentences that illustrate the left-hand distributive property and horizontal folds represent examples of right-hand distributivity.

The distributive law of multiplication over subtraction can be taught in a way similar to that just described for the multiplication-addition principle. Have students fold grid paper as before or place pegs in a pegboard or rubber bands around pegs of a geoboard, but this time one of the smaller arays is to be removed from the original one.

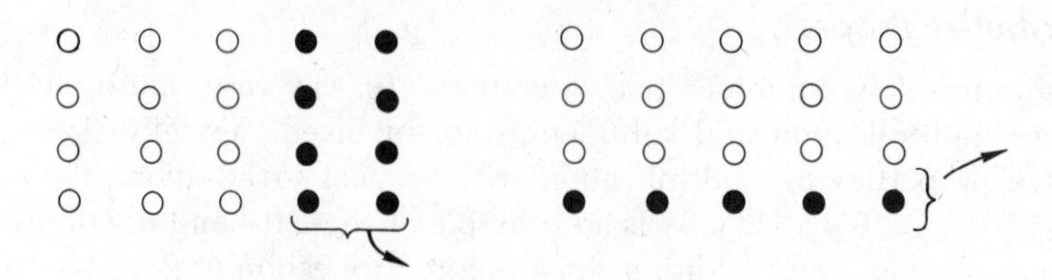

The first model suggests $4 \times 3 = (4 \times 5) - (4 \times 2)$, or $4 \times (5-2) = (4 \times 5) - (4 \times 2)$, and the second one shows $3 \times 5 = (4 \times 5) - (1 \times 5)$, or $(4-1) \times 5 = (4 \times 5) - (1 \times 5)$. To provide further experience with the distributive property, place several partitioned arrays on an overhead-projector transparency for students to write the associated mathematical sentences and some sentences for which they must draw the corresponding array model.

Cuisenaire rods also provide a good model for the distributive property. The following indicates how the rods can be used to show $3 \times (4+2) = (3 \times 4) + (3 \times 2)$:

4	2
4	2
4	2

4
4
4

2
2
2

$$3 \times (4+2) \qquad = \qquad (3 \times 4) + (3 \times 2)$$

The distributive property aids in mental computation and provides an opportunity for students to break down tasks into simpler components. Multiplications such as these are commonly a part of the third-grade curriculum:

$$\begin{aligned} 8 \times 13 &= 8 \times (10+3) \\ &= (8 \times 10) + (8 \times 3) \\ &= 80 + 24 \\ &= 104 \end{aligned}$$

Research gives supportive evidence that emphasizing the distributive property promotes comprehension, transfer, and retention.[3]

Introducing Division

The early childhood teacher is concerned with helping a student conceptualize division and discover some properties of the operation and with providing instruction on some early computational procedures in division. The final division algorithm for whole numbers, however, does not occur until fourth or fifth grade. In introductory work, the three notations $a \div b$, $b\overline{)a}$, and $\frac{a}{b}$ should be used. They may appear in different contexts, but a student should learn that all of them mean division. In the sentence $a \div b = c$, a is the dividend, b is the divisor, and the answer c is the quotient. In many third-grade textbooks, the simpler terms of factors and product are given to the parts of a division sentence, terms which are descriptive when a division equation is related to multiplication. The sentence $a \div b = c$ means $c \times b = a$ and consequently a in both equations can be called the product and b and c, factors. This terminology is analogous to that of sum and addends in a subtraction sentence.

Approaches

A student needs experience in several approaches to division.

1. *the number-of-subsets interpretation of quotient, or measurement approach*

   This interpretation of quotient is used in a child's beginning work in division. Examples of this approach are "Fifteen pennies have the same value as how many nickels?" and "Sandy has twenty cookies to give to her friends. If each friend receives four cookies, how many of her friends will receive cookies?" To solve $20 \div 4 = \square$ by the measurement approach, a set of twenty elements is separated into subsets of four.

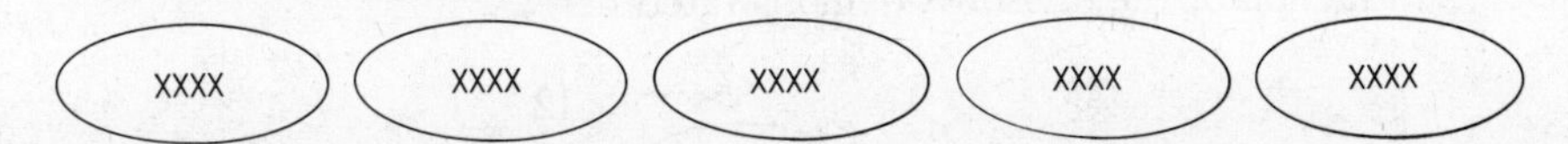

   There are five such subsets; thus $20 \div 4 = 5$. It is helpful to review with the students that the separation of a set into nonequivalent subsets is associated with subtraction and not division.

2. *number-per-subset interpretation of quotient, or partitive approach*

   Studies show that young children have more difficulty with the partitive approach than the measurement approach to division.[4] Some examples of the partitive interpretation are "If three children share twelve paper dolls equally, how many dolls will each child get?" and "If you share twenty crayons with your friends, how many crayons will each of you have?" The following diagram shows how $12 \div 3 = \square$ is solved by this method.

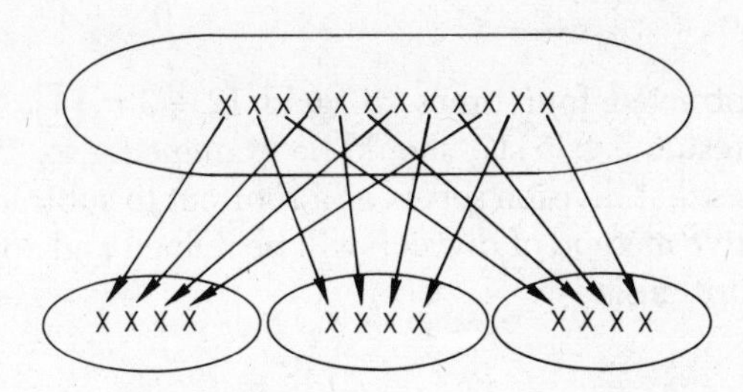

   The elements of the twelve-element set are placed in the three subsets in a "one for this set, one for this set, and so on" fashion until the set of twelve is exhausted. Since four elements occur in each subset, the student should conclude $12 \div 3 = \boxed{4}$. The manner in which partitive situations must be solved may account for the increased difficulty students have with this approach over the measurement one.

3. *the missing-factor method*

After children have examined division with sets, they should relate it to multiplication in a way analogous to the addition-subtraction pattern. The step in solving $10 \div 2 = \square$ are

$$10 \div 2 = \square, \square \times 2 = 10, \boxed{5} \times 2 = 10, 10 \div 2 = \boxed{5}.$$

The quotient 5 is the missing factor in the related multiplication sentence $5 \times 2 = 10$.

Provide readiness for this approach by giving examples such as "In an array there are ten dots and two columns of dots. How many rows are there?" and "How many sets of 3 are contained in 15?" Also, in examining a sentence such as $3 \times 6 = 18$, cover the product sometimes and the factors at other times so that children learn to solve the equation when it is in the form $3 \times 6 = \square$, $3 \times \square = 18$, and $\square \times 6 = 18$.

4. *division as repeated subtraction*

Further breadth in division can be supplied by having students relate division to subtraction. The procedure for solving $12 \div 3 = \square$ by repeated subtraction of the divisor is demonstrated here:

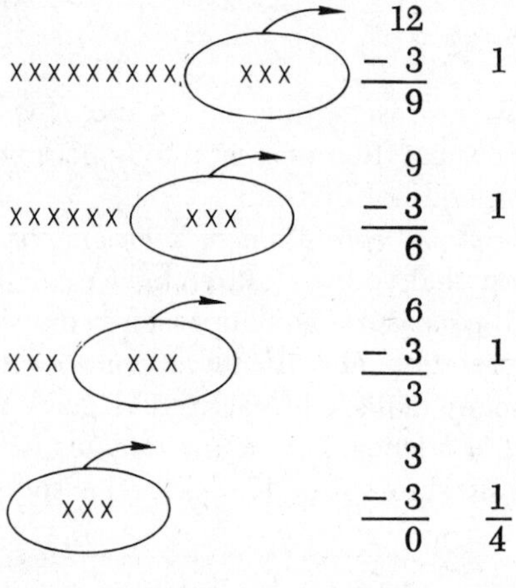

XXXXXXXXX (XXX) $\quad \begin{array}{r} 12 \\ -\ 3 \\ \hline 9 \end{array} \quad 1$

XXXXXX (XXX) $\quad \begin{array}{r} 9 \\ -\ 3 \\ \hline 6 \end{array} \quad 1$

XXX (XXX) $\quad \begin{array}{r} 6 \\ -\ 3 \\ \hline 3 \end{array} \quad 1$

(XXX) $\quad \begin{array}{r} 3 \\ -\ 3 \\ \hline 0 \end{array} \quad \begin{array}{r} 1 \\ \hline 4 \end{array}$

Three is subtracted four times to yield $12 \div 3 = \boxed{4}$. Divisions in which remainders result, e.g., $5\overline{)17}$, should be examined also. The preceding problem illustrates that division serves as a shortcut to subtraction in many cases. The subtractive method of division will be refined and applied in subsequent division algorithms.

Teaching Suggestions

The interpretations of division just described should be included in beginning activities in division. Some ideas follow:

1. *Have students enact a situation in which they serve as elements of a set to be separated into equivalent subsets.* For instance, line up six children in front of the class and ask them to break up into groups of three. Have other

members of the class tell the number of groups of three there are in six. Repeat this activity with other numbers. Return to one of the problems the students have enacted and introduce the associated division sentence. In the example with six children, write $6 \div 3 = 2$ on the chalkboard, explaining that this is a division sentence read as "six divided by three equals two."

At a later time, the students should also enact the number-per-subset approach. Place rope circles on the floor and have the children stand inside the circles so that the same number is in each circle. Say, for example, to the class, "We have six students who have formed three equivalent subsets. How many students are in each subset?" and record $6 \div 3 = 2$ on the chalkboard.

2. *In a small-group or class activity, have students use plastic counters to explore division.* Distribute counters in envelopes so that not all children get the same number of counters. For example, give each student 12, 15, 18, 21, 24, or 27 counters. Instruct the class to make stacks of three, and call on different students to tell the number of counters and the number of stacks of three they have and, if possible, state the division sentence while other students record the equations on the chalkboard.
3. *Involve the children in a feltboard activity on division.* For example, place twelve "scoops" of ice cream and six "cones" on the feltboard. Select a child to put the scoops of ice cream on the cones so that each cone has the same number of scoops. The process involved is suggested here:

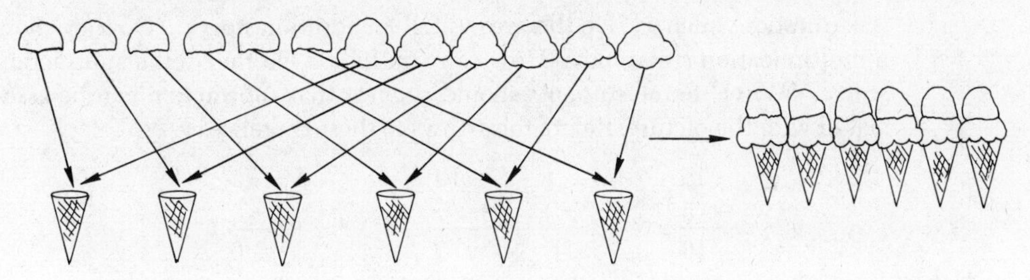

Make the sentence $12 \div 6 = 2$ on the feltboard, stressing that a set of twelve separated into six subsets has two in each subset. Create other situations of the number-per-subset interpretation of quotient with the felt materials, and ask students to write the division sentences represented.

4. *Use Cuisenaire rods as an embodiment of the number-of-subsets interpretation of quotient.* Give an open sentence such as $8 \div 2 = \square$, and let the class members suggest ways to solve it with the rods. They should find how many 2 rods can be placed in a train to exactly fit the length of the 8 rod.

8			
2	2	2	2

The model shows $8 \div 2 = \boxed{4}$. Explain to the students this division can also be written as $\begin{array}{r}4\\ 2\overline{)8}\end{array}$. Then introduce division situations with a remainder, e.g., $3\overline{)14}$, using the new notation. Instruct them to find the number of 3 rods that match the rods representing 14. When they have made a train of 3 rods alongside 14, point out that four 3 rods are shorter than the rods for 14 but another 3 rod would make the train of 3 rods too long. Ask them to find the missing rod that completes the train of 3 rods. The students' work should look like this:

10				4
3	3	3	3	2

Have a student tell the number of threes in 14 and the number left over. Introduce the term remainder and record $\begin{array}{r} 4, \text{R2} \\ 3\overline{)14} \end{array}$ on the chalkboard, emphasizing that 4 is the quotient and 2, the remainder. Give the children other problems such as $4\overline{)24}$ and $8\overline{)26}$ to solve using the rods.

5. *Guide students in solving division sentences on the number line.* Show an arrow diagram such as

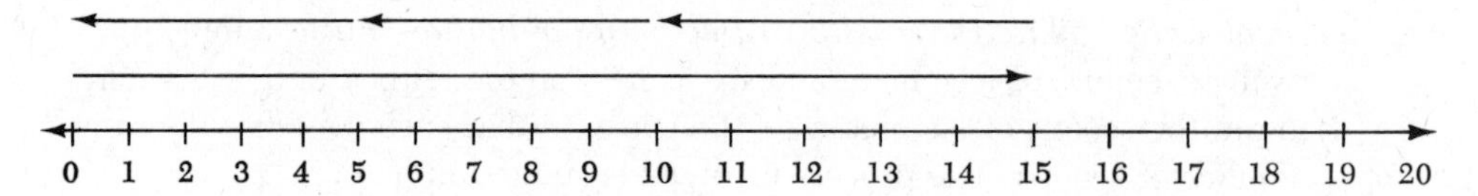

Ask questions such as "Do the arrows tell an addition story?" "Do they show a multiplication situation?" "How can you tell?" "Do they tell a subtraction story?" "Why?" Some students should suggest that subtraction can be associated with the picture. Relate the arrows to these sentences:

$$\begin{array}{r} 15 \\ -\ 5 \\ \hline 10 \end{array} \qquad \begin{array}{r} 10 \\ -\ 5 \\ \hline 5 \end{array} \qquad \begin{array}{r} 5 \\ -5 \\ \hline 0 \end{array}$$

Tell them these three subtraction sentences show there are three fives in fifteen and that means $15 \div 5 = 3$. Give students a worksheet of number-line problems for which they write repeated-subtraction and division sentences.

6. *Have children work in small groups to develop their understanding of division.* Supply each student with a felt-tip pen and each group with a box of about a hundred plastic counters and a set of approximately twenty laminated cards each with a division problem such as $15 \div 3 = \square$, $9\overline{)18}$, or $4\overline{)17}$ on it. Instruct the children to take turns drawing a card from the stack, solving the division sentence on the card by withdrawing the correct number of counters and separating the set of counters into subsets, and writing the quotient and remainder, if one, on the card with the pen. Each student should read the division sentence aloud and explain to the members of his group his procedure for solving it. After a card is completed and checked by the other members of the group, the child solving the problem should place the card face up in a discard pile. You may need to demonstrate the procedure before the students begin.

7. *It may be fitting for some children to explore division on the Minicomputer.* Model the procedure for them or give them a handout showing the process. To solve $9 \div 3 = \square$, for example, place counters on the ones board to enter the dividend 9 and loops to represent the divisor 3. Move the counters until the same number occurs in each loop, this number being the quotient and the counters left over representing the remainder. The exchange of counters is in a left-to-right direction as it was for subtraction on the Minicomputer. These are the steps for solving $9 \div 3 = \square$:

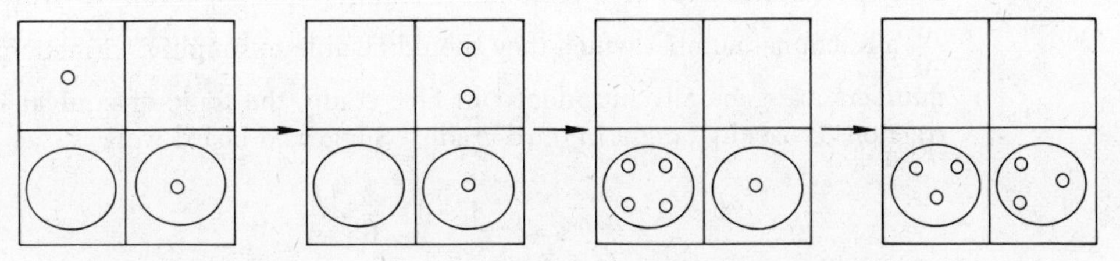

Since there are three counters in each divisor loop, $9 \div 3 = \boxed{3}$.

8. *On the chalkboard or an overhead projector transparency, draw multiplication-division pictures and ask students to suggest a multiplication and a division sentence for each picture.* In the example

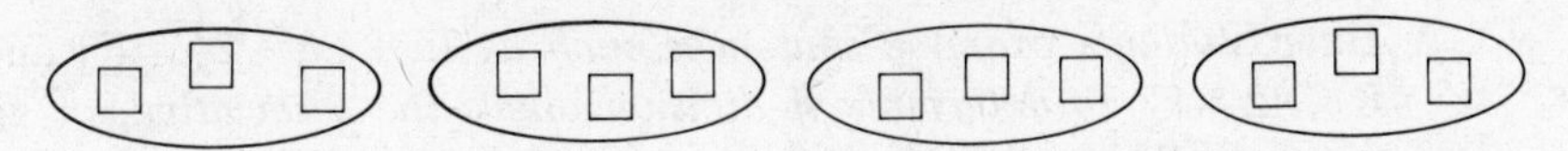

some students may say there are four sets of three, or twelve objects. Others may see a set of twelve squares separated into subsets of three of which there are four. Record the sentences corresponding to these respective situations on the transparency: $4 \times 3 = 12$, $12 \div 3 = 4$. Tell the students that these equations are equivalent and that division problems can be solved by writing the related multiplication sentence. Demonstrate the process.

$20 \div 4 = \square$	$24 \div \square = 6$	$\square \div 3 = 6$
$\square \times 4 = 20$	$6 \times \square = 24$	$6 \times 3 = \square$
$\boxed{5} \times 4 = 20$	$6 \times \boxed{4} = 24$	$6 \times 3 = \boxed{18}$
$20 \div 4 = \boxed{5}$	$24 \div \boxed{4} = 6$	$\boxed{18} \div 3 = 6$

Provide follow-up exercises that require students to write multiplication and division sentences associated with set models and solve division equations by solving the related multiplication sentence.

Properties of Division

Certain properties of division can be explored as students participate in activities such as the foregoing ones; however, for the most part, division patterns of whole numbers can best be examined at the symbolic level using series of related division sentences. Some students may be able to treat these principles in depth while others may do little with them. Help pupils understand that most of the properties of multiplication do not hold for division but that division principles are very similar to those for subtraction.

Division properties and considerations for teaching them follow:

1. *Have students investigate whether or not division is commutative.* They may use sets or Cuisenaire rods to compare examples such as $6 \div 3 = \square$ and $3 \div 6 = \square$. The first quotient is 2 because there are two subsets of 3 in a set of 6. The second quotient is not a whole number since only part of a set of 6 will fit in a set of 3. Remind the children that $3 \div 6$ can be written $\frac{3}{6}$, a fractional number which they should be able to simplify. (Fractional numbers are generally introduced in first grade; the topic of equivalent fractions is usually taught in third grade.) Suggested board work is

$$6 \div 3 = 2 \qquad 3 \div 6 = \frac{3}{6} = \frac{1}{2}$$

$$6 \div 3 \neq 3 \div 6.$$

   Verbalize that changing the order of the dividend and divisor, in general, changes the quotient. Challenge students to think of an exception to this rule. When $a = b$ (a and b whole numbers), $a \div b$ does equal $b \div a$ provided a and b are not 0.
2. *Have students examine sentences such as* $12 \div (6 \div 2) = \square$ and $(12 \div 6) \div 2 = \square$ *to determine if division obeys the associative, or grouping, property.* The quotients in the two examples are different; thus division is not associative.
3. *Guide students to discover the property of* one *in division.* This property is illustrated with the open sentences $1 \div 1 =$ _____, $2 \div 1 =$ _____, $3 \div 1 =$ _____, . . ., $9 \div 1 =$ _____. Have students complete them and urge the pupils to make the generalization that any number divided by one is the number itself.
4. *Children should explore examples in which the dividend and divisor are the same:* $1 \div 1 =$ _____, $2 \div 2 =$ _____, $3 \div 3 =$ _____, . . ., $9 \div 9 =$ _____. Verbalize that any number (different from 0) divided by itself is one. Extend the pattern to an example such as $2847 \div 2847 = \square$.
5. *A limited distributive property holds for division with respect to addition and to subtraction.* It is limited in that division is distributive on the right with respect to addition and subtraction but not on the left. Students should perform computations such as these:

$$\begin{aligned}(8 + 4) \div 2 &= 12 \div 2 \\ &= 6 \\ (8 \div 2) + (4 \div 2) &= 4 + 2 \\ &= 6\end{aligned} \qquad \begin{aligned}(8 - 4) \div 2 &= 4 \div 2 \\ &= 2 \\ (8 \div 2) - (4 \div 2) &= 4 - 2 \\ &= 2\end{aligned}$$

   The pairs of sentences name the same number. It may be helpful to model each sentence by joining or separating sets as the operations indicate. For instance, $(8 + 4) \div 2$ would be represented

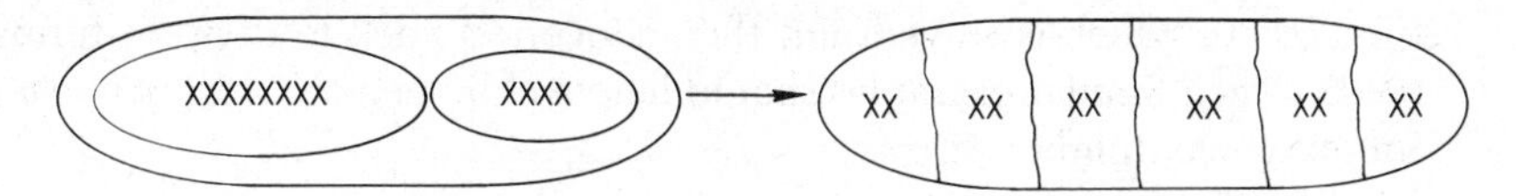

and $(8 \div 2) + (4 \div 2)$ by this procedure:

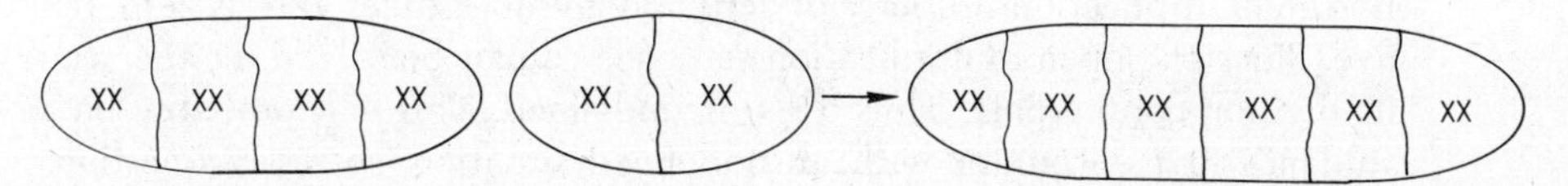

The diagrams indicate that $(8+4) \div 2 = (8 \div 2) + (4 \div 2)$. Have students contrast the result in the preceding equations with examples such as these:

$$12 \div (4+2) = 12 \div 6 = 2 \qquad (12 \div 4) + (12 \div 2) = 3 + 6 = 9$$

Since $2 \neq 9$, $12 \div (4+2) \neq (12 \div 4) + (12 \div 2)$. Also, $12 \div (4-2) \neq (12 \div 4) - (12 \div 2)$. Thus only the right-hand distributive property of division holds with respect to addition and to subtraction.

Cuisenaire rods also offer a good model for the distributive property of division with respect to addition and to subtraction. The equation $(4 + 6) \div 2 = (4 \div 2) + (6 \div 2)$ is suggested with the rods this way:

The distributive property for division is important in later work with addition and subtraction of fractional numbers. It also helps a student relate division facts, e.g.,

$$\begin{aligned} 54 \div 6 &= (30 + 24) \div 6 \\ &= (30 \div 6) + (24 \div 6) \\ &= 5 + 4 \\ &= 9 \end{aligned}$$

and helps him understand the long division algorithm.

6. *There are three cases involving zero in division of whole numbers.* They are (a) $0 \div a = \square$ for any whole number $a \neq 0$, (b) $a \div 0 = \square$ for whole numbers $a \neq 0$, and (c) $0 \div 0 = \square$. To present division with zero in a systematic way, review the students on the missing-factor approach to division and the zero property of multiplication. Then guide them through examples such as these:

a) $0 \div 5 = \square$
$\square \times 5 = 0$
$\boxed{0} \times 5 = 0$
$0 \div 5 = \boxed{0}$

Here $0 \div 5 = \square$ is changed to its equivalent multiplication equation

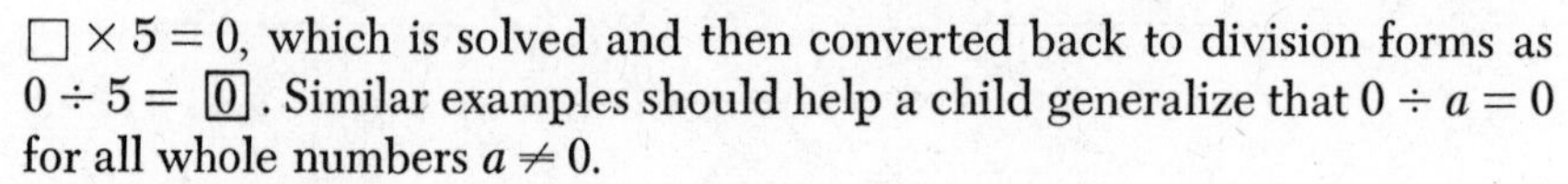

$\square \times 5 = 0$, which is solved and then converted back to division forms as $0 \div 5 = \boxed{0}$. Similar examples should help a child generalize that $0 \div a = 0$ for all whole numbers $a \neq 0$.

b) $5 \div 0 = \square$

$\square \times 0 = 5$

By the multiplication property of zero, any number times zero is zero, not five. Since the open multiplication sentence cannot be solved, neither can its division counterpart. Thus $5 \div 0$ is undefined. This will probably be a student's first encounter with an undefined situation, so many questions concerning it may ensue.

c) $0 \div 0 = \square$

$\square \times 0 = 0$

By the multiplication property of zero, any number times zero is zero, so $0 \times 0 = 0$, $1 \times 0 = 0$, $2 \times 0 = 0$, and so on. Thus, the solution to $\square \times 0 = 0$ and hence $0 \div 0 = \square$ is any number. Since the quotient is not unique, $0 \div 0$ is also considered to be undefined.

7. *Another pattern in division involves interchanging divisor and quotient to produce two related division sentences, e.g.,* $10 \div 2 = 5$ *and* $10 \div 5 = 2$. This property can be taught after a child has learned the measurement and partitive interpretations of division. The set model

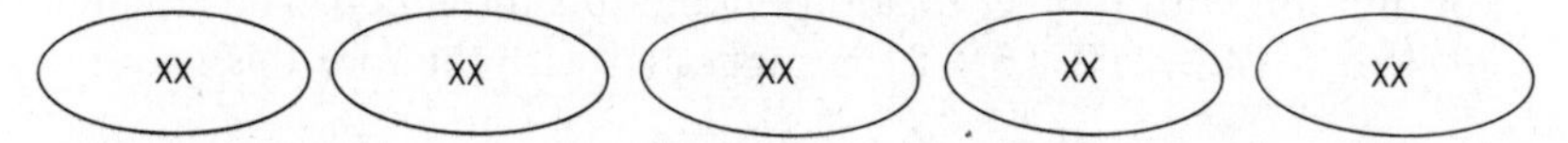

represents $10 \div 2 = 5$ if the quotient means the number of subsets and $10 \div 5 = 2$ if the number-per-subset interpretation of quotient is used. Students should understand that if they know a division sentence of the form $a \div b = c$, b and c whole numbers different from 0, they can write another in the form $a \div c = b$.

8. *Have students solve related open division sentences in which the dividend is multiplied or divided by a nonzero number but the divisor remains the same.*

Examples are:

$6 \div 3 =$ ____, $12 \div 3 =$ ____, $18 \div 3 =$ ____, $24 \div 3 =$ ____

and pairs such as:

$3\overline{)6}$ $3\overline{)24}$, $2\overline{)8}$ $2\overline{)16}$, $5\overline{)30}$ $5\overline{)15}$, and $6\overline{)48}$ $6\overline{)6}$.

Have the children perform the computations first; then question them on how the quotients change and how that result corresponds to the change in the dividend. If they cannot generalize that multiplying or dividing the dividend by a (nonzero) number produces the same effect in the quotient, supply more examples. Extend the pattern to include exercises such as "If $12 \div 6 = 2$, then $(12 \times 9) \div 6 =$ ____" and "If $72 \div 8 = 9$, then $(72 \div 3) \div 8 =$ ____." The students should use the division pattern to mentally find the quotients.

Understanding this property of division will greatly assist the student in performing division algorithms. For example, he should be able to conclude that $\begin{array}{r} 400 \\ 3\overline{)1200} \end{array}$ since $\begin{array}{r} 4 \\ 3\overline{)12} \end{array}$. (The dividend 12 has been multiplied by 100 so the quotient 4 must be multiplied by 100 also.) If a child can reason this way, he is ready to extend his division experience to a problem such as $3\overline{)1247}$. In chapter 10, methods for division at higher levels are discussed.

9. *Another pattern in division involves multiplying or dividing the divisor by a nonzero number and keeping the dividend constant.* Pairs of examples to explore are:

$$2\overline{)18} \quad 6\overline{)18}, \; 3\overline{)27} \quad 9\overline{)27},$$
$$4\overline{)16} \quad 2\overline{)16}, \text{ and } 8\overline{)24} \quad 2\overline{)24}.$$

Call on various students to state the quotients. Question the class members on how the divisors in each pair are related and how the quotients in the same pair compare. Some students should see that multiplying or dividing the divisor by a (nonzero) number produces the opposite effect in the quotient. Extend the pattern to a problem such as "If $72 \div 9 = 8$, then $72 \div 18 = \_\_\_\_\_$," which the students must compute mentally. Have them explain how $72 \div 9 = 8$ served as a hint in solving $72 \div 18 = \_\_\_\_\_$.

10. *If both the dividend and divisor are multiplied by the same nonzero number, the quotient remains the same.* Several related sentences and proper questioning should help students discover this pattern. Some sample problems include:

$$2\overline{)12} \quad 4\overline{)24},$$
$$3\overline{)6} \quad 9\overline{)18}, \text{ and}$$
$$8\overline{)24} \quad 2\overline{)6}.$$

In the first pair, 2 and 12 have been doubled, but the quotient in both cases is 6. In the second example, dividend and divisor in the first sentence have been multiplied by 3, and in the last pair the dividend 24 and divisor 8 have been divided by 4 to produce $2\overline{)6}$. Assess the students' understanding of the pattern by having them complete an example such as "If $8 \div 2 = 4$, then $(8 \times 20) \div (2 \times 20) = \_\_\_\_\_$" without paper and pencil or calculator.

This property of division justifies the rule for dividing numbers with ending zeros. For example, $18{,}000 \div 3{,}000 = 6$ since $(18 \times 1000) \div (3 \times 1000) = 18 \div 3 = 6$. Such considerations provide readiness for the final division algorithm.

Summarizing the Basic Facts of Multiplication and Division

The one hundred basic facts of multiplication are $0 \times 0 = 0$ through $9 \times 9 = 81$, and the basic division facts are those which can be associated with basic multiplication facts. For instance, $12 \div 6 = 2$ and $48 \div 6 = 8$ are basic facts of division since $2 \times 6 = 12$ and $8 \times 6 = 48$ are basic multiplication facts, but $30 \div 2 = 15$ is not since $15 \times 2 = 30$ is not a basic fact of multiplication. You should begin to summarize the basic multiplication facts after your students have gained experience in multiplication at the enactive, iconic, and symbolic levels of learning.

When summarizing the facts, present them in an orderly fashion. You may, for example, have students generate all the basic facts in which two is a factor or give all the facts having a certain product, e.g., $8 = 1 \times 8$, $8 = 8 \times 1$, $8 = 2 \times 4$, $8 = 4 \times 2$. As these facts are found, have students compose a summary table for multiplication similar to that discussed for addition in the previous chapter. For the more difficult products, construct open sentences guiding students to use the properties and facts they already know to discover new facts.

$$\begin{aligned} 5 \times 8 &= (2 + \underline{\quad\quad}) \times 8 \\ &= (2 \times \underline{\quad\quad}) + (3 \times 8) \\ &= 16 + \underline{\quad\quad} \\ &= \underline{\quad\quad} \end{aligned}$$

$$\begin{aligned} 7 \times 9 &= (\underline{\quad\quad} + 5) \times 9 \\ &= (2 \times 9) + (5 \times \underline{\quad\quad}) \\ &= \underline{\quad\quad} + 45 \\ &= \underline{\quad\quad} \end{aligned}$$

If they know the facts in which 2 and 3 are factors and the distributive property and can add, they should be able to complete the blanks in the first example. In the second example, the facts for twos and fives help the students determine a basic fact in which seven is a factor. Have the students record new products on their composite tables. The completed composite table looks like this:

×	0	1	2	3	4	5	6	7	8	9
0	0	0	0	0	0	0	0	0	0	0
1	0	1	2	3	4	5	6	7	8	9
2	0	2	4	6	8	10	12	14	16	18
3	0	3	6	9	12	15	18	21	24	27
4	0	4	8	12	16	20	24	28	32	36
5	0	5	10	15	20	25	30	35	40	45
6	0	6	12	18	24	30	36	42	48	54
7	0	7	14	21	28	35	42	49	56	63
8	0	8	16	24	32	40	48	56	64	72
9	0	9	18	27	36	45	54	63	72	81

The table can be used to read division facts also. To find $63 \div 9$, for example, locate 63 under the 9 at the top and read the missing factor, or quotient, 7 at the left. Notice the eye action in reading division facts on the table is the reverse to that in reading multiplication facts—another reminder of the inverse relationship of multiplication and division.

When the composite table is completed, it should be used for examining patterns and reinforcing properties of multiplication and division. Let students spend time describing the patterns they see, and then guide them on those they overlooked. These are some considerations:

1. How many basic multiplication facts are there in the table?
2. In the first row and column all the products are zero. No where else in the table is a product zero shown. Why are the products in the first row and column zero and why are those the only zero products?
3. What is unusual about the second row and second column? How do they compare with the top and left side of the table? What is the product when 1 is a factor?
4. Why are the first row and first column, second row and second column, . . ., tenth row and tenth column the same?
5. Across from 1, the products increase by one from one product to the next, and across from 2 they increase by two from one to the next. What happens in the row across from 3, from 4, and so on? What can you say, then, about finding all the products having factor 2, 3, 4, . . ., 9? (You count by twos, threes, fours, etc.)
6. Show how to read the table for finding the product 6×9, for example.
7. Show how to find a quotient such as $18 \div 6$ on the table.
8. Find $0 \div 1, 0 \div 2, 0 \div 3, \ldots, 0 \div 9$ on the table. What is the quotient in each case? What do you think the quotient will always be when the dividend is zero and divisor is different from zero?
9. Can you find $6 \div 0$ on the table? What do you think this says about $6 \div 0$?
10. Can you find $0 \div 0$ on the table? How many times is $0 \div 0$ shown? What does this say about $0 \div 0$?

The table can also be used to introduce the perfect squares 0, 1, 4, 9, . . ., 81, which occur along the main diagonal, and to have students explore multiplication of even and odd whole numbers. (See chapter 14.)

Practice in the Basic Facts

After a student has formed the concepts of multiplication and division and has gained experience in properties of the operations, you should promote mastery by providing interesting practice activities. Some children will require more drill than others. Many of the games and activities for practicing the basic addition and subtracton facts suggested in the preceding chapter can easily be adapted for multiplication and division facts. Other ideas include:

1. *Have students make a set of Napier's rods.* They will need eleven strips of poster board, one of which serves as an index rod with the factors 0 through 9 listed. Each of the other ten rods has one of the factors 0, 1, 2, . . ., 9 entered at the top and the products of that factor and those on the index rod given next. A diagonal is drawn to separate tens and ones in the products. The index rod and the ten strips of products appear this way:

When the index rod and the other ten strips are lined up in order, the arrangement resembles the composite multiplication table discussed earlier..

As students construct these rods, they are actually recording basic multiplication facts. Also, a child can refer to the appropriate strips if he forgets a product. For example, 7×9 is found on the rods by placing the index rod by the 9 rod and reading the product 63 on the 9 rod across from 7 on the index rod.

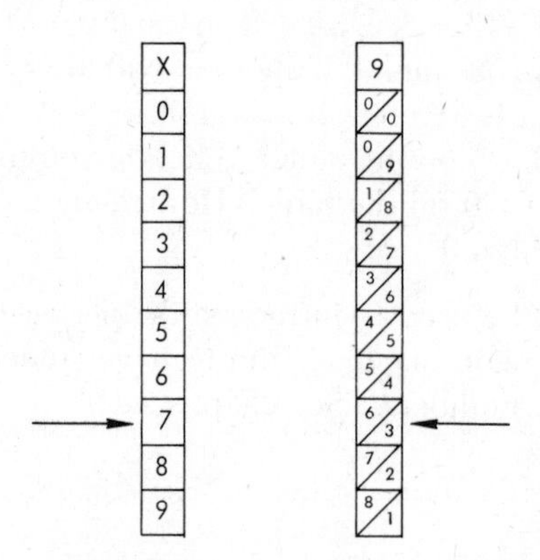

Products of multi-digit factors can also be found with the rods (see chapter 10).

Encourage students to look for a set of Napier's rods in novelty shops. You may plan an activity combining social studies and mathematics using the rods or let mathematically talented students conduct independent work on the history of the rods.

2. *Introduce multiplication and division fact teams, e.g.,*

$$30 \div 5 = 6 \qquad 6 \times 5 = 30$$
$$30 \div 6 = 5 \qquad 5 \times 6 = 30$$

As seatwork, give students examples representing one equation in a fact team and have them generate the other equations.

3. *Have children practice multiplication facts in which the blank occurs in different positions.* Give them a handout of a series of open sentences to solve. Include examples such as:

$$\begin{array}{llll} 6 \times 3 & = \_\_\_\_ & 18 \div 2 & = \_\_\_\_ \\ 4 \times \_\_\_\_ & = 4 & 54 \div \_\_\_\_ & = 9 \\ \_\_\_\_ \times 8 & = 0 & \_\_\_\_ \div 5 & = 4 \\ \_\_\_\_ \times \_\_\_\_ & = 24 & \_\_\_\_ \div \_\_\_\_ & = 1. \end{array}$$

You may have to remind the students to write division sentences as multiplication sentences if they become confused, e.g., $54 \div \_\_\_\_ = 9$ is equivalent to $9 \times \_\_\_\_ = 54$.

4. *Give students flash cards with problems involving zero in multiplication and division.* Examples include

$$\begin{array}{r} 3 \\ \underline{\times\, 0} \end{array}, \quad 0 \times 0, \quad 0 \div 8, \quad 7 \div 0, \quad 0 \div 0, \quad \text{and } 2\overline{)0}.$$

Have the children sort the cards into two sets, one designated as "Names for 0" and the other as "Not Names for 0." It is also good to include addition and subtraction expressions such as $6 - 6$, $10 - 0$, $0 - 4$, $0 + 0$, and $5 + 0$.

In a similar way, children may be asked to sort flash cards dealing with the basic facts of multiplication and division into sets labeled as "Names for 6," "Names for 18," and so on. Such an exercise provides practice in generating all pairs of whole-number factors of a number as well as drill in the basic multiplication and division facts.

5. *Revise the "Hint" game suggested in chapter 7 to include statements for practicing division properties.* Write examples on the chalkboard and have students mentally compute the answer.

If $64 \div 8 = 8$, then $64 \div 16 = \_\_\_\_$
If $28 \div 7 = 4$, then $(28 \times 5) \div 7 = \_\_\_\_$
If $6 \div 3 = 2$, then $54 \div 27 = \_\_\_\_$
If $16 \div 2 = 8$, then $(16 \times 7) \div (2 \times 7) = \_\_\_\_$

Question the students on how the hint helped them complete the open sentence. For instance, a child answering the fourth problem may say, "16 and 2 have both been multiplied by 7 so the answer is still 8."

6. *Design practice activities in which students compute basic facts while working with the associative and commutative properties, e.g.,*

$$\begin{aligned} 9 \times 4 &= (3 \times \_\_\_\_) \times (2 \times 2) \\ &= (3 \times 2) \times (\_\_\_\_ \times 2) \\ &= \_\_\_\_ \times 6 \\ &= \_\_\_\_. \end{aligned}$$

Such an exercise is a meaningful drill in that the student is relating basic multiplication facts.

Closing

When guiding students in exploring multiplication and division basic facts, teachers should be careful to appeal to the learners' previous understandings of addition and subtraction. The properties of multiplication and addition are similar as are those of division and subtraction. Individual differences will occur among students on the amount of time needed to form the concepts of multiplication and division, to discover properties of the operations, and to master the basic facts. The teacher should provide activities on these operations at various levels. For instance, on a given day, one group of children may work independently on certain properties of division or multiplication, another group solve multiplication or division sentences on the Minicomputer, a group engage in practice activities on the basic facts of the operations, another group solve open sentences which relate the more difficult basic multiplication facts to easier ones, and still another group model basic facts of the operations with counters or blocks, the number line, or Cuisenaire rods. It is with this latter group the teacher may need to concentrate her time.

Notes

1. Fredricka K. Reisman, *Diagnostic Teaching of Elementary School Mathematics* (Chicago: Rand McNally College Publishing Company, 1977), p. 187.
2. Margaret A. Hervey, "Children's Responses to Two Types of Multiplication Problems," *The Arithmetic Teacher* 13 (April 1966): 291–92.
3. Marilyn N. Suydam and J. Fred Weaver, *Using Research: A Key to Elementary School Mathematics*, (Columbus, Ohio: ERIC Information Analysis Center for Science, Mathematics and Environmental Education, 1975), p. 7–8.
4. Suydam and Weaver, *Using Research: A Key to Elementary School Mathematics*, p. 7–8.

Assignment

1. Why is the array approach not a good way to model products having one factor zero?
2. Why is the set-of-equivalent-disjoint-sets approach not a good model for a problem such as $0 \times 5 = \square$? What strategies can be used for teaching facts in which zero is the first factor?
3. Write an activity for teaching the Cartesian-product interpretation of multiplication. Include ideas on how to assess the students to determine if they are ready to use the Cartesian-

product approach to multiplication and suggestions on how you intend to meet the needs of slow and gifted learners in the class.

4. Use a number line to teach a child the commutative property of multiplication.
5. Teach a student the distributive property of multiplication over addition using an array. After teaching the activity, tell how you would modify it before teaching it again.
6. Study the approaches used in teaching multiplication and division facts involving zero in several second- and third-grade mathematics textbooks. What examples are used and how many pages are given to the concepts?
7. Design an integrated mathematics and social studies lesson in which Napier's rods are used.
8. Why are there only 90 basic division facts when there are 100 each of basic addition, subtraction, and multiplication facts?

Laboratory Session

Preparation: Study chapter 8 of this textbook.

Objectives: Each student should be able to:

1. use plastic counters, Cuisenaire rods, arrays on grid paper, the number line, and the Minicomputer to model the basic facts of multiplication and division of whole numbers and
2. illustrate the commutative, associative, and identity properties of multiplication, the left- and right-hand distributive property of multiplication over addition and over subtraction, and the right-hand distributive property of division over addition.

Materials: plastic counters, Cuisenaire rods, grid paper, yarn loops, Minicomputer boards (two for each student)

Procedure: Students should form groups of three or four and solve these problems:

1. Use plastic counters and loops to
   a) solve $4 \times 3 = \square$,
   b) solve $10 \div 2 = \square$,
   c) solve $4 \times 0 = \square$,
   d) solve $15 \div \square = 3$,
   e) show $6 \times 2 = 2 \times 6$,
   f) show $4 \times 1 = 4$ and $1 \times 4 = 4$,
   g) demonstrate $(12 + 8) \div 4 = (12 \div 4) + (8 \div 4)$.
2. Use Cuisenaire rods to
   a) find the solution to $3 \times 6 = \square$,
   b) solve $14 \div 7 = \square$,
   c) show $5 \times 2 = 2 \times 5$.
   What is the name of this property?
   d) demonstrate $2 \times (5 \times 3) = (2 \times 5) \times 3$.
   What is the name of this property?
   e) show $(8 + 6) \div 2 = (8 \div 2) + (6 \div 2)$.
   What is this property called?

f) demonstrate $3 \times (5 + 2) = (3 \times 5) + (3 \times 2)$.
What is the name of this property?
g) demonstrate all the basic multiplication facts with a product of 12.

3. Use arrays on grid paper to
   a) solve $5 \times 6 = \square$,
   b) show $5 \times 6 = 6 \times 5$,
   c) demonstrate $4 \times (5 + 2) = (4 \times 5) + (4 \times 2)$,
   d) show $(6 - 3) \times 8 = (6 \times 8) - (3 \times 8)$,
   e) show $7 \times 1 = 7$ and $1 \times 7 = 7$.
4. Draw arrow diagrams on the number line to
   a) solve $2 \times 6 = \square$,
   b) solve $6 \div 3 = \square$,
   c) show $4 \times 3 = 3 \times 4$,
   d) show $5 \times 1 = 5$ and $1 \times 5 = 5$.
5. Use the Minicomputer to compute:
   a) $6 \times 5 = \square$, b) $20 \div 4 = \square$,
   c) $3 \times 7 = \square$, d) $3\overline{)17}$.

Bibliography

Banks, J. Houston. *Elementary-School Mathematics, A Modern Approach for Teachers.* Boston: Allyn and Bacon, Inc., 1966.

Davidson, Jessica. *Using the Cuisenaire Rods—A Photo/Text Guide for Teachers.* New Rochelle, New York: Cuisenaire Company of America, Inc., 1969.

Grossnickle, Foster E., and Reckzeh, John. *Discovering Meanings in Elementary School Mathematics,* 6th ed. New York: Holt, Rinehart and Winston, Inc., 1973.

Hervey, Margaret A. "Children's Responses to Two Types of Multiplication Problems." *The Arithmetic Teacher* 13 (April 1966): 288–92.

Jungst, Dale G. *Elementary Mathematics Methods: Laboratory Manual.* Boston: Allyn and Bacon, Inc., 1975.

Reisman, Fredricka K. *Diagnostic Teaching of Elementary School Mathematics.* Chicago: Rand McNally College Publishing Company, 1977.

Suydam, Marilyn N., and Weaver, J. Fred. *Using Research: A Key to Elementary School Mathematics.* Columbus, Ohio: ERIC Information Analysis Center for Science, Mathematics and Environmental Education, 1975.

Extending Addition and Subtraction of Whole Numbers

In chapter 7 methods and materials for introducing addition and subtraction were discussed, and attention was given in chapter 6 to ideas for teaching place value and grouping by tens. If a child has learned the basic facts of addition and subtraction and understands place value, he is ready to add and subtract larger whole numbers, the subject of the present chapter. To teach higher level computational procedures, the same materials used for teaching place value are serviceable: a pocket chart, base-ten blocks, paper strips and squares, bean sticks, Cuisenaire rods, an abacus, money, and the Minicomputer.

The view of mathematics educators on the importance of developing computational skills is evidenced by Trafton and Suydam's statement of the position of the Editorial Panel of *The Arithmetic Teacher*. Points presented are:

1. Computational skill is one of the important, primary goals of a school mathematics program.
2. All children need proficiency in recalling basic number facts, in using standard algorithms with reasonable speed and accuracy, and in estimating results and performing mental calculations, as well as an understanding of computational procedures.
3. Computation should be recognized as just one element of a comprehensive mathematics program.
4. The study of computation should promote broad, long-range goals of learning.

5. Computation needs to be continually related to the concepts of the operations and both concepts and skills should be developed in the context of real-world applications.
6. Instruction on computational skills needs to be meaningful to the learner.
7. Drill-and-practice plays an important role in the mastery of computational skills, but strong reliance on drill-and-practice alone is not an effective approach to learning.
8. The nature of learning computational processes and skills requires purposeful, systematic, and sensitive instruction.
9. Computational skills need to be analyzed carefully in terms of effective sequencing of the work and difficulties posed by different types of examples.
10. Certain practices in teaching computation need thoughtful reexamination.[1]

Early childhood educators should be thoughtful of these points since it is in early childhood that many of the ideas that make computational processes meaningful begin.

Higher Level Addition Tasks

Extending addition from the basic facts and place value includes instruction in finding simple sums, adding by endings, adding numbers represented by multi-digit numerals in the case when no regrouping (carrying) is required, and adding numbers named by multi-digit numerals in the case when regrouping is needed.

Simple Sums

Simple, or easy, sums are sums of multiples of ten, one hundred, one thousand, and so on. Examples are $20 + 40 = 60$ and $700 + 800 = 1500$. This instructional task usually begins in first grade. An introductory lesson can be planned with a pocket chart, Cuisenaire rods, base-ten blocks, or beansticks. To guide students to solve $20 + 40 = \square$ with Cuisenaire rods, for example, tell them the white rod represents 10 (remember the rods can be used to represent different numbers in different discussions) and establish with them that the other rods would represent 20, 30, 40, . . ., 100. Let them suggest how 20 and 40 would be added using the rods. They should form a train of the 20 and 40 rods and then find the rod of equivalent length.

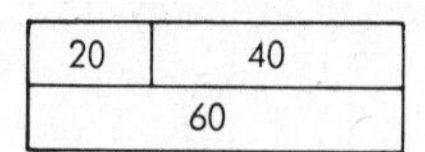

Record and have them record the fact in both horizontal and vertical notation: $20 + 40 = 60$, $\begin{array}{r} 20 \\ +\ 40 \\ \hline 60 \end{array}$. More explorations can be made with other materials. The sentence $30 + 90 = 120$ is illustrated on an abacus as

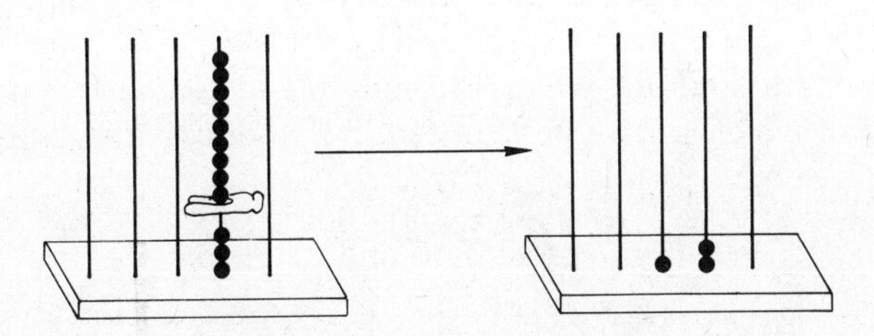

and with money (dimes and dollars) as

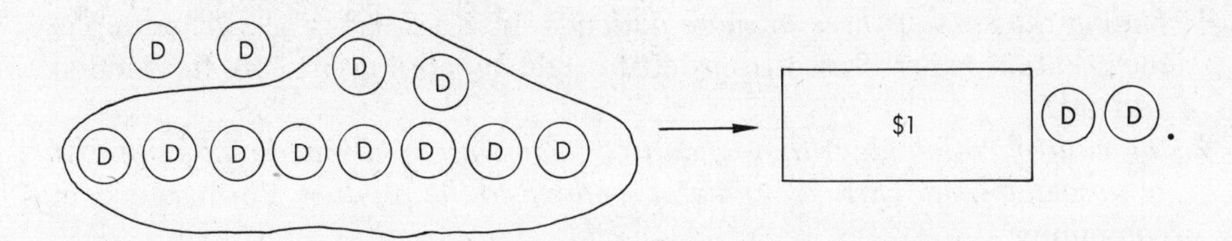

Have students represent each addend (notice a clothespin separates the represented addends on the abacus), exchange ten tens for one hundred, and record and verbalize the resulting addition sentences.

After supplying introductory experiences with concrete and pictorial materials, have students complete open sentences such as

$$\begin{aligned} 40 + 30 &= 4 \text{ tens} + \underline{\quad\quad} \text{ tens} \\ &= \underline{\quad\quad} \text{ tens} \\ &= \underline{\quad\quad} \end{aligned} \qquad \text{and} \qquad \begin{aligned} 800 + 500 &= \underline{\quad\quad} \text{ hundreds} + 5 \underline{\quad\quad} \\ &= \underline{\quad\quad} \text{ hundreds} \\ &= \underline{\quad\quad}. \end{aligned}$$

Bring to the children's attention the number of zeros in each addend and in the sum. Guide them to discover that these sums can be found quickly by computing a basic fact of addition and attaching the same number of zeros as in each addend.

After multiplication is introduced in second grade, precocious learners may benefit from examining the mathematical explanation of why the sum in these special addition sentences has the same number of zeros as each addend. Have these children complete sentences and give reasons for each step as suggested in this example:

$$\begin{aligned} 50 + 30 &= (5 \times \underline{10}) + (\underline{3} \times 10), \text{ place value} \\ &= (5 + \underline{3}) \times 10, \text{ distributive property} \\ &= \underline{8} \times 10, \text{ basic addition fact} \\ &= \underline{80}, \text{ place value} \end{aligned}$$

Most students will have very little trouble computing simple sums. All children need to be skillful in this task before attempting to find more difficult sums. Furthermore, being able to compute these sums is important in estimating sums when calculating with the final addition algorithm.

Addition, No Regrouping Required

Beginning work with addition of numbers named by multi-digit numerals includes finding the sum of numbers represented by a two-digit and one-digit numeral, by a two-digit numeral and one which is a multiple of ten, and by two two-digit numerals neither of which is a multiple of ten. Representative examples are

$$\begin{array}{r} 23 \\ +\underline{\ 6} \end{array}, \quad \begin{array}{r} 23 \\ +\underline{40} \end{array}, \quad \text{and} \quad \begin{array}{r} 23 \\ +\underline{46} \end{array}.$$

The first case $\begin{array}{r} 23 \\ +\underline{\ 6} \end{array}$ is an example in which the student adds by endings. That is, the sum can be found by using the single-digit numeral and the ones digit, or ending, of the two-digit numeral. Addition by endings is important in

1. *finding the sum of three or more addends.* In $7 + 9 + 5 = \_\_$, for example, the addition is performed from left to right by adding 5 to 16, the hidden addend.
2. *the multiplication algorithm when two factors are multiplied and a number represented by a "carried" numeral is added to the product.* For instance, in computing

$$\begin{array}{r} 2 \\ 86 \\ \times\ \underline{4} \\ 344 \end{array}$$

4 and 8 are multiplied and 2 is added to the product 32.

3. *adding costs, e.g., 49¢ and 6¢, of items.*[2]

To teach addition by endings, construct activities in which the child counts forward from the number represented by the two-digit numeral as far as the one-digit numeral suggests. For example, write on the chalkboard exercises such as those indicated here:

$$\begin{array}{r} 4 \\ +\underline{3} \end{array} \quad \begin{array}{r} 14 \\ +\underline{\ 3} \end{array} \quad \begin{array}{r} 24 \\ +\underline{\ 3} \end{array} \quad \begin{array}{r} 34 \\ +\underline{\ 3} \end{array} \ldots \begin{array}{r} 94 \\ +\underline{\ 3} \end{array}$$

$$\begin{array}{r} 7 \\ +\underline{8} \end{array} \quad \begin{array}{r} 17 \\ +\underline{\ 8} \end{array} \quad \begin{array}{r} 27 \\ +\underline{\ 8} \end{array} \quad \begin{array}{r} 37 \\ +\underline{\ 8} \end{array} \ldots \begin{array}{r} 97 \\ +\underline{\ 8} \end{array}$$

Have students count three forward from 14, 24, 34, . . . , 94 and eight forward from 17, 27, 37, . . . , 97 to arrive at the higher decade sums. Question them on how the tens digit in the two-digit addend compares to the tens digit in the sum. In the first row of examples, the number of tens in the sum is the same as that in the two-digit addend because the sum of the ones is less than ten. The second set of examples illustrates that the number of tens in the sum is one more than that in the two-digit addend when the sum of the ones is at least ten.

If students have difficulty with this process, let them show the addition on a number line or with money. A model with dimes and pennies suggesting the sums in the first set of sentences in the preceding examples looks like this:

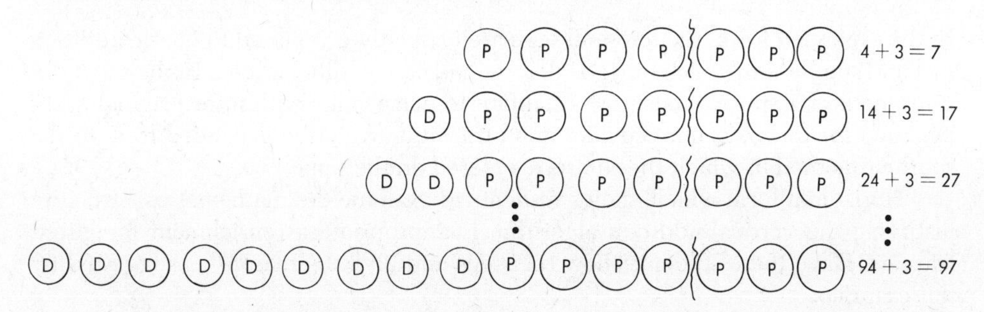

Also, provide oral exercises such as "I'm thinking of a number 6 more than 28. What is the number?" The one-digit numeral need never be more than 8 since 8 is the largest possible numeral to be "carried."

Let us now examine procedures for teaching how to add two two-digit addends in which no regrouping is required. Similar techniques would apply for all other additions involving no regrouping. The example $\begin{array}{r}23\\ +\underline{46}\end{array}$ can be solved with base-ten blocks by representing each addend with the blocks, uniting the units and recording the sum, and uniting the tens blocks and recording the sum.

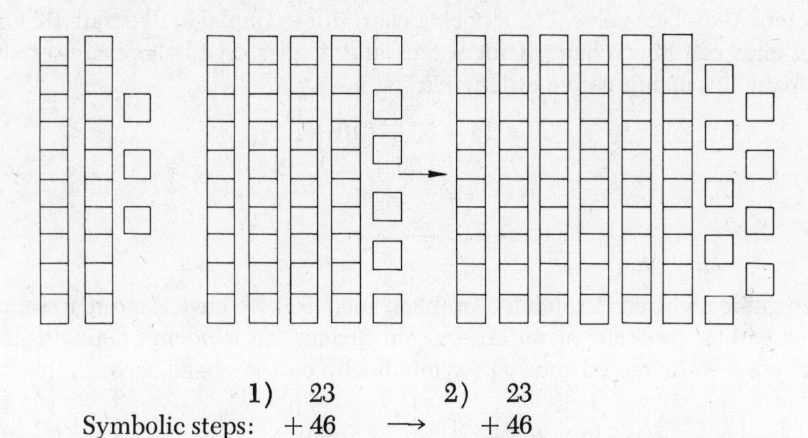

$$\text{Symbolic steps:}\quad 1)\ \begin{array}{r}23\\ +\,46\\ \hline 9\end{array} \quad\longrightarrow\quad 2)\ \begin{array}{r}23\\ +\,46\\ \hline 69\end{array}$$

Two symbolic methods which can be taught prior to introducing the final algorithm for addition are the expanded-notation and separate-sums methods. These processes should help a student understand the final algorithm and should also serve as remediation devices for children having trouble with the final algorithm. An illustration of these methods and the final algorithm for addition follow:

Expanded notation	Separate sums	Final algorithm
$\begin{array}{rl}23 = & 20+3\\ +\,46 = & \underline{40+6}\\ & 60+9=69\end{array}$	$\begin{array}{rl}23 & \\ +\,46 & \\ \hline 9 & (3+6=9)\\ 60 & (20+40=60)\\ \hline 69 & \end{array}$	$\begin{array}{r}23\\ +\,46\\ \hline 69\end{array}$

In the expanded form, the tens and ones are explicitly given, and the basic addition fact and simple sum which must be determined can readily be seen. In the separate-sums method, a student must give attention to place value within the original problem, and for this reason it provides good transition from the expanded form to the shortcut method in which the sums are entered on the same line.

Horizontal form which allows students to examine the mathematical structure justifying the vertical addition algorithm is appropriate as enrichment for gifted learners. Have these students show the steps and give the reasons as suggested by this example:

$$\begin{aligned} 23 + 46 &= (20 + 3) + (40 + 6), \text{ expanded notation (place value)} \\ &= (20 + 40) + (3 + 6), \text{ commutative and associative properties of addition} \\ &= 60 + 9, \text{ simple sum, basic fact of addition} \\ &= 69, \text{ place value.} \end{aligned}$$

Addition, Regrouping Required

This type of instructional task is usually introduced in second grade. Help students get ready for the regrouping (carrying) step by reviewing them on grouping by tens and place value. Use a pocket chart, for example, to illustrate that 5 tens and 14 ones can be exchanged for 6 tens and 4 ones. Symbolic exercises should accompany this manipulative experience.

$$\begin{aligned} 50 + 14 &= 50 + (10 + \_\_) \\ &= (50 + \_\_) + 4 \\ &= \_\_ + 4 \\ &= \_\_ \end{aligned}$$

To guide children in solving a problem such as $\begin{array}{r} 37 \\ +49 \\ \hline \end{array}$, have them represent each addend with a concrete material, e.g., an abacus. As students demonstrate the process concretely, record the steps symbolically on the chalkboard.

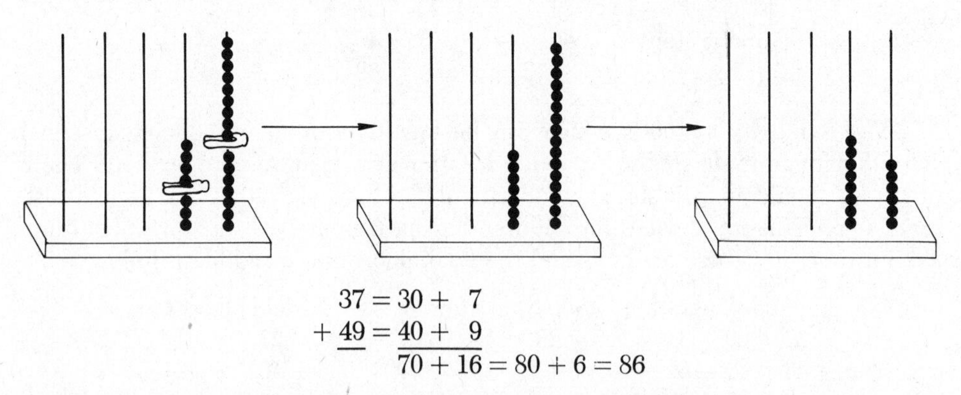

$$\begin{array}{rl} 37 = & 30 + \ \ 7 \\ +\underline{49} = & \underline{40 + \ \ 9} \\ & 70 + 16 = 80 + 6 = 86 \end{array}$$

Notice the subordinate skills and understandings involved in this symbolic task: writing numerals in expanded notation, computing basic addition facts and simple sums, regrouping ones and tens (associative property), and writing numerals in positional notation.

As suggested in chapter 3, a hand-held calculator can be used to teach the process in algorithms of arithmetic. Students should understand that ones, tens, and hundreds are added separately in addition algorithms. A task card such as

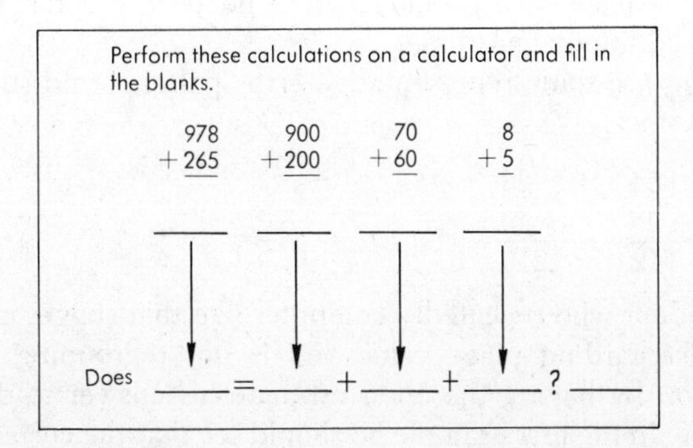

is appropriate in preliminary work with the expanded form of addition. The separate-sums method also provides readiness for the final algorithm for addition.

$$\begin{array}{rl} & 37 \\ + & \underline{49} \\ & 16\ (7+9=16) \\ & \underline{70}\ (30+40=70) \\ & 86 \end{array} \longrightarrow \begin{array}{r} 37 \\ +\ \underline{49} \\ 86 \end{array}$$

Both the expanded and separate-sums forms require a student to write down much of his thinking and, in so doing, he should learn the process involved in the shortcut form. Help the children understand that the final form is simply an abbreviated version of the separate-sums method.

As students perform addition of numbers represented by multi-digit numerals, emphasize that they estimate the sums by rounding each addend and computing the resulting simple sum. In the immediately preceding example, 37 and 49 would be rounded to 40 and 50, respectively, and the real sum should be close to 40 + 50, or 90. A child who thinks in this manner can thus determine the answer 86 is reasonable. Teaching children to estimate answers deserves emphasis in third grade. Estimation procedures, however, are to be used in addition to, not instead of, actual checking of the computation for accuracy.

You will need to exercise your own discretion as to when individual students are ready to advance to the final method for addition. Gifted students will need to spend little time with the introductory symbolic methods. Once students understand the final algorithm, they need extensive practice experience. Practice exercises are readily available in elementary mathematics textbooks.

Computational Errors in the Final Addition Algorithm

Some errors students make in computation are careless and fit no systematic pattern. Many, however, are due to the child's performing an algorithm incorrectly, a situation that sometimes occurs when learning an algorithm for a new operation interferes with previously learned algorithms. Other errors are caused by lack of mastery of the basic facts and confusion on place value and properties of the operation involved. Third-grade teachers especially should be alert to common computational errors in the final addition algorithm.

The following are some representative error patterns and suggestions for remediating them.

1. $$\begin{array}{r} 74 \\ +29 \\ \hline 913 \end{array} \qquad \begin{array}{r} 54 \\ +28 \\ \hline 712 \end{array} \qquad \begin{array}{r} 86 \\ +49 \\ \hline 1215 \end{array}$$

A student who repeatedly computes like that suggested in these examples is disregarding place value and is not regrouping correctly. Begin remediation by having the child estimate his answer to determine if it is reasonable. In the first example he should see that the correct sum is close to 100.

$$\begin{array}{rcr} 74 & \longrightarrow & 70 \\ +29 & \longrightarrow & +30 \\ \cline{1-1} \cline{3-3} & & 100 \end{array}$$

To help the student correct his error, have him model the problem with base-ten blocks, beansticks, or some other manipulative or pictorial device and record his work in expanded notation, separate-sums form, or even the final algorithm. If you use expanded or separate-sums form, refer to the method as you guide him to condense that form to the final algorithm. As the student uses the final method, emphasize that only one digit at a time can be recorded in each column in the sum and that if the sum in one place is ten or more, one ten must be added in the next place to the left. Have him record the auxiliary numeral at the top. It may be helpful to present problems to the student on grid paper on which he enters the sums so that one digit is in each square needed or to have him turn his notebook paper ninety degrees from its regular position so that his computation is shown by this arrangement:

$$\begin{array}{c|c|c} & 7 & 4 \\ & 2 & 9 \\ \hline 1 & 0 & 3 \end{array}$$

2. $$\begin{array}{r} 54 \\ +\ 2 \\ \hline 76 \end{array} \qquad \begin{array}{r} 32 \\ +\ 3 \\ \hline 65 \end{array} \qquad \begin{array}{r} 47 \\ +22 \\ \hline 69 \end{array}$$

The student who performs this computation is adding the number named by the tens place and the one-digit addend. His work, however, is correct in the case when numbers represented by two two-digit numerals are added. The error may be due to interference from learning the multiplication algorithm and points to the need for reviewing previously learned processes.

Begin remediation by having the student estimate the sum, in which case he should conclude 76 in the first example and 65 in the second one are not reasonable answers. Then ask him to complete open sentences in horizontal notation so that he sees that the one-digit addend is added only to the ones in the two-digit addend. Referring to the horizontal notation, guide the student through the vertical algorithm, first by drawing a line to separate tens and ones and then in the final form.

$$\begin{aligned} 54 + 2 &= (50 + \underline{4}) + 2 \\ &= 50 + (\underline{4} + \underline{2}) \\ &= 50 + \underline{6} \\ &= \underline{56} \end{aligned} \quad \longrightarrow \quad \begin{array}{c|c} \text{Tens} & \text{Ones} \\ 5 & 4 \\ & 2 \\ \hline 5 & 6 \end{array} \quad \longrightarrow \quad \begin{array}{r} 54 \\ +\ 2 \\ \hline 56 \end{array}$$

3. $\begin{array}{r} 240 \\ +103 \\ \hline 300 \end{array} \qquad \begin{array}{r} 400 \\ +348 \\ \hline 700 \end{array} \qquad \begin{array}{r} 706 \\ +203 \\ \hline 909 \end{array}$

Errors in computing when the addends have zeros can probably easily be alleviated by reviewing the student on the additive-identity property. Design exercises such as

$$\begin{array}{r} 3 \\ +\underline{0} \end{array} \qquad \begin{array}{r} 0 \\ +\underline{8} \end{array} \qquad \begin{array}{r} 0 \\ +\underline{0} \end{array}$$

asking the child to model them with concrete objects or marks on paper. The student could be confusing the zero property of multiplication with the identity property of addition in which case it would be good to also include examples such as these:

$$\begin{array}{r} 0 \\ \times\underline{0} \end{array} \qquad \begin{array}{r} 4 \\ \times\underline{0} \end{array} \qquad \begin{array}{r} 0 \\ \times\underline{6} \end{array}$$

Other systematic errors and ways to remediate them are given in Robert B. Ashlock's book, *Error Patterns in Computation,* published by Charles E. Merrill, 1300 Alum Creek Drive, Columbus, Ohio 43216.

Sequencing Addition of Whole Numbers

In the present chapter and chapter 7, addition tasks have been discussed from the basic facts to the final algorithm. The following list summarizes how addition instructional tasks which usually occur in grades one through three can be sequenced:

1. the basic facts of addition (which implies prerequisite learning such as forming subsets, demonstrating one-to-one correspondences, determining the cardinal number of a set, uniting sets, and so on, has already taken place)
   Examples: $0 + 1 =$ ____, $8 + 3 =$ ____, $9 + 9 =$ ____
2. numeration facts of addition (place value and expanded notation)
   Example: $48 = 40 +$ ____

3. simple sums

   Examples: $40 + 50 = \_\_\_\_$, $900 + \_\_\_\_ = 1300$

4. the sum of a two-digit or three-digit addend (not a multiple of 10 or 100) and a one-digit addend, no regrouping required

   Examples: $\begin{array}{r} 42 \\ +\ 5 \\ \hline \end{array}$ $\begin{array}{r} 814 \\ +\ 3 \\ \hline \end{array}$

5. the sum of a two-digit or three-digit addend (not a multiple of 10 or 100) and a two-digit or three-digit addend (a multiple of 10 or 100), no regrouping needed

   Examples: $\begin{array}{r} 32 \\ +\ 40 \\ \hline \end{array}$ $\begin{array}{r} 734 \\ +\ 20 \\ \hline \end{array}$ $\begin{array}{r} 249 \\ +\ 300 \\ \hline \end{array}$

6. the sum of two two-digit addends (neither a multiple of 10), a three-digit addend (not a multiple of 100) and a two-digit addend (not a multiple of 10), or two three-digit addends (neither multiple of 100), no regrouping required

   Examples: $\begin{array}{r} 53 \\ +\ 26 \\ \hline \end{array}$ $\begin{array}{r} 372 \\ +\ 26 \\ \hline \end{array}$ $\begin{array}{r} 424 \\ +\ 135 \\ \hline \end{array}$

7. renaming such as $40 + 12 = \_\_\_\_ + 2$ and $60 + 16 = 70 + \_\_\_\_$

8. the sum of a two-digit addend (not a multiple of 10) and a one-digit addend or a three-digit addend (not a multiple of 100) and a one-digit addend, regrouping required

   Examples: $\begin{array}{r} 47 \\ +\ 9 \\ \hline \end{array}$ $\begin{array}{r} 238 \\ +\ 6 \\ \hline \end{array}$

9. the sum of two two-digit addends (one a multiple of 10), a three-digit (not a multiple of 100) and a two-digit addend (a multiple of 10), or two three-digit addends (one a multiple of 100), regrouping required

   Examples: $\begin{array}{r} 76 \\ +\ 90 \\ \hline \end{array}$ $\begin{array}{r} 824 \\ +\ 90 \\ \hline \end{array}$ $\begin{array}{r} 563 \\ +\ 600 \\ \hline \end{array}$

10. the sum of two two-digit addends, one regrouping needed

    Example: $\begin{array}{r} 47 \\ +\ 26 \\ \hline \end{array}$

11. the sum of a three-digit addend (not a multiple of 100) and a two-digit addend (not a multiple of 10) or two three-digit addends (neither a multiple of 100), one regrouping necessary

    Examples: $\begin{array}{r} 635 \\ +\ 92 \\ \hline \end{array}$ $\begin{array}{r} 473 \\ +\ 308 \\ \hline \end{array}$

12. the sum of two two-digit addends, two regroupings needed

    Example: $\begin{array}{r} 83 \\ +\ 59 \\ \hline \end{array}$

13. addition of a number represented by a three-digit numeral (not a multiple of 100) and one named by a two-digit numeral (not a multiple of 10) or two numbers, each named by three-digit numerals, two regroupings required

    Examples: $\begin{array}{r} 754 \\ +\ 86 \\ \hline \end{array}$ $\begin{array}{r} 396 \\ +\ 842 \\ \hline \end{array}$

14. addition of numbers represented by two three-digit numerals, three regroupings required

$$\begin{array}{r} 478 \\ +\underline{863} \end{array}$$

Example:

This sequence can be extended to include larger addends and addition of more than two numbers. As students advance through the steps, encourage understanding of the algorithmic procedures by having them enact the processes with objects or pictures as they begin new symbolic methods. Identifying sequential learning tasks and presenting topics at various levels of maturity is important in realizing mathematical objectives and is in accordance with the learning theory discussed in chapter 1.

Higher Level Subtraction Tasks

Extending subtraction from the basic facts involves giving instruction in computing simple differences, subtraction of numbers represented by multi-digit numerals in the case when no regrouping is required, and subtraction of numbers named by multi-digit numerals when renaming is necessary. Both decomposition and equal-additions methods can be used in problems in which renaming must be done. Beginning work with subtraction algorithms should involve the students in demonstrating the process with materials such as paper strips and squares, base-ten blocks, an abacus, or beansticks. The final subtraction algorithm is usually presented in third grade.

Simple Differences

Simple, or easy, differences are differences of two numbers, each of which is a multiple of ten, of one hundred, of one thousand, and so on. Students can begin computing them soon after they have been introduced to basic subtraction facts and place value. Children should understand, for example, that $50 - 30 = 20$ when they remove three of the tens beansticks from a set of five tens sticks and see that two tens sticks are left.

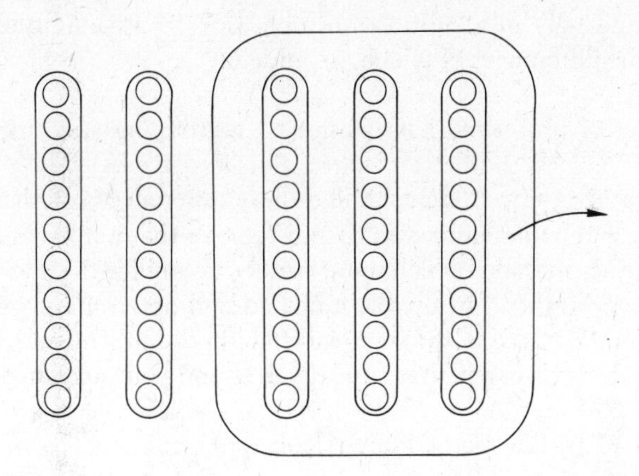

Children also need symbolic exercises to guide their thinking in computing simple differences. Examples are

$$\begin{aligned} 70 - 30 &= \_\_\_\_ \text{ tens} - 3 \text{ tens} \\ &= (7 - \_\_\_\_) \text{ tens} \\ &= \_\_\_\_ \text{ tens} \\ &= \_\_\_\_ \end{aligned} \qquad \text{and} \qquad \begin{aligned} 800 - 200 &= \_\_\_\_ \text{ hundreds} - \_\_\_\_ \text{ hundreds} \\ &= (\_\_\_\_ - 2) \text{ hundreds} \\ &= \_\_\_\_ \text{ hundreds} \\ &= \_\_\_\_. \end{aligned}$$

When students complete examples such as these, have them examine the number of zeros in the minuend, subtrahend, and difference. They should discover that these differences can be determined by finding the difference in a basic subtraction fact and attaching the same number of ending zeros as in either minuend or subtrahend.

After multiplication is introduced, superior learners can delve more deeply into the mathematical justification of the pattern in simple differences. These children can complete open sentences and give reasons for each step as suggested by this example:

$$\begin{aligned} 800 - 500 &= (8 \times \underline{100}) - (\underline{5} \times 100), \text{ place value} \\ &= (8 - \underline{5}) \times 100, \text{ distributive property of multiplication over subtraction} \\ &= \underline{3} \times 100, \text{ basic subtraction fact} \\ &= \underline{300}, \text{ place value.} \end{aligned}$$

Subtraction, No Renaming Required

In second grade most students are ready to perform subtractions requiring no regrouping, or borrowing, step. Subordinate skills include writing numerals in expanded notation, using place value, computing differences in basic facts and in simple subtractions, and recognizing the rearrangement property of addition and subtraction. A student needs experience in the rearrangement pattern so that he can understand why in a subtraction such as $\begin{array}{r} 87 \\ -\underline{24} \end{array}$ two subtractions are actually performed, the differences of which are added.

Examples of problems that require no regrouping step are $\begin{array}{r} 28 \\ -\underline{\ 6} \end{array}$, $\begin{array}{r} 47 \\ -\underline{20} \end{array}$, and $\begin{array}{r} 53 \\ -\underline{21} \end{array}$. One way to guide students to find these differences is to have them represent the minuend with concrete materials and remove the number dictated by the subtrahend. Another method, which more closely resembles the symbolic process, is to represent the subtrahend below the minuend and determine how many more ones, tens, and so on there are in the minuend than in the subtrahend. The work the students would do with paper strips and squares and your accompanying board work for solving $\begin{array}{r} 53 \\ -\underline{21} \end{array}$ by this method might look like this:

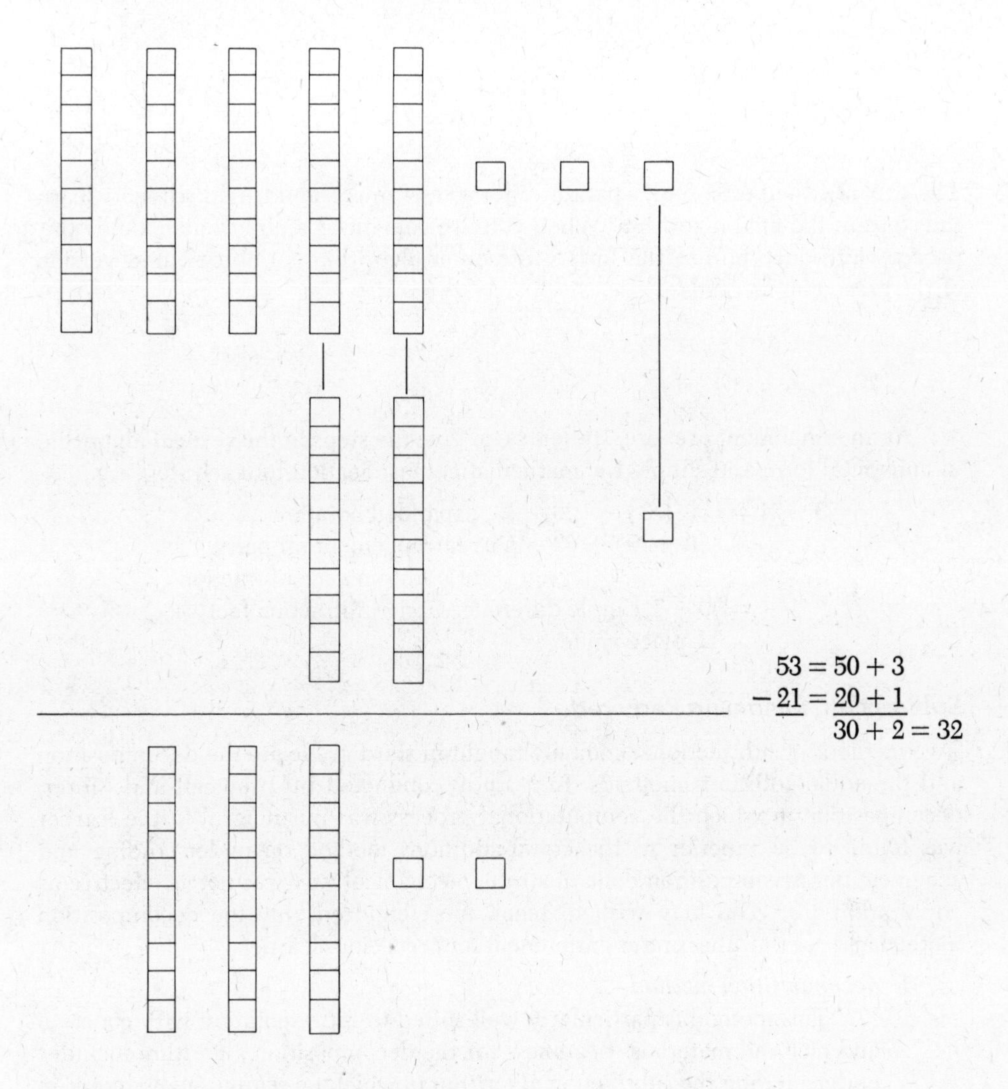

Actually correct notation in the example is

$$\begin{array}{rl} 53 = & 50 + 3 \\ -\underline{21} = & \underline{-(20 + 1)} \end{array} \quad \text{or} \quad \begin{array}{rl} 53 = & 50 + 3 \\ -\underline{21} = & \underline{-20 - 1} \end{array}$$

but these forms may be more confusing than helpful to most young students. Emphasize that subtraction should begin in the ones place. Children may not understand why this right-to-left direction is important until problems requiring regrouping occur. As an aside, you may wish to illustrate in a more advanced problem involving regrouping of the minuend why the ones are subtracted first.

In the separate-differences algorithm, another symbolic method for performing subtraction, the fact that the ones and tens are subtracted separately becomes evident in the form itself.

$$
\begin{array}{r}
53 \\
-\underline{21} \\
2\ (3-1=2) \\
\underline{30}\ (50-20=30) \\
32
\end{array}
$$

The expanded-notation and separate differences forms should help students learn the steps in the final algorithm. When you are convinced a child understands this process, introduce him to the final subtraction algorithm, an abbreviated version of the preceding methods.

$$
\begin{array}{r}
53 \\
-\underline{21} \\
32
\end{array}
$$

As an enrichment exercise, students can give the steps in the vertical algorithm in horizontal form and supply the mathematical justification for each step, e.g.,

$$
\begin{aligned}
53-21 &= (50+3)-(20+1), \text{ expanded notation} \\
&= (50-20)+(3-1), \text{ rearrangement property of addition and subtraction} \\
&= 30+2, \text{ simple difference, basic subtraction fact} \\
&= 32, \text{ place value.}
\end{aligned}
$$

Subtraction, Renaming Required

Two methods of subtraction commonly taught in third grade are the decomposition and the equal-additions methods. In a study conducted by Brownell and Moser, decomposition in which the computational process was meaningful to the learner was found to be superior to the equal-additions method on understanding and accuracy, but decomposition done in a rote, mechanical way was not as effective as equal additions.[3] You may wish to teach most children only the decomposition method and use the other one as enrichment for precocious learners.

1. *decomposition method*

   This method is particularly well suited to representation with concrete and pictorial materials. Readiness for the decomposition algorithm includes understanding the subtraction algorithm in which no regrouping is required and regrouping a ten with ones, a hundred with tens, etc. To give students experience in the regrouping step, have them trade in one of the dimes in 4 dimes and 6 pennies, for example, for pennies and tell the resulting number of dimes and pennies. Supply associated symbolic exercises.

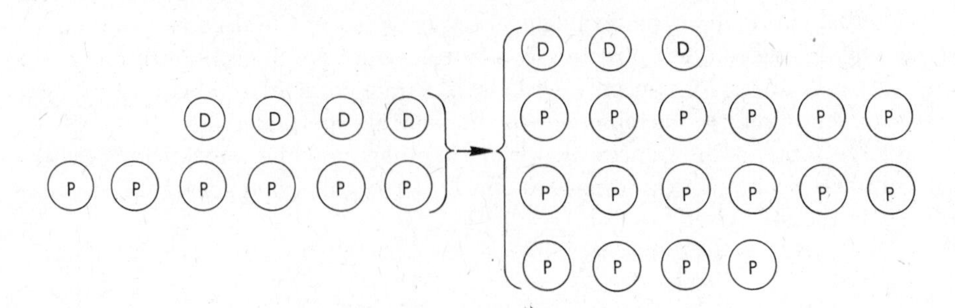

4 dimes, 6 pennies = 3 dimes, _____ pennies
$46 = \_\_\_\_\_ + 6$
$= 30 + \_\_\_\_\_$

Examples in which regrouping is done in the hundreds should also be included.

$$\begin{aligned} 347 &= 300 + \_\_\_\_\_ + 7 \\ &= 300 + 30 + \_\_\_\_\_ \\ &= \_\_\_\_\_ + 130 + 17 \end{aligned}$$

When students are ready to begin the decomposition algorithm, have them physically present the algorithm as you record the symbolic process on the chalkboard. The steps in solving $\begin{array}{r} 65 \\ -\underline{37} \end{array}$ with the aid of an abacus follow:

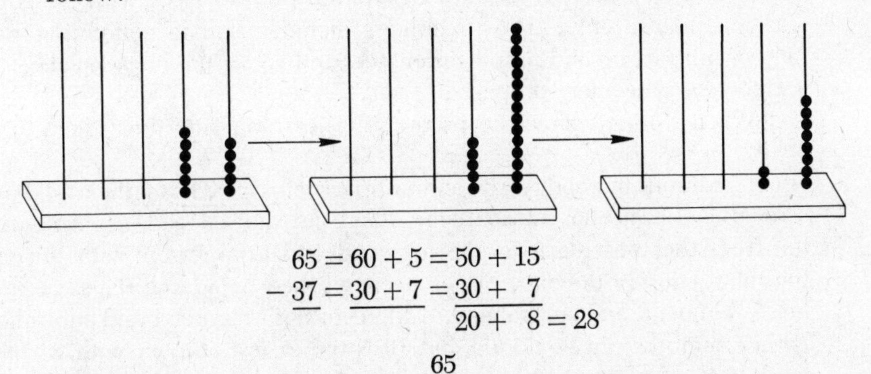

$$\begin{array}{rcl} 65 = 60 + 5 &=& 50 + 15 \\ -\underline{37} = \underline{30 + 7} &=& \underline{30 + \ \ 7} \\ && 20 + \ \ 8 = 28 \end{array}$$

You may prefer to use the form $\begin{array}{r} 65 \\ -\underline{37} \\ 28 \end{array}$ instead of expanded notation to record the work. After students learn to decompose the minuend and regroup one time, they can begin subtraction with two regroupings, e.g.,

$$\begin{array}{rcl} 523 = 500 + 20 + 3 &=& 400 + 110 + 13 \\ -\underline{274} = \underline{200 + 70 + 4} &=& \underline{200 + \ \ 70 + \ \ 4} \\ && 200 + \ \ 40 + \ \ 9 = 249 \end{array} \qquad \text{or} \quad \begin{array}{r} 523 \\ \underline{274} \\ 249 \end{array}$$

A calculator is useful in reinforcing the fact that a multi-digit subtraction can be accomplished by separately subtracting the face values in each place value. For example, prepare a worksheet or task cards such as this one:

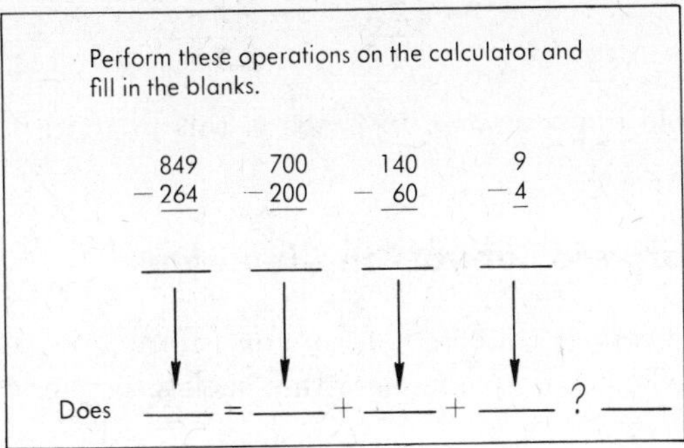

Perform these operations on the calculator and fill in the blanks.

$$\begin{array}{r} 849 \\ -\underline{264} \end{array} \quad \begin{array}{r} 700 \\ -\underline{200} \end{array} \quad \begin{array}{r} 140 \\ -\underline{\ \ 60} \end{array} \quad \begin{array}{r} 9 \\ -\underline{4} \end{array}$$

Does _____ = _____ + _____ + _____ ? _____

When students exhibit understanding of the subtraction process, introduce them to the short form for subtraction. Let them use some auxiliary numerals, or crutches, if they need them. As a student becomes more skillful in the short form, encourage him to omit the auxiliary numerals and use the most efficient form.

$$\begin{array}{r} {}^{4}\not{5}\,{}^{11}\not{2}\,3 \\ 2\,7\,4 \\ \hline 2\,4\,9 \end{array} \quad \longrightarrow \quad \begin{array}{r} 5\,2\,3 \\ 2\,7\,4 \\ \hline 2\,4\,9 \end{array}$$

2. *equal-additions method*

This method for subtraction is based on the property of subtraction which asserts that increasing (or decreasing) the minuend and subtrahend by the same number leaves the difference unchanged. (See chapter 7.) This subtraction pattern should be reviewed prior to work with the equal-additions algorithm. Further readiness includes skill in computing basic facts of subtraction and simple differences and understanding of place value and the rearrangement property.

When students have the prerequisite learning, introduce them to the equal-additions method. Write an example such as $\begin{array}{r} 762 \\ -\underline{345} \end{array}$ on the chalkboard or overhead-projector transparency. Point out that instead of borrowing a ten from the tens place in the minuend and grouping it with the ones the subtraction in the ones can be accomplished by adding the same number to minuend and subtrahend. Add 10 to both minuend and subtrahend (other numbers can be added, but 10 is the easiest number with which to work because of place value) and show the students, or let them suggest, that the 10 added in the minuend should be combined with the 2 ones and the 10 added in the subtrahend should be grouped with the 4 tens. You may present your work in expanded notation or in the short form.

$$\begin{array}{rl} 762 = 700 + 60 + 2 + 10 = & 700 + 60 + 12 \\ -\underline{345} = \underline{300 + 40 + 5 + 10} = & \underline{300 + 50 + \ \ 5} \\ & 400 + 10 + \ \ 7 = 417 \end{array} \qquad \begin{array}{r} 7\,6\,{}^{1}2 \\ {}^{5} \\ -3\,\not{4}\,5 \\ \hline 4\,1\,7 \end{array}$$

The equal-additions method also works well in cases in which more than one addition is made and in problems in which the minuend contains zeros.

$$\begin{array}{r} \not{5}\,{}^{1}\not{2}\,{}^{1}3 \\ {}^{3}\,{}^{8} \\ -\not{2}\,\not{7}\,4 \\ \hline 2\,4\,9 \end{array} \qquad \begin{array}{r} 6\,{}^{1}0\,{}^{1}3 \\ {}^{3}\,{}^{5} \\ -\not{2}\,\not{4}\,8 \\ \hline 3\,5\,5 \end{array}$$

Students need ample opportunities to practice this method in order to master it.

Computational Errors in Subtraction Algorithms

Often systematic errors in the final form for subtraction can be corrected by returning to a less efficient symbolic algorithm such as expanded notation, separate differences, or the use of more auxiliary numerals. If one of these symbolic methods fails to correct the error, students should use manipulative or pictorial devices to model the computation. Also, encourage children to estimate the answer to determine if the computed answer is reasonable. You may notice, too, that a student's mistake is not in the algorithmic process but in computing basic facts of subtraction in which case intensive practice in the facts is in order.

Some representative error patterns and suggestions for remediating them follow:

1. $$\begin{array}{r} 873 \\ -\,289 \\ \hline 616 \end{array} \qquad \begin{array}{r} 73 \\ -\,48 \\ \hline 35 \end{array} \qquad \begin{array}{r} 219 \\ -\,143 \\ \hline 136 \end{array}$$

   Computation such as this done repeatedly indicates the child is subtracting the smaller number from the larger in every place even when he should regroup (decomposition method) or add a number to minuend and subtrahend (equal-additions method). The student may correct his error by simply being reminded that the bottom number is always subtracted from the top number. If more help is needed, have him use expanded notation or, if you want him to use the decomposition method, represent the minuend and subtrahend with base blocks or an abacus so that in the first example he can see that a ten must be "borrowed" from the 7 tens and one of the hundreds in 800 should be regrouped with the 6 tens. Have him verbalize the process as he works with the materials and performs the subtractions.

2. $$\begin{array}{r} \overset{5}{\cancel{6}}\overset{13}{\cancel{4}}7 \\ -\,285 \\ \hline 3512 \end{array} \qquad \begin{array}{r} \overset{6}{\cancel{7}}\overset{12}{\cancel{3}}2 \\ -\,364 \\ \hline 368 \end{array} \qquad \begin{array}{r} \overset{1}{\cancel{2}}\overset{16}{\cancel{7}}9 \\ -\,86 \\ \hline 1813 \end{array}$$

   The student who shows this work is using the decomposition method but is regrouping even when it is not needed. (Notice his work in the second example is correct since two regroupings are necessary.) Begin remediation by having the student round the minuend and subtrahend in the first example to the nearest hundred, compute the simple difference 600 − 300, and determine that a difference close to 300 is a reasonable answer. Then ask him to represent the minuend and subtrahend with physical objects and go through the steps in subtracting the ones, tens, and hundreds, explaining each time whether or not he must regroup. Next give him several examples to decide if regrouping is needed or not and have him draw a loop around each column that requires a "borrowing" step, e.g.,

   $$\begin{array}{r} 5\,\boxed{2}\,6 \\ -\,2\,\boxed{5}\,3 \\ \hline \end{array}$$

   When the student demonstrates he knows when to regroup, have him perform the computations in the examples.

3. $$\begin{array}{r} 6\overset{7}{\cancel{8}}{}^{1}2 \\ -\,2\overset{4}{\cancel{3}}7 \\ \hline 435 \end{array} \qquad \begin{array}{r} \overset{7}{\cancel{8}}{}^{1}2 \\ -\,\overset{4}{\cancel{3}}7 \\ \hline 35 \end{array} \qquad \begin{array}{r} \overset{7}{\cancel{8}}\overset{10}{\cancel{1}}{}^{1}4 \\ -\,\overset{3}{\cancel{2}}\overset{7}{\cancel{6}}5 \\ \hline 439 \end{array}$$

The student who does this work is using the decomposition and equal-additions algorithms in the same problem. Decide on one method for the student to review and practice. In the case of decomposition, point out, possibly with the help of physical materials, that regrouping is done in the minuend and the subtrahend remains unchanged. If you decide the child should use the equal-additions method, review the subtraction pattern of increasing the minuend and subtrahend by the same amount and then have the student apply it in practice exercises.

Sequencing Subtraction of Whole Numbers

The following list summarizes some of the preceding material and suggests a sequence for teaching subtraction of whole numbers from the basic facts to the final algorithm (decomposition):

1. basic facts of subtraction
   Examples: $9 - 3 = 6$, $16 - 9 = 7$
2. simple differences
   Examples: $120 - 80 = 40$, $900 - 400 = 500$
3. difference in a two-digit or three-digit minuend and a one-digit subtrahend, no regrouping required
   Examples: $\begin{array}{r} 47 \\ -\ \underline{\ 2} \end{array} \qquad \begin{array}{r} 836 \\ -\ \underline{\ \ 4} \end{array}$
4. difference in a two-digit minuend and a two-digit subtrahend (multiple of 10) and the difference in a three-digit minuend and a two-digit subtrahend (multiple of 10) or a three-digit subtrahend (multiple of 100), no regrouping required
   Examples: $\begin{array}{r} 47 \\ -\underline{20} \end{array} \qquad \begin{array}{r} 280 \\ -\ \underline{\ 50} \end{array} \qquad \begin{array}{r} 463 \\ -\underline{200} \end{array}$
5. review of rearrangement property
   $(7 + 5) - (2 + 1) = (7 - 2) + (5 - 1)$
   Examples: $(40 + 8) - (20 + 3) = (40 - 20) + (8 - 3)$
6. difference in a two-digit minuend and a two-digit subtrahend (not a multiple of 10) and the difference in a three-digit minuend and a two-digit subtrahend (not a multiple of 10) or a three-digit subtrahend (not a multiple of 100), no regrouping required
   Examples: $\begin{array}{r} 67 \\ -\underline{24} \end{array} \qquad \begin{array}{r} 865 \\ -\ \underline{\ 24} \end{array} \qquad \begin{array}{r} 726 \\ -\underline{514} \end{array}$
7. renaming such as
   $$\begin{aligned} 64 &= \overline{60} + 4 \\ &= \overline{50} + 14 \end{aligned}$$
8. difference in a number represented by a two-digit numeral and one represented by a one-digit numeral, regrouping necessary
   Example: $\begin{array}{r} 53 \\ -\ \underline{\ 8} \end{array}$

9. difference in two numbers each named by two-digit numerals, regrouping required

   Example: $\begin{array}{r} 62 \\ -\underline{38} \end{array}$

10. renaming such as

$$\begin{aligned} 342 &= 300 + 40 + 2 \\ &= 300 + 30 + 12 \\ &= 200 + 130 + 12 \end{aligned}$$

11. difference in a three-digit minuend and a two-digit or three-digit subtrahend, one regrouping necessary

    Examples: $\begin{array}{r} 758 \\ -\underline{29} \end{array}$ $\quad \begin{array}{r} 546 \\ -\underline{70} \end{array}$ $\quad \begin{array}{r} 637 \\ -\underline{254} \end{array}$

12. difference in a three-digit minuend and a two-digit or three-digit subtrahend, two regroupings required

    Examples: $\begin{array}{r} 614 \\ -\underline{87} \end{array}$ $\quad \begin{array}{r} 703 \\ -\underline{268} \end{array}$

    Include problems with more than three digits when the students are ready.

If you teach the equal-additions method also, you should review the subtraction property $a - b = (a + n) - (b + n)$, where a, b, and n are whole numbers and $a \geq b$, and then present examples in this order: those in which 10 must be added to minuend and subtrahend, examples in which 100 is added to each number, and those in which both 100 and 10 must be added to minuend and subtrahend. Have children check their computation by redoing the subtraction or by writing the related addition problem and determining if the difference plus subtrahend yields the minuend. Encourage them to be neat in their work.

Closing

After students learn place value and the basic facts of addition and subtraction, they are ready to explore simple sums and differences. Later addition and subtraction are extended to multi-digit situations. Manipulative and pictorial materials should be used in demonstrating a new procedure. Expanded notation offers readiness for more efficient algorithms and serves as a remediation device for students who have trouble with the final algorithms. Decomposition in subtraction can be explained in regrouping problems with the use of physical materials, but the equal-additions method has proven to be effective with many children also. As students advance to more complex problems, they should be allowed to suggest methods of solution and should not be forced to use only one procedure. When computing, students should be encouraged to estimate the answer and check their work for accuracy. Furthermore, when children can compute with the addition and subtraction algorithms, they need the opportunity to apply them to real-life situations. (See chapter 15.)

Notes

1. Paul R. Trafton and Marilyn N. Suydam, "Computational Skills: A Point of View," *The Arithmetic Teacher* 22 (November 1975): 528.
2. Leonard M. Kennedy, *Guiding Children to Mathematical Discovery*, 2nd ed. (Belmont, Calif.: Wadsworth Publishing Company, Inc., 1975), p. 167.
3. Marilyn Suydam and Fred Weaver, "Research on Learning Mathematics," in Joseph N. Payne, ed., *Mathematics Learning in Early Childhood*, Thirty-seventh Yearbook (Reston, Virginia: National Council of Teachers of Mathematics, 1975), pp. 59–60.

Assignment

1. Suppose you are teaching third grade. Design a laboratory activity in which addition and subtraction algorithms are reinforced and extended. Provide for at least four centers in the activity. Describe the materials at each center and your plan for management.
2. a) Write a teaching activity for the regrouping step in subtraction.
   b) Outline a lesson on the subtraction decomposition method to be taught on the day following the activity you planned in part a).
3. a) Develop an activity for teaching the equal-additions method of subtraction.
   b) Teach the lesson you planned in (a) to an individual child, group, or entire class.
   c) What modifications of your activity do you suggest after having taught it?
4. Develop a teaching activity in which a calculator is used to teach the process of adding separately the ones, tens, and so on in multi-digit addends. Refer to the task card on addition on the calculator given in this chapter.
5. Read these articles in *The Arithmetic Teacher* and list points you consider useful for your future teaching:
   a) "Addition for the Slow Learner" by Gabriel J. Batarseh, December 1974, pp. 714–15,
   b) "Low-stress Subtraction" by Barton Hutchings, March 1975, pp. 226–32,
   c) "Diagnosing and Remediating Systemic Errors in Addition and Subtraction Computations" by L. S. Cox, February 1975, pp. 151–57,
   d) "The Odometer in the Addition Algorithm" by Donald E. Hall and Cynthia T. Hall, January 1977, pp. 18–21.

Laboratory Session

Preparation: Study chapter 9 of this textbook.
Objectives: Each student should
1. use base-ten blocks, plastic counters, and an abacus to represent common addition and subtraction algorithms,
2. relate the representations of algorithms with physical objects to steps in computing at the symbolic level,
3. compute using standard addition and subtraction algorithms, and
4. identify systematic errors children make in computation and give suggestions for correcting the errors.

Materials: abacus, base-ten blocks, plastic counters of at least three different colors
Procedure: Students should work in small groups to solve and discuss these problems:

1. Demonstrate the steps in solving each problem with base-ten blocks, and record the steps using expanded notation.

$$\begin{array}{r} 73 \\ +\,\underline{65} \end{array} \qquad \begin{array}{r} 608 \\ +\,\underline{297} \end{array} \qquad \begin{array}{r} 475 \\ -\,\underline{152} \end{array} \qquad \begin{array}{r} 2506 \\ -\,\underline{678} \end{array}$$

2. Solve each of the following with an abacus. Record your work using separate sums.

$$\begin{array}{r} 235 \\ +\,\underline{143} \end{array} \qquad \begin{array}{r} 463 \\ +\,\underline{288} \end{array}$$

3. Use an abacus to find each difference. Record your final answers.

$$\begin{array}{r} 324 \\ -\,\underline{103} \end{array} \qquad \begin{array}{r} 2001 \\ -\,\underline{674} \end{array}$$

4. Let one color of the plastic counters represent ones; another color, tens; another color, hundreds. Use the counters to represent each problem, and record your work in the final form. Use the decomposition method to find the last difference.

$$\begin{array}{r} 37 \\ +\,\underline{12} \end{array} \qquad \begin{array}{r} 658 \\ +\,\underline{269} \end{array} \qquad \begin{array}{r} 354 \\ -\,\underline{132} \end{array} \qquad \begin{array}{r} 403 \\ -\,\underline{186} \end{array}$$

5. a) Demonstrate the subtraction pattern which justifies the equal-additions method by representing each of these equations with two sets of plastic counters.

$$\begin{array}{l} 11 - 7 = 4 \\ (11 + 1) - (7 + 1) = 4 \\ (11 + 2) - (7 + 2) = 4 \\ (11 + 3) - (7 + 3) = 4 \end{array}$$

   b) Use the equal-additions method to solve $\begin{array}{r} 5406 \\ -\,\underline{1578}. \end{array}$

6. Assume that a child systematically makes the error given in each representative example. Identify each error and illustrate how you would use the materials of this laboratory activity to help the student correct his error.

$$\begin{array}{r} 87 \\ +65 \\ \hline 142 \end{array} \qquad \begin{array}{r} 43 \\ +78 \\ \hline 1111 \end{array} \qquad \begin{array}{r} 356 \\ -187 \\ \hline 231 \end{array} \qquad \begin{array}{r} \overset{4}{\not 5}\overset{1}{4} \\ -23 \\ \hline 211 \end{array}$$

Bibliography

Ashlock, Robert B. *Error Patterns in Computation.* Columbus, Ohio: Charles E. Merrill Publishing Company, 1976.

Grossnickle, Foster E., and Reckzeh, John. *Discovering Meanings in Elementary School Mathematics,* 6th ed. New York: Holt, Rinehart and Winston, Inc., 1973.

Jungst, Dale G. *Elementary Mathematics Methods: Laboratory Manual.* Boston: Allyn and Bacon, Inc., 1975.

Kennedy, Leonard M. *Guiding Children to Mathematical Discovery,* 2nd ed. Belmont, California: Wadsworth Publishing Company, Inc., 1975.

Payne, Joseph N., ed. *Mathematics Learning in Early Childhood.* Thirty-seventh Yearbook. Reston, Virginia: National Council of Teachers of Mathematics, 1975.

Trafton, Paul R., and Suydam, Marilyn N. "Computational Skills: A Point of View." *The Arithmetic Teacher* 22 (November 1975): 528–37.

10

Extending Multiplication and Division of Whole Numbers

Beginning work with multiplication and division in which students learn various meanings and properties of the operations, ways to represent the operations concretely and pictorially, and the basic facts of the operations were discussed in chapter 8. In this chapter multiplication and division are extended to include simple products and quotients and common algorithms which serve as readiness for the final multiplication and division algorithms. Many of the devices useful for introducing multiplication and division and for teaching addition and subtraction algorithms are also appropriate in work with multiplication and division algorithms. These devices include arrays, sets of objects, base-ten blocks, a pocket chart, an abacus, money, and Cuisenaire rods.

Higher Level Multiplication Tasks

Finding simple products and multiplying two numbers, one represented by a two-digit numeral and the other by a one-digit numeral, can occur soon after a student has conceptualized the operation of multiplication and has begun to practice the basic facts of multiplication. Motivation for learning how to find more difficult products can often be supplied by story situations such as "If your father bought four dozen eggs at the store, how many eggs did he buy?" and "Fifty quarters are equivalent to how many nickels?"

Simple Products

Simple, or easy, products involve basic facts of multiplication and place value. Examples of simple products are $3 \times 20 = 60$ and $30 \times 40 = 1200$. When students have learned that $3 \times 2 = 6$ and understand place value, they may find 3×20 by entering 20 on the abacus three times.

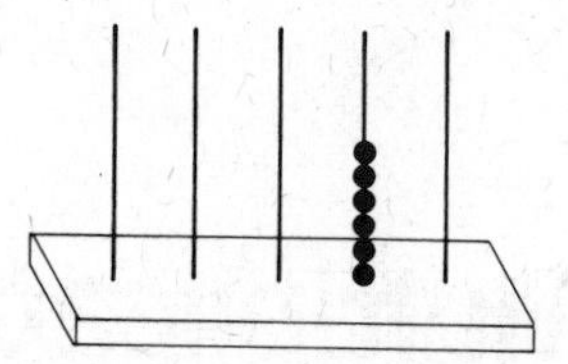

They should see that the product is six tens, or 60.

When students are ready (i.e., when they have learned the basic multiplication facts and the commutative and associative properties of multiplication and know place value), you may guide their thinking in finding simple products by having them record the steps as they work with concrete or pictorial materials and by presenting them with open sentences such as these:

$$\begin{aligned} 3 \times 80 &= 3 \times (8 \times \underline{\quad}) \\ &= (3 \times \underline{\quad}) \times 10 \\ &= \underline{\quad} \times 10 \\ &= \underline{\quad} \end{aligned} \qquad \begin{aligned} 60 \times 40 &= (6 \times \underline{\quad}) \times (\underline{\quad} \times 10) \\ &= (6 \times 4) \times (\underline{\quad} \times 10) \\ &= \underline{\quad} \times 100 \\ &= \underline{\quad} \end{aligned}$$

By questioning the students on the number of ending zeros in the factors and the product, help them generalize how to find the products. In $60 \times 40 = 2400$, each factor has a zero and the product has two zeros. When pupils see that the product can be computed by determining the product in the basic multiplication fact involved and attaching the sum of the numbers of zeros in the factors, they can quickly compute products such as 7×800, 70×40, and 40×200. You can emphasize this pattern by underlining the zeros in the factors and placing blanks where zeros go in the product, e.g., $7\underline{0} \times 4\underline{0} = 28\underline{0}\underline{0}$. Students should also contrast this pattern with that of addition and subtraction with ending zeros. Examples are $70 + 40 = 110$ and $70 - 40 = 30$.

Multiplication of a Two-digit Factor and a One-Digit Factor

Children need to understand expanded notation and the distributive property of multiplication over addition and subtraction before the process involved in multiplying $\begin{array}{r} 42 \\ \times \underline{\ 3} \end{array}$ is meaningful to them. Such a product can be investigated on grid paper. You may let students explore how the product 3×42 can be found, or you may wish to structure the activity by having them outline a 3×42 array on grid paper and separate the array vertically to form two smaller arrays in which the number of squares can be easily determined. Students may offer alternative suggestions on partitioning the 3×42 array, but guide them to understand that the

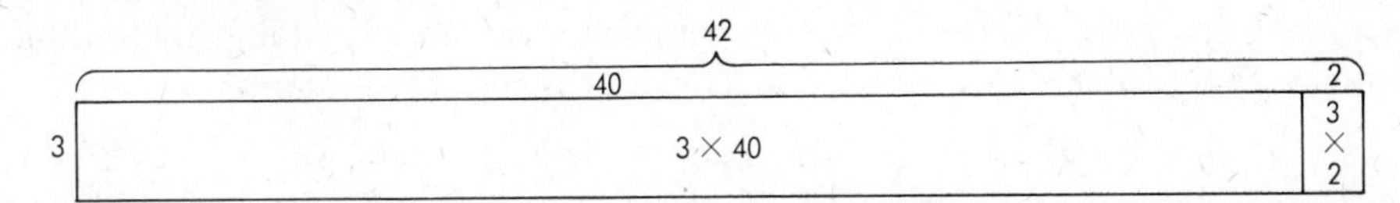

partition allows them to use multiplication skills they already have. Represent the multiplication indicated using expanded notation, pointing out that the two products must be added since the arrays associated with them must be joined.

$$\begin{array}{rl} 42 = & 40 + 2 \\ \times \underline{\ 3} & \underline{\quad\quad 3} \\ & 120 + 6 = 126 \end{array}$$

The basis for the vertical algorithm can be seen when the products are written in horizontal notation. Many students, especially the more capable ones, profit from such work.

$$\begin{aligned} 3 \times 42 &= 3 \times (40 + 2) \\ &= (3 \times 40) + (3 \times 2) \\ &= 120 + 6 \\ &= 126 \end{aligned}$$

An alternative form for presenting the symbolic work is the separate-products method. This method takes into account place value within the original problem and can be used to provide transition from the expanded form to the final vertical algorithm. In the example that follows, notice that each partial product corresponds to one of the smaller arrays in the array model given earlier.

$$\begin{array}{rl} 42 & \\ \times \underline{\ 3} & \\ 6 & (3 \times 2 = 6) \\ \underline{120} & (3 \times 40 = 120) \\ 126 & \end{array}$$

In this method, as in the expanded form, students are recording the steps in the thinking process required in the final algorithm. Once students understand this process, they can use the short form for multiplication. Appeal to their work with the separate-products form, however, when introducing the final algorithm.

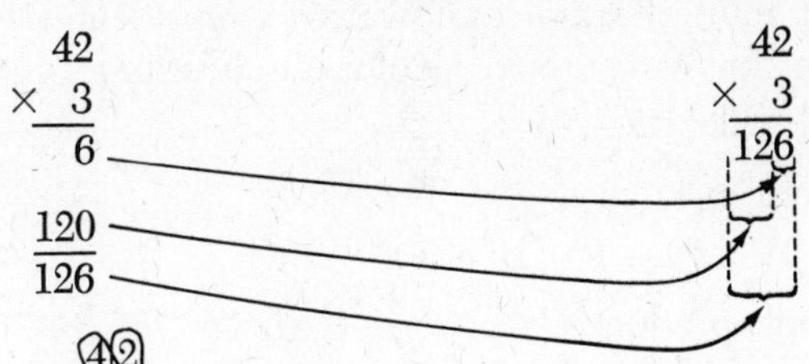

Visual cues such as $\times$ 42 over 3 (circled) often reduce errors in computation and are especially helpful to learning disabled children.

These same methods can be utilized when teaching multiplication of a two-digit factor and one-digit factor when regrouping, or carrying, is required. Before introducing the final algorithm, however, review students on adding by endings (examined in chapter 9). Having students work through the examples

$$(7 \times 8) + 6 = 56 + 6 = 62 \quad \text{and} \quad (6 \times 7) + 5 = 42 + 5 = 47$$

should help them remember to multiply before they add in the second step in finding the products $\begin{array}{r} 89 \\ \times \underline{\ 7} \end{array}$ and $\begin{array}{r} 79 \\ \times \underline{\ 6} \end{array}$, respectively.

Some students may enjoy using Napier's rods discussed in chapter 8 to perform the multiplications. To find the product 8 × 42, for example, the child should place the 4 and 2 rods to the right of the index rod, look across from 8 on the index rod, and add along the diagonals in the two squares.

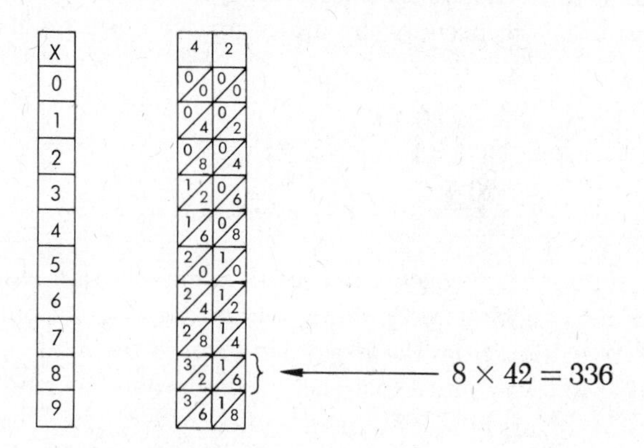

Sequencing Multiplication

The following summary suggests a hierarchy of multiplication tasks culminating in the final multiplication algorithm. Such a sequence can be used to order presentation of material, prepare diagnostic tests to determine a child's level of concept and skill development in multiplication, and to remediate a student who is having difficulty working at a certain level in the sequence. Students should

1. learn the basic facts of multiplication which involves understanding various meanings of multiplication and ways to represent them concretely and pictorially and knowing properties of multiplication,
   Example: $7 \times 6 = 42$
2. use 10, 100, etc. as factors,
   Examples: $12 \times 10 = 120$, $10 \times 100 = 1000$
3. compute simple products,
   Example: $40 \times 20 = 800$
4. use the distributive property to multiply two numbers, one represented by a two-digit numeral and the other by a one-digit numeral,

Example: $$\begin{aligned} 4 \times 17 &= 4 \times (10 + 7) \\ &= (4 \times 10) + (4 \times 7) \\ &= 40 + 28 \\ &= 68 \end{aligned}$$

5. use expanded notation and vertical form to multiply a number represented by a two-digit numeral by a number represented by a one-digit numeral,

Example: $$\begin{array}{r} 47 \\ \times \underline{\ \ 2} \end{array} \quad \begin{array}{r} 40 + \ \ 7 \\ \underline{\qquad\ \ 2} \\ 80 + 14 = 94 \end{array}$$

6. use partial products to multiply a number represented by a two-digit numeral by a number represented by a one-digit numeral,

Example: $$\begin{array}{r} 27 \\ \times \underline{\ \ 3} \\ 21 \\ \underline{60} \\ 81 \end{array}$$

7. multiply a number represented by a two-digit numeral by a number represented by a one-digit numeral using the final multiplication algorithm,

Example: $$\begin{array}{r} 45 \\ \times \underline{\ \ 6} \\ 270 \end{array}$$

Before step 6 in the sequence, the student needs to have learned the final addition algorithm, a common objective for the last part of second grade, and be able to add by endings prior to the "carrying" step in the final algorithm in step 7.

Higher Level Division Tasks

Since division is generally not introduced until third grade, many students at the early childhood level do very little in division beyond the basic facts. Others, however, can explore properties of division, compute simple quotients, and divide larger numbers.

Simple Quotients

If a division sentence can be associated with a multiplication sentence involving a simple product, the quotient in the division sentence is called a simple quotient. For instance, $80 \div 4 = 20$ and $1200 \div 30 = 40$ are simple quotients since $20 \times 4 = 80$ and $40 \times 30 = 1200$ are simple products. Students should be able to relate division to multiplication and to compute simple products before you, as teacher, introduce simple quotients. The following two examples illustrate how this prerequisite learning is applied to finding simple quotients:

$60 \div 2 = \square$	$800 \div 40 = \square$
$\square \times 2 = 60$	$\square \times 40 = 800$
$\boxed{30} \times 2 = 60$	$\boxed{20} \times 40 = 800$
$60 \div 2 = \boxed{30}$	$800 \div 40 = \boxed{20}$

Question the students on the number of zeros in the dividend, divisor, and quotient to guide them in discovering the pattern in finding simple quotients: compute the quotient in the basic division fact involved and then attach to this quotient the number of zeros found by subtracting the number of zeros in the divisor from the number of ending zeros in the dividend. Such a generalization allows the pupils to quickly determine the quotients in examples such as $40\overline{)1200}$ and $60\overline{)540}$.

Division on the Number Line

The number line is used in early work with division when students are exploring the basic facts of division. As their experience in division widens, the number line is useful in teaching children to relate division to sentences of the form $a = (q \times b) + r$, where a, q, b, and r are whole numbers, $b \neq 0$, and $0 \leq r < b$. In the sentence, b is the divisor; a, the dividend; q, the quotient; and r, the remainder. The quotient must be the maximum number which can be multiplied with the divisor to yield a product less than or equal to the dividend. The form $a = (q \times b) + r$ serves as a check in long division. The role of the quotient and remainder in this new look at division is illustrated in the following number-line model of 29 divided by 3.

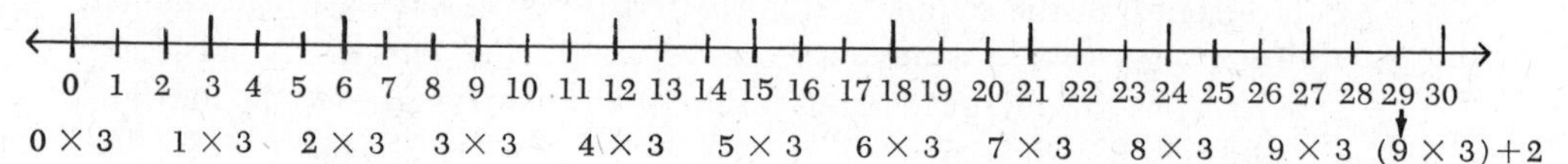

The student should mark off multiples of 3 and find that 9×3 is the maximum multiple of 3 contained in 29 and two more steps beyond 9×3 are needed to reach 29. Hence 9 is the quotient, 2 is the remainder, and $29 = (9 \times 3) + 2$. Even though 29 can be named in other ways, e.g., $(8 \times 3) + 5$ and $(7 \times 3) + 8$, these expressions do not yield a possible quotient or remainder in the division $3\overline{)29}$ since neither 8×3 nor 7×3 is the maximum multiple of 3 contained in 29 and since both 5 and 8 are greater than the divisor, 3.

Division Using the Distributive Property

Beginning work with the basic facts of division involves the distributive property (right-hand) of division with respect to addition and subtraction. This property is also important in division of larger numbers. Guide students to divide $3\overline{)36}$, for example, by using the number-per-subset interpretation of quotient and manipulative materials such as beansticks, base-ten blocks, Cuisenaire rods, or rectangular strips and squares cut out of grid paper. Have students to represent 36, the dividend, with the concrete material and to make three loops with string or yarn to represent the divisor, 3. Instruct them to place the tens and ones representations in the three loops so that the same number of each is in each set. The children's work with rectangular strips and squares and your accompanying board work are suggested here:

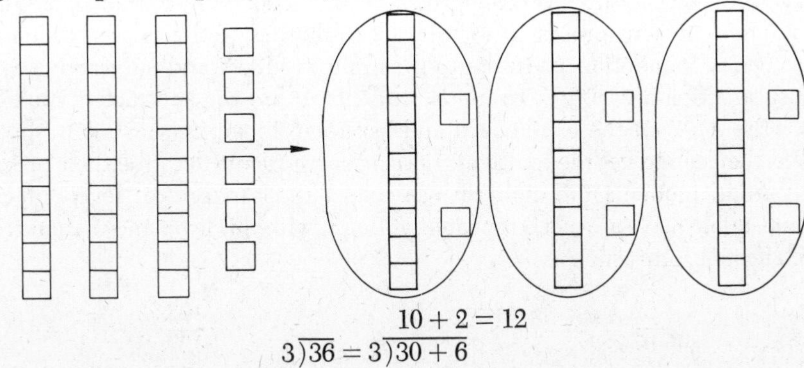

$$\begin{array}{r} 10 + 2 = 12 \\ 3\overline{)36} = 3\overline{)30 + 6} \end{array}$$

You may also have the mathematically mature students record the steps in the division in horizontal form which shows clearly how the distributive property justifies the division process in the preceding example.

$$\begin{aligned} 36 \div 3 &= (30 + 6) \div 3 \\ &= (30 \div 3) + (6 \div 3) \\ &= 10 + 2 \\ &= 12 \end{aligned}$$

Children also need experience in division problems in which the dividend must be regrouped, e.g., $3\overline{)42}$. This example can be modeled in the same way as that just shown except one ten in the dividend must be exchanged for ten ones.

The completed symbolic work is $\begin{array}{r} 10 + \ \ 4 = 14 \\ 3\overline{)42} = 3\overline{)30 + 12} \end{array}$

Emphasize that the original division problem is broken down into easier components: the student regroups 42 and thinks, "3 divides into 30 ten times and into 12 four times." This thinking process will be applied in the conventional distributive long-division algorithm which occurs after third grade.

Division by Repeated Subtraction of the Divisor

This method can be used when teaching the basic facts of division as well as with larger numbers. The symbolic work is shown in the example

$$\begin{array}{rl} 2 & \\ 13\overline{)38} & \\ \underline{13} & 1 \\ 25 & \\ \underline{13} & \underline{1} \\ 12 & 2 \end{array}$$

but students who have difficulty should show the subtractive process and the partial quotients with counters, marks on paper, or some other device.

Subtractive Method

Many students will discover that they need not always subtract only the divisor. A more efficient process is to subtract multiplies of the divisor which implies they need a foundation in computing simple products and subtracting with the final subtraction algorithm. To guide students to use the subtractive method, you may write $3\overline{)97}$ on the chalkboard and ask them to subtract as many threes at a time as they can to get the quotient. Let various students describe their procedures. The partial quotients the children use reflect their individual levels of comprehension. Some may subtract 3 repeatedly. Begin with this level and then move to the more efficient forms such as

$$\begin{array}{rr} 32 & \\ 3\overline{)97} & \\ \underline{30} & 10 \\ 67 & \\ \underline{30} & 10 \\ 37 & \\ \underline{30} & 10 \\ 7 & \\ \underline{6} & \underline{2} \\ 1 & 32 \end{array} \qquad \text{and} \qquad \begin{array}{rr} 32 & \\ 3\overline{)97} & \\ \underline{90} & 30 \\ 7 & \\ \underline{6} & \underline{2} \\ 1 & 32, \end{array}$$

the latter method representing the most refined form of the subtractive method. Supply other examples and guide students to progress to the more efficient forms.

When they have completed the division, have pupils check their work for accuracy. In our example,

$$\begin{array}{r} 32 \\ \underline{\times\ 3} \\ 96 \end{array} \qquad \begin{array}{r} 96 \\ \underline{+\ 1} \\ 97 \end{array}$$

shows the quotient and remainder determined are correct.

Research is not conclusive as to which is better to teach children—the subtractive or the distributive form of division. In one study, both methods had some advantages. The subtractive method was found to be especially effective for low-ability students and helped in transfer to new but similar learning situations while students taught the distributive form did better on problem solving. In another study on comparison of the two approaches, results on application problems favored the subtractive method while results on a retention test favored the distributive form.[1] You, then, should exercise your own discretion in choosing the form or forms you teach your students. If one method proves to be ineffective with a child, try another one. Also, alternative forms may serve as enrichment for gifted students.

Pyramid Method

Instead of writing the partial quotients to the side as in the subtractive method, they can be entered above the dividend in pyramid fashion, thus giving attention to place value within the original problem. The thinking process is the same as that in the subtractive form.

$$\begin{array}{r} 18 \\ \hline 8 \\ 10 \\ 4\overline{)75} \\ 40 \\ \hline 35 \\ 32 \\ \hline 3 \end{array}$$

Sequencing Division

Some third-grade students will be able to progress through all of the following steps; many, however, will not. These division tasks provide a foundation for the higher level division algorithms. Students should:

1. Learn the basic facts of division. This involves exploration of various interpretations and properties of division (the distributive property, in particular) of whole numbers as well as practicing the facts.

   Example: $72 \div 8 = 9$

2. Compute special quotients.

   Example: $80 \div 40 = 2$

3. Divide using the distributive property.

   Example: $4\overline{)48} = \begin{array}{r} 10 + 2 = 12 \\ 4\overline{)40 + 8} \end{array}$

4. Use repeated subtraction to solve division problems with or without remainders.

   Example:

$$9\overline{)41} \longrightarrow \begin{array}{rr} 41 & \\ -\ 9 & 1 \\ \hline 32 & \\ -\ 9 & 1 \\ \hline 23 & \\ -\ 9 & 1 \\ \hline 14 & \\ -\ 9 & 1 \\ \hline 5 & 4 \end{array} \longrightarrow \begin{array}{r} 4,\ \mathrm{r}\,5 \\ 9\overline{)41} \end{array}$$

5. Use the subtractive method in division by a one-digit divisor.

   Example:

$$\begin{array}{rr} 12 & \\ \hline 6\overline{)74} & \\ 60 & 10 \\ \hline 14 & \\ 12 & 2 \\ \hline 2 & 12 \end{array}$$

6. Use the pyramid form when dividing by a one-digit divisor.
   Example:

```
     21
     --
      1
     20
  9)195
    180
    ---
     15
      9
     --
      6
```

When his understanding breaks down, have the student go back to a prior symbolic method or represent the division process with concrete devices or diagrams.

Closing

Early childhood teachers are responsible for providing the foundation work in multiplication and division algorithms. The algorithms presented in this chapter are not the only ones students may discover. These, however, should help you offer guidance and relate properties of multiplication and division to algorithmic procedures to make these procedures meaningful to the learner. Such an approach is a problem solving one rather than a rote memorization of a routine.

Students should not be rushed from one algorithm to a more refined one and should be allowed to use objects or draw pictures when they need them to proceed through some of the steps in the algorithm or to check their computation. The pocket calculator can be used to help children understand the process in multiplication and division algorithms and in checking computational work. Once children understand the algorithms at the symbolic level, facilitate retention by giving opportunities for practice and application.

Notes

1. Marilyn N. Suydam and J. Fred Weaver, *Using Research: A Key to Elementary School Mathematics* (Columbus, Ohio: ERIC Information Analysis Center for Science, Mathematics and Environmental Education, 1975), pp. 7–5, 7–6.

Assignment

1. Make a set of Napier's rods and find these products with the rods:

   4×23 8×726

2. a) Examine two or three current second- and third-grade textbooks, and outline types of multiplication tasks presented beyond the basic facts. What manipulative or pictorial materials do the textbooks suggest to use along with the symbolic methods?
   b) Review several third-grade textbooks and outline how division is extended beyond the basic facts at third-grade level. What materials are suggested as representations of the algorithmic procedures?
3. Write a guided-discovery lesson plan on multiplication of a two-digit factor by a one-digit factor where "carrying" is required.
4. How do you think a tape recorder could be used in teaching multiplication and division algorithms?

Laboratory Session

Preparation: Study chapter 10 in this textbook.

Objectives: Each student should
1. illustrate how rectangular strips and squares can be used to develop multiplication and division algorithms at the early childhood level,
2. relate the steps in the representation of the algorithm to steps in the symbolic solution, and
3. use symbolic methods to solve multiplication and division problems.

Materials: 10×10 square arrays, strips of ten squares, and individual squares for each student

Procedure: Pairs of students should work and discuss these problems:
1. Model each problem with the strips and squares and record the steps symbolically using expanded notation.

   $3 \times 32 = \square$ $\begin{array}{r} 27 \\ \times \underline{6} \end{array}$

2. Use the strips of ten and unit squares to illustrate the distributive method for computing each of the following. Record your work using the distributive method of division.

   $4\overline{)48}$ $4\overline{)52}$
3. Using the subtractive method, compute $7\overline{)96}$. Represent each step in the process with the strips and squares.

Bibliography

Grossnickle, Foster E., and Reckzeh, John. *Discovering Meanings in Elementary School Mathematics,* 6th ed. New York: Holt, Rinehart and Winston, Inc., 1973.

Suydam, Marilyn N., and Weaver, J. Fred. *Using Research: A Key to Elementary School Mathematics.* Columbus, Ohio: ERIC Information Analysis Center for Science, Mathematics and Environmental Education, 1975.

11

Fractional Numbers

The mathematical definition of a fractional number is any number which can be represented in the form $\frac{a}{b}$, where a and b are whole numbers and $b \neq 0$. These numbers are needed in measurement and in comparison of equivalent parts of an object to the whole object.

Decimal fractions are especially important with the increased attention given to the metric system, but common fractions will continue to have a place in the early childhood curriculum. The early childhood educator is concerned with teaching these fractional-number topics:

1. writing fractional numbers represented by concrete and pictorial models
2. representing proper fractional numbers with objects or pictures
3. writing improper fractional numbers when given the representation
4. representing improper fractional numbers
5. demonstrating when fractional numbers are equivalent
6. comparing fractional numbers
7. generating classes of equivalent fractional numbers

Several structured materials are useful in teaching fractional numbers. They include paper folding, Cuisenaire rods, geometric regions, sets, fraction charts, and the number line. Candy boxes with the dividers that lift out serve as an inexpensive aid. Numerals for fractional numbers can be placed underneath the dividers. Relating fractional numbers to the study of measurement helps children understand how fractional numbers can be used. Some teachers are highly successful in teaching fractional-number concepts though experiences in cooking. Also, fractional numbers should be used in informal ways such as "On the school bus, you sit on half of the seat." and "If the three of you share this apple, each of you will receive one third of it."

Regardless of the material chosen to introduce fractional numbers, emphasize these points: a unit, or whole, object must be subdivided into equivalent parts; the numerater in the fraction tells the number of equivalent parts considered; and the denominator is associated with the number of equivalent parts in the unit object. Another basic concept students should understand is that fractional numbers have meaning only when related to the unit object. For example, $\frac{3}{4}$ of one object is not the same size as $\frac{3}{4}$ of an object of different size:

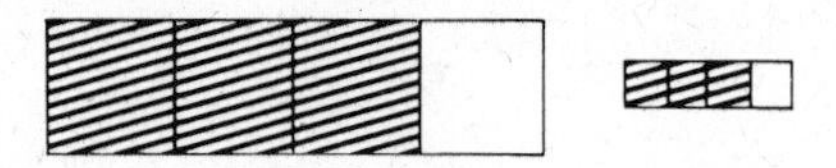

Beginning Work with Fractional Numbers

Most first and second graders can deal with the meaning and symbolism of fractional numbers. Suggestions for introductory activities follow:

1. Provide an exercise on part-to-whole relationships. Give the students several rectangular strips of paper all the same size, and tell them to fold one of the strips so that two parts of the same size are formed. Ask them to mark the fold and color one of the two parts. Then instruct them to take another strip, fold it into four equal-sized parts, mark the folds, and color one of the four parts. Note with the children that there is more than one way the paper can be folded, e.g.,

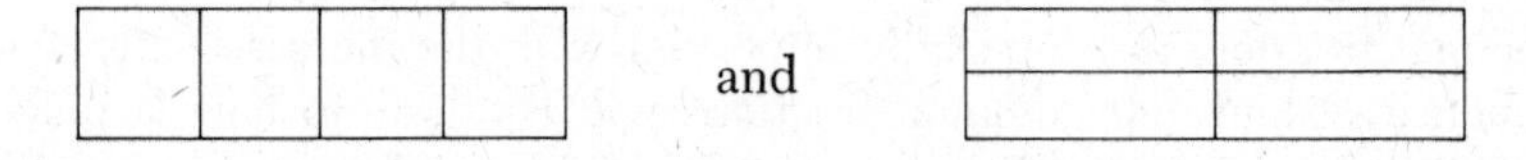

   Direct the students' thinking: "How many parts in the two-part region make a whole strip? How many of the two parts did you color? How many parts of the four-part strip make a whole strip? How many of these parts did you color? Which is larger—one of the parts of the two-part region or one of those in the four-part region? How can you tell? Will two of the parts of the four-part region exactly match one part of the two-part region? Hold up your two strips to show your answer is correct." Emphasize that a certain number out of a certain number of parts is being considered each time and that all of the parts are the same size.
2. After a readiness activity such as the previous one, plan an activity in which you introduce the notation for a fractional number. For example, place a circular felt cut-out on the feltboard and cover it with a red region and blue region, each one half of the unit region. Guide the students' thinking this way: "The whole region is covered by how many parts? Are the two parts

the same size? (Have a student demonstrate that they are the same size.) One of the two parts is blue and one of them is red. This means one half of the region has been covered with blue felt and one half has been covered with red felt. We write one half this way: $\frac{1}{2}$. (Show the notation on the feltboard.)

Blue | Red $\frac{1}{2}$ blue, $\frac{1}{2}$ red

What does the 1 in the symbol $\frac{1}{2}$ mean? What does the 2 in $\frac{1}{2}$ mean?" Introduce $\frac{1}{3}$, $\frac{2}{3}$, and $\frac{3}{4}$ by a similar procedure. At this time, you may introduce the terms numerator, which tells how many equivalent parts are considered, and denominator, which tells the number of parts in each unit object.

3. Children need to learn correct representations for fractional numbers. A task card or worksheet with exercises such as this one may help:

   Place a "✓" by each picture which shows $\frac{1}{2}$ and an "x" by each picture which does not represent $\frac{1}{2}$.

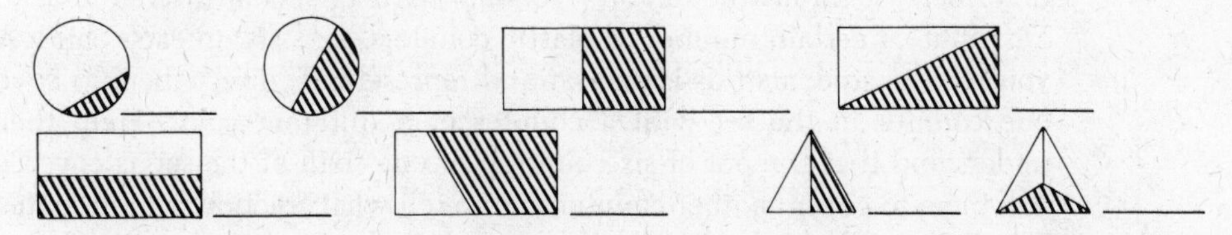

4. Have students make a fraction chart. Give each child a piece of construction paper and several strips of paper, each the same length as the width of the construction paper. Ask students to glue one strip across the top of the construction paper and write "1" on the strip. Then instruct them to fold, mark, and label another strip to indicate halves and glue it below the "1" strip. Continue with thirds through sixths. (You may need to help them fold thirds and fifths.) Their work should take this form:

1					
$\frac{1}{2}$	$\frac{1}{2}$				
$\frac{1}{3}$	$\frac{1}{3}$	$\frac{1}{3}$			
$\frac{1}{4}$	$\frac{1}{4}$	$\frac{1}{4}$	$\frac{1}{4}$		
$\frac{1}{5}$	$\frac{1}{5}$	$\frac{1}{5}$	$\frac{1}{5}$	$\frac{1}{5}$	
$\frac{1}{6}$	$\frac{1}{6}$	$\frac{1}{6}$	$\frac{1}{6}$	$\frac{1}{6}$	$\frac{1}{6}$

Call on different students to read the fractional numerals $\frac{1}{2}$, $\frac{1}{3}$, $\frac{1}{4}$, $\frac{1}{5}$, and $\frac{1}{6}$ as you point to them. Continue the activity with "How many halves make a whole?" (Record $\frac{2}{2} = 1$ on the chalkboard.) "How many thirds make a whole?" (Record $\frac{3}{3} = 1$.) "How many fourths are equivalent to the unit strip?" (Record $\frac{4}{4} = 1$.) "Which is larger—a half, third, or fourth of a unit object?" "Show me three fourths of the unit object." "Point to three fifths on your chart." "Which is larger—three fourths or three fifths?" Let them compare other pairs of fractional numbers and write the relationships.

5. Play a game with Cuisenaire rods. Each pair of students needs a set of rods. Hold up a rod, e.g., the brown one, and say, "If this rod represents 1, find the rod representing $\frac{1}{4}$." Let a student explain why the red one would be one fourth of the brown rod. (He should see that four red rods are the same length as a brown rod, so one red rod is one fourth of a brown one.) Hold up the dark green rod, tell the children it represents one unit, and ask them to find the $\frac{1}{2}$ rod, the $\frac{1}{3}$ rod, and the $\frac{2}{3}$ rod. Ask them what fractional number would be associated with the white rod in this case. Provide several more examples.
6. When students understand fractional parts of continuous objects as just described, begin instruction on fractional parts of sets of discrete objects. Distribute a certain number of plastic counters, e.g., six, to each child. As you tell the students this is one set and represents 1, direct them to cover one counter in the set with a counter of a different color. Help them understand that one out of six counters, or one sixth of the set, is covered. Ask them to cover another counter and to tell what fraction is represented. Continue until all six are covered, recording the fractional numeral each time. Point out that $\frac{6}{6}$ means all the set is covered, and write $\frac{6}{6} = 1$ on the chalkboard. Now ask the children to remove all the top counters so that none of the set of six is covered and to suggest what fractional number represents 0 out of 6 counters. Record $\frac{0}{6} = 0$ on the chalkboard. Conduct a similar exercise with eight counters. Emphasize that $\frac{6}{6}$ and $\frac{8}{8}$ are new names for 1 and $\frac{0}{6}$ and $\frac{0}{8}$ can be renamed as 0. Let them suggest other names for 1 and 0.
7. Provide pictorial and symbolic exercises such as these:

a)

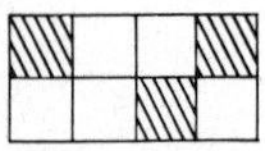

1) _____ out of _____ parts are shaded. The fraction of parts shaded is _____.
2) _____ out of _____ parts are not shaded. The fraction of parts unshaded is _____.

b)

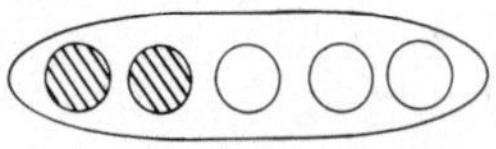

1) The set has _____ members.
2) _____ out of _____ members are shaded. How much of the set is shaded? _____
3) _____ out of members are unshaded. How much of the set is unshaded? _____

c) 1) _____ out of _____ parts are shaded. What fractional number is represented by the shaded parts? _____

2) _____ out of _____ parts is unshaded. Write the fraction for the unshaded parts of a region.

_____

d) Represent each fractional number in the given unit object.

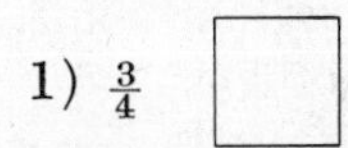

1) $\frac{3}{4}$ 2) $\frac{5}{6}$ 3) $\frac{1}{3}$

(Note: Let the pupils make free-hand subdivisions so the work will not be too tedious.)

8. An improper fraction is one in which the numerator is greater than or equal to the denominator. Examples such as $\frac{6}{6}$ and $\frac{8}{8}$ have already been considered. To introduce numbers represented by fractions having the numerator greater than the denominator, give the students strips of paper of equal length to fold. Instruct them to fold a strip into halves. Then ask the students to suggest how $\frac{3}{2}$ can be represented. Some may say that another strip must be folded into halves, the two strips placed one after the other, and then three parts shaded. A few may suggest to fold the strip in half and let each of the resulting parts represent one unit. Then the improper fraction $\frac{3}{2}$ can be shown on a single strip of paper. Have students shade $\frac{3}{2}$ as you diagram their work on the chalkboard.

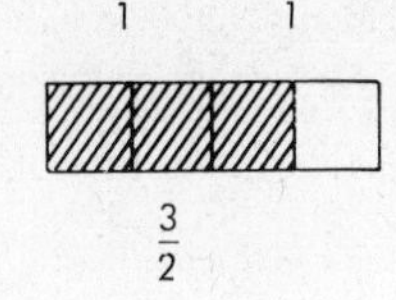

Ask, "Is $\frac{3}{2}$ less than 1 or greater than 1?" "Is $\frac{3}{2}$ less than 2 or greater than 2?" "Can we say that $\frac{3}{2}$ is between 1 and 2?" "How much more than 1 is $\frac{3}{2}$?" "Can you think of another name for $\frac{3}{2}$?" Some may point out that $\frac{3}{2}$ means $\frac{2}{2} + \frac{1}{2}$, $\frac{1}{2} + \frac{1}{2} + \frac{1}{2}$, $1 + \frac{1}{2}$, or $2 - \frac{1}{2}$. You may also introduce the mixed numeral $1\frac{1}{2}$ at this time: $\frac{3}{2} = 1\frac{1}{2}$. Let children represent other improper fractional numbers such as $\frac{5}{4}$. More capable students should be encouraged to represent larger improper fractional numbers such as $\frac{10}{3}$ and $\frac{15}{4}$.

9. Relate fractional numbers to measurement. When students have been introduced to centimeters, decimeters, and meters, question the students on their relationships: How many centimeters measure the same distance as one decimeter? (10) One centimeter is what part of one decimeter? ($\frac{1}{10}$) An object one meter long is how many decimeters long? (10) One decimeter is what part of a meter? ($\frac{1}{10}$) How many centimeters measure the same length as one meter? (100) What part of a meter is one centi-

meter? ($\frac{1}{100}$) Decimal fractions 0.1 and 0.01 can be introduced this way. Such notation is gradually being introduced in the late primary grades.

Measurement of capacity in cooking also lends itself to the study of fractional numbers. Students read the quantities listed in a recipe, measure them, and consequently discover equivalences and perform operations on fractional numbers. As children answer questions such as "If the recipe calls for $\frac{1}{4}$ cup of sugar, how much sugar will we need if we double the recipe?" "One-half cup measures the same amount as how many eighths?" "If we pour $\frac{3}{4}$ cup of milk into our measuring cup, how much more milk do we need to have a cup?" "If you add $\frac{1}{2}$ cup of flour to $\frac{1}{4}$ cup of flour, how much flour will you have?" they are dealing with many aspects of fractional numbers.

Determining half past and quarter past an hour in telling time on a clockface should be related to fractional numbers. Also, the money system provides a model for certain fractional numbers: How many pennies is a nickel worth? A penny is worth what fractional part of a nickel? A dime has the same value as how many nickels? A nickel is worth what part of a dime? and so on.

10. Use the number line to graph and compare fractional numbers. The two units between 0 and 2 can be used in a beginning activity. Have the children subdivide each unit into two parts of equal length and label the points as a certain number of halves: $\frac{0}{2}, \frac{1}{2}, \frac{2}{2}, \frac{3}{2}, \frac{4}{2}$. Next instruct the students to subdivide the units into fourths and label the points.

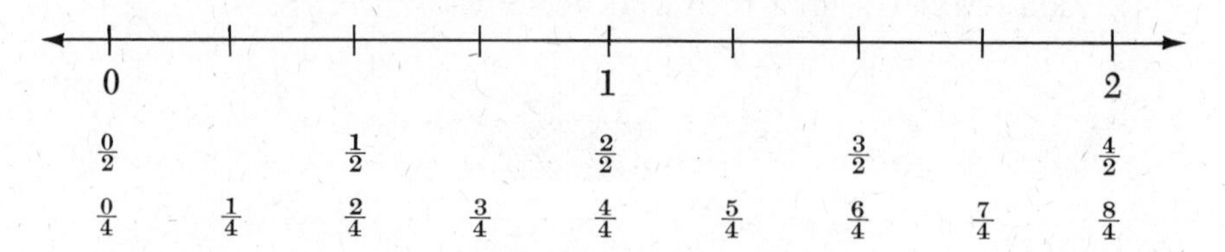

Extend the labeling to eighths and other fractional parts when students are ready. Then have them use the number line to solve problems such as "Which is larger—$\frac{5}{4}$ or $\frac{3}{2}$? $\frac{8}{8}$ is equal to how many halves? $\frac{8}{4}$ is another name for what whole number? Order the fractional numbers $\frac{1}{2}, \frac{1}{4}, \frac{3}{8}, \frac{5}{8}$ from smallest to largest."

Equivalence and Comparison of Fractional Numbers

Some of the activities just described serve as readiness for fractional number equivalences and inequalities. Cuisenaire rods, sets, paper folding, and the number line are useful in helping children understand when fractional numbers are equivalent. The equivalence $\frac{1}{4} = \frac{2}{8} = \frac{3}{12}$ can be seen when the white rod is placed over the purple rod; the red rod, over the brown rod; and the light green rod, over the rods which represent 12.

1 4 2 8 3 10 2

In each case a train of four of the smaller rods exactly matches the longer rod. The fact that any fractional number has many names should be related to the analogous situation with whole numbers, each of which can be named in many ways.

Equivalent fractional numbers can also be modeled with sets. In the first of the following pictures, $\frac{3}{9}$ of the set is shaded; if a loop is placed around subsets of 3, $\frac{1}{3}$ of the set is shaded. Thus $\frac{3}{9} = \frac{1}{3}$.

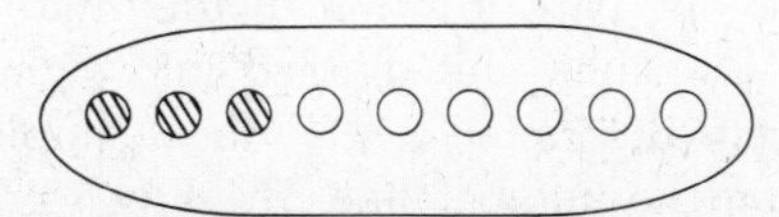

A similar model can be presented at the concrete level by placing counters of one color over a subset of counters of another color and putting loops around equivalent subsets. As students find equivalent fractional numbers, they should record them and seek patterns in naming them.

Paper folding can be used in introductory work with equivalent fractional numbers. Give each student five strips of paper labeled A, B, C, D, and E.* The strips should be the same size. Instruct the pupils to fold strip A into halves, mark the folds, and shade one half of the strip. Then have them fold strip B into fourths, place B below A, shade the same part of B as they did for A, and, by direct comparison, tell the number of fourths that match one half of strip A. Then help them fold strip C into sixths, place C below B and A, shade the correct number of sixths, and determine the number of sixths that are equivalent to $\frac{1}{2}$ and $\frac{2}{4}$. The students should repeat the folding process for eighths on strip D and tenths on strip E. Record and have students record the equivalence $\frac{1}{2} = \frac{2}{4} = \frac{3}{6} = \frac{4}{8} = \frac{5}{10}$. Urge them to extend the pattern without folding paper. Writing the equalities $\frac{2}{4} = \frac{1 \times 2}{2 \times 2}$, $\frac{3}{6} = \frac{1 \times 3}{2 \times 3}$, $\frac{4}{8} = \frac{1 \times 4}{2 \times 4}$, and $\frac{5}{10} = \frac{1 \times 5}{2 \times 5}$ should help all students see the pattern for generating sets of equivalent fractions.

The number line provides a model for equivalent fractional numbers and for the ordering of fractional numbers. Young children can easily determine that $\frac{3}{4}$ is greater than $\frac{1}{3}$ by directly comparing three fourths of a region and one third of the same region or one of the same size, but numbers which are closer in value can more easily be compared using the number line. You might provide task cards such as this one:

1. Use the number line to complete these sentences so the fractional numbers will be equivalent.

   a) $\frac{1}{3} = \frac{\square}{6}$ b) $\frac{12}{6} = \frac{4}{\square}$ c) $\frac{\square}{2} = \frac{6}{4}$

2. Make a dot to represent each number on the number line and write >, <, or = in each ○.

---

* Braille paper works well for blind students. The creases in it are more pronounced than those in ordinary paper.

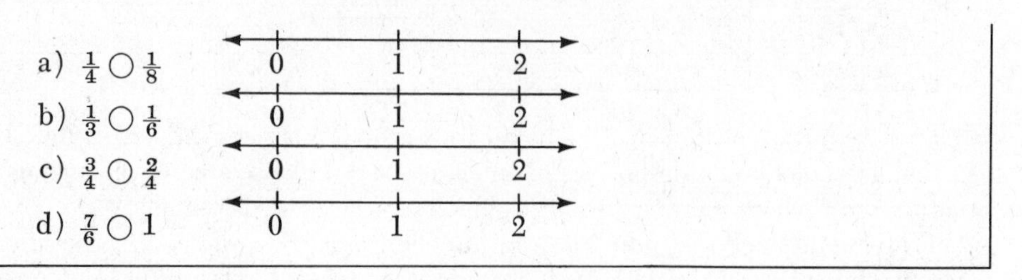

Help students discover patterns in comparing fractional numbers. For instance, $\frac{1}{4}$ is greater than $\frac{1}{8}$ because one part of a unit divided into four equivalent parts is larger than one part of the same unit divided into eight equivalent parts. This thinking should help them determine $\frac{2}{7} < \frac{2}{5}$, for example, without drawing the graphs of the numbers on the number line. Brighter students may discover that renaming the fractional numbers so that they have common denominators allows for easy comparison.

After students have worked with equivalent fractional numbers at the concrete and iconic levels, use examples such as these to teach the symbolic pattern for naming equivalent fractional numbers:

$$\frac{1}{3} = \frac{1 \times \square}{3 \times \square} = \frac{2}{6}, \qquad \frac{1}{3} = \frac{1 \times \square}{3 \times \square} = \frac{3}{9}$$

$$\frac{1}{3} = \frac{1 \times 10}{3 \times 10} = \frac{\square}{\triangle}, \qquad \frac{1}{3} = \frac{\square}{12}, \ \frac{1}{3} = \frac{5}{\square}$$

These exercises should help the students write a set of fractional numerals naming numbers equivalent to $\frac{1}{3}$: $\{\frac{1}{3}, \frac{2}{6}, \frac{3}{9}, \frac{4}{12}, \frac{5}{15}, \ldots, \frac{1 \times k}{3 \times k}, \ldots\}$, where k is a counting number. Children should generate sets of fractions equivalent to $\frac{1}{4}$, $\frac{1}{2}$, $\frac{2}{3}$, and so on and should graph sets of equivalent fractional numbers on the number line. Sample problems may be:

1. Write two more fractions in each set and graph each set with a dot on the number line:

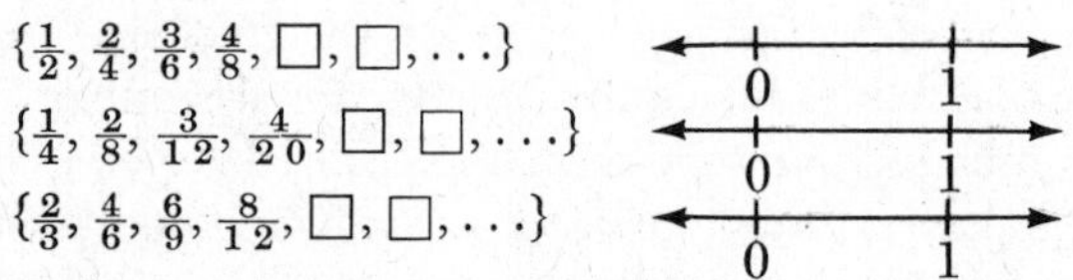

2. Make up your own example:

This experience is valuable in helping students to later rename fractions so they will have common denominators.

Closing

As students work with various materials in exploring fractional number concepts, the teacher should unify these experiences so learners can abstract the meanings and processes involved. Teachers should appeal to the students' prior understanding of whole-numbers operations to do much of the work with fractional numbers. Young children should understand why there is a need for fractional numbers and that a fractional number compares equivalent parts of an object to the whole object. They should be actively involved in associating concrete and pictorial models with fractional numerals, representing fractional numbers, demonstrating and writing equivalent fractional numbers, and comparing fractional numbers. Common fractions having denominators 10 and 100 are especially important in a modern mathematics program in that these fractions relate to decimal fractions and the metric system of measurement.

Assignment

1. Write a teaching activity for introducing fractional numbers to first graders. Use materials suggested in this chapter or create your own. Assume the class is mainstreamed and contains an exceptional child. Specify the exceptionality and allow for it in your lesson plan.
2. Develop a laboratory activity on equivalent fractional numbers appropriate for third grade.
3. Read and describe principal ideas or activities given in these articles in *The Arithmetic Teacher:*
   a) "Fun with Fractions for Special Education" by Ruth S. Jacobson, October 1971, pp. 417–21.
   b) "Some Thoughts on Piaget's Findings and the Teaching of Fractions" by R. J. Duquette, April 1972, pp. 273–75.

Laboratory Session

Preparation: Study chapter 11 of this textbook.

Objectives: Students should use manipulative and pictorial materials to

1. represent fractional numbers,
2. demonstrate equivalence of fractional numbers, and
3. compare fractional numbers.

Materials: plastic counters of three different colors, Cuisenaire rods, strips of paper

Procedure: Students should form groups of two or three to solve these problems:

1. Fold strips of paper to
   a) represent $\frac{3}{8}$, b) represent $\frac{8}{3}$, c) show $\frac{5}{4} = \frac{10}{8}$.
2. Use Cuisenaire rods to
   a) represent $\frac{6}{7}$, b) show $\frac{3}{2} = \frac{9}{6}$.

3. Cover a subset or subsets of a set of plastic counters to
   a) represent $\frac{4}{7}$, b) demonstrate $\frac{9}{12} = \frac{3}{4}$.
4. Use the number line to
   a) represent $\frac{9}{5}$,
   b) show $\frac{3}{4} = \frac{6}{8} = \frac{9}{12}$,
   c) graph the set $\{\frac{3}{4}, \frac{6}{8}, \frac{9}{12}, \ldots\}$,
   d) determine if $\frac{3}{4}$ is $<$, $=$, or $>$ $\frac{5}{6}$.

Bibliography

Heddens, J. W. *Today's Mathematics,* 2nd ed. Chicago: Science Research Associates, Inc., 1971.

Kramer, Klaas. *Teaching Elementary School Mathematics,* 3rd ed. Boston: Allyn and Bacon, Inc., 1975.

May, Lola June. *Teaching Mathematics in the Elementary School.* New York: The Free Press, 1974.

12

Geometry

There are two areas of geometry—metric and nonmetric. This chapter focuses on the nonmetric aspect; metric geometry (measurement) is discussed in the next chapter. Inductive and deductive strategies can be used in teaching geometry. The informal inductive approach is employed at the elementary level, whereas the deductive method is used in junior high school and high school geometry when the student is advancing to the intellectual stage of formal operations.

There is no general agreement on what geometric concepts should be taught and in what sequence they should be taught at the early childhood level. Piaget's research, however, provides guidelines for selecting and sequencing topics in geometry and is taken into account in this chapter. Some topics appropriate in the early childhood years are:

1. topological concepts,
2. Euclidean plane and solid shapes,
3. properties of Euclidean shapes,
4. points, line segments, lines, and rays,
5. congruence,
6. symmetry,
7. polygons,
8. angles, and
9. coordinate geometry.

Topological Concepts

A child's first concepts appear to be topological in nature. A baby sees his mother's face getting larger as she moves closer to him and changing shape as she turns to the right or left. He views these changes as physical distortions since he has not yet developed perspective. Such distortions in size and shape are part of geometry called *topology*. Objects are topologically equivalent when one can be transformed to the other by a squeezing, bending, or stretching action. To understand the difference in topology and Euclidean geometry, consider a rubber band placed in the shape of a square. It can be changed by stretching or squeezing to make other shapes which are topologically equivalent but different Euclidean figures. The shapes are topologically equivalent. The same is true for a sphere, cylinder, cone,

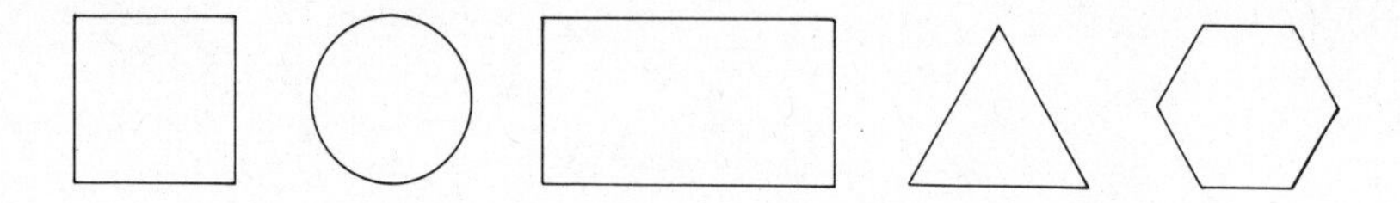

prism, and pyramid. (Visualize these transformations with a ball of clay.) Cuts or tears, however, produce figures which are not equivalent topologically, e.g.,

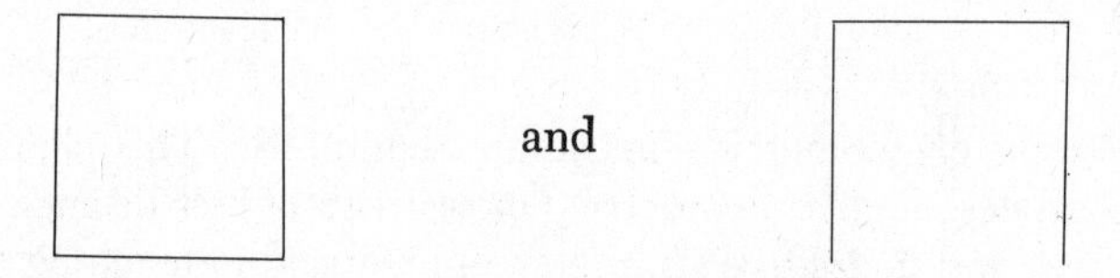

A two- or three-year-old child may draw a circle when asked to copy a square since he does not yet view objects in space as having a fixed shape. Developmentally, then, according to Piaget, a child is ready to learn about topological concepts before fixed Euclidean shapes. For this reason, beginning activities in geometry should deal with topology. As children's discrimination skills develop, they should be introduced to Euclidean shapes. Important concepts in topology include:

1. *proximity*

   Proximity is the most elementary topological spatial relation. The concepts of "near" and "far" should be taught along with classification and comparison of objects. Use terms related to near and far in everyday experiences: "Elyssa is nearer to the window than Marty is. Susan is nearer to me than she is to Mark. Which door in the room is nearest to Kevin? Which door is farthest from Kevin?" Provide structured experiences using a feltboard and duplicated handouts. For example, create a scene such as a zoo and ask, "Which animal is nearest to the tiger?" "Is the monkey nearer to the giraffe or to the hippopotamus?" "What animal is farthest from the black bear?" In another activity, you might arrange several plastic counters on the table and, as you point to various counters, ask, "Which counter is nearest to this yellow one?" "Which red counter is nearest to this blue one?"

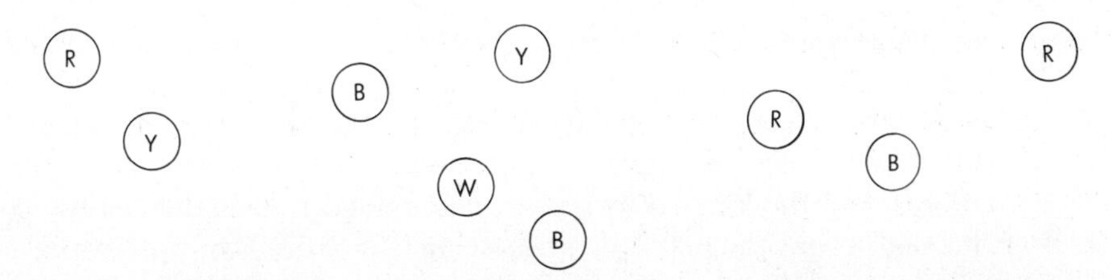

"Which red counter is farthest from this blue one?" "Can you move the white counter so that it is farther from this blue one than this yellow one?"

Children should also relate the nearness of an object to perspective. Take them outside and have them select two objects which are approximately the same size and shape, e.g., two children who are about the same size. Situate the two children on opposite sides of the playground. Have the other children stand close to one of the children and compare how distance affects the way the two of them look. Have the children who are observing walk toward the other child and describe how their perception of him changes as they get closer to him.

2. *open and closed curves*

Open and closed curves are different topologically, and young children recognize this distinction before they can distinguish between Euclidean shapes such as triangles and squares. You may begin instruction by making a closed figure with popsicle sticks and having children copy it. Ask them to select a point on the figure, move their fingers around the shape, and return to the starting place. Tell them that figures which can be made by starting and stopping at the same point are called *closed* figures. Now rearrange the popsicle sticks to form an open curve, and ask the children to copy it with their sticks and trace it with their fingers. Let them tell how the two curves are different. Tell them the figure having different starting and stopping points is an *open* curve. This tactile-visual experience aids the student in conceptualizing open and closed curves. Let the children make other examples of open and closed curves with the popsicle sticks. A similar activity can be conducted in which children make open and closed curves with pipe cleaners or with yarn strips on small, individual feltboards.

Further exploration of open and closed curves can be done using pieces of rubber cut from balloons. Give one piece to each child and ask him to draw an open and a closed curve on the piece of rubber with a felt-tip pen. Ask the children to stretch or shrink their balloon pieces and describe what happens to the curves. Pose these questions: Can you stretch or shrink the piece of rubber to make one closed curve look like another closed curve? Can you stretch or shrink the piece to make one open curve look like another open curve? Can you change a closed curve to an open one by stretching or shrinking the balloon piece? Can you change an open curve to a closed one by a stretching or shrinking action? Such considerations should help the children understand that open curves are topologically equivalent as are closed curves, but open and closed curves are not equivalent to each other.

Open and closed curves can also be investigated on the geoboard. As children create examples on the geoboard, lead discussion on the various shapes they make. For closed curves, ask questions about the number of sides, the number of pegs inside, on, and outside the curve, the number of corners, and the kinds of curves as determined by whether all the corners "jut out" (convex curve) or at least one "juts in" (concave curve). As a follow-up activity, have them examine letters of the alphabet for examples of open and closed curves. Also provide pictorial exercises in which children must recognize and draw open and closed curves. Examples are:

a) Tell whether each curve is open or closed. Write the correct word below each figure.

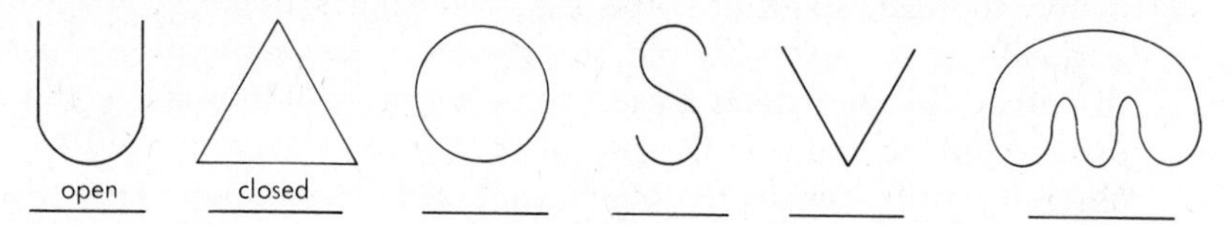

b) Draw three closed curves different from those in part a).

c) Draw three open curves different from those in part a).

When the children can distinguish between open and closed curves, guide them to find examples of them in their environment. Key rings, for instance, must be closed to hold the keys but hooks must be open to catch fish.

3. *inside, on, and outside*

These concepts are topological notions since the same points remain inside, outside, and on a closed curve after it has been squeezed or stretched. In a beginning activity, you might build a "corral" (closed curve) with some toy horses inside, toy animals such as a cat and dog outside the "fence," and a "cowboy" on the "fence." Help the children understand that the horses are *inside* the corral, the cat and dog are *outside* the corral, and the cowboy is *on* the fence. Ask them what happens when the gate is opened so that they are reminded that open and closed curves serve different purposes in real life situations. Practice (or evaluation) exercises may include coloring the inside of a closed curve one color, outside of the curve in another color, and the closed curve itself in a third color. Also, as children generate sets and show subsets with loops, you can ask them to name objects inside the loop and outside the loop.

Concepts of inside, outside, and on should be reinforced in play activities such as hopscotch and bean-bag toss. Have children answer questions such as "Is the bean bag inside, outside, or on the square of the playing board?"

Experiences in distinguishing between the interior of a closed curve and the closed curve itself provide readiness for the study of area and perimeter. It is important to teach the concepts of inside, on, and outside early so that second and third graders will understand that perimeter is a property of a closed curve whereas area is a property of the interior of a closed curve.

4. *order*

The ordering activities suggested in chapter 5 help a child understand order. Young children need experiences in arranging beads on a string, pegs on a pegboard, and blocks in a certain order in a row, reversing the order, and arranging the beads, pegs, or blocks in the same order but in a circular pattern. Bring out the idea of "betweenness" when teaching order. On the pegboard, for example, point to a peg and tell the children it is between the green and yellow pegs.

Activities dealing with sets and identification of shape, color, and size of blocks can contribute to a child's understanding of order. For example, ask each group of four children to share a set of attribute blocks equally. Guide them, if necessary, to think of a systematic way to share the blocks, e.g., each child could take all of one color or shape. Next ask the children to arrange the blocks in their sets in an orderly way. Place a manila folder between the child and his blocks, and take away a block. Then remove the folder and have the student identify the block you have taken. Go to another child and repeat the procedure with him.

Euclidean Geometry

Euclidean geometry involves fixed shapes, sizes, lengths of sides, number of sides, and number of angles. The transition from topology to Euclidean geometry usually begins in kindergarten.

Euclidean Shapes

Very young children become aware of various Euclidean shapes as they place a shape in a matching hole of puzzles which may range in difficulty from one hole per puzzle to several different shapes in the same puzzle. Such materials should be available in the room or center for children's experimentation. A sequence for presenting shapes is to have children to match, recognize, identify, and, finally, draw shapes. In an activity on matching, recognizing, and identifying shapes, you might, with the children sitting at a table, spread several shapes (circular, triangular, square, and rectangular) out on the table, hold up a shape, and call on a child to select the matching shapes. Introduce the name of the shape. Then repeat this procedure with the remaining shapes. Next, distribute to each child an envelope of paper shapes in various colors and sizes. Instruct the children to take all the shapes out of their envelopes and make sets of all the square shapes, all the circular shapes, and so on. Continue the activity by asking them to identify each shape, i.e., they verbally respond with the correct label when you ask, "What shape is this?" To conclude the activity, you might have the children paste the shapes on paper to make an art design.

Tactile experience helps some children, blind ones in particular, distinguish among the shapes. In the activity just described, you could have children move around each shape with their fingers and answer questions such as "Which shape is round? Which shapes have corners?" Another possibility would be to give each child shapes cut out of sandpaper or made from wire and let the children describe what they feel (whether the shape is round or has corners and sides).*

Activities involving body movement have proven effective with children with learning disabilities or behavior disorders. These children may learn Euclidean shapes by swinging circles and other shapes with their arms or joining hands in a group to form various shapes. You might also sketch a shape in chalk on the floor and have a child or several children lie down on the shape outline. All children may benefit from such kinesthetic experiences.

When children can match, recognize, and identify shapes, they should draw the shapes when you specify the name. By first grade, most students can match, recognize, identify, and draw the common Euclidean shapes. Another strategy for presenting shapes is to introduce one shape each day and let children describe it, learn its name, find examples of it in their environment, and draw it. In addition to circles, squares, rectangles, and triangles, children should work with rhombi (diamonds) and ellipses (ovals). Let them compose their own "shape" booklets by cutting out pictures of the shapes from magazines and pasting them on the appropriate page of their booklets.

To help children compare shapes using the number of corners as a criterion, give each one five yarn loops and shapes cut out of poster board. Have them classify the shapes as suggested here:

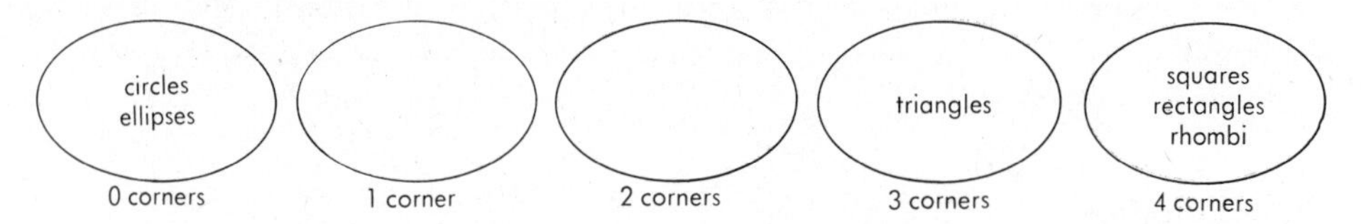

Notice two empty sets are generated. This activity is best conducted with a small group working in a large, clear area.

After children have had some experience in identifying and drawing the shapes, provide further experience in a laboratory setting. For example, set up four centers on tables in the classroom, and explain the task involved at each center. At center 1, have for each pair of children a bag of shapes of various sizes and textures and a chart headed by the four shapes. Children work in pairs and alternate feeling a shape in the bag, pointing to the matching shape on the chart, withdrawing the

* Wire shapes can be purchased from the American Printing House for the Blind.

shape, and placing it in the correct column on the chart. You should check their completed work on the chart. Ask them to then return the shapes to the bag. At center 2, have available for each child in the group shapes drawn on index cards and pipe cleaners or strips of yarn. Each child selects a set of cards and copies the shapes with the pipe cleaners or yarn strips. Children then discuss with a partner the number of sides and corners each shape has and identify each shape. At center 3, children practice drawing the shapes. Have strips of paper in a box in the middle of the table. Each child selects a strip of paper and performs the task designated. Sample tasks are "Draw 2 squares (□)" and "Draw 3 circles (○)." Each child draws the shapes on the strip of paper, writes his name, if possible, on the paper, and puts the paper in a collection box at the center. Later you can examine their work. At center 4, have students play "I'm thinking of a block" problem-solving game. Supply attribute blocks of one color on the table for each pair of children to use. One child selects a block and covers it with his hands while the other child's eyes are closed. Then the"guesser" must ask questions about the shape's attributes such as "Is it round?" "Does it have three corners?" "Does it have four sides?" "Does it have square corners?" until he guesses the block. The children then exchange roles.

After children can identify two-dimensional shapes, introduce them to three-dimensional ones such as cubes, spheres, cylinders, and cones. The extent to which you can involve them in the study of solid shapes depends, as usual, on the mathematical maturity of the children and the other topics of the curriculum which must be treated. Before names of the solids are introduced, have the children examine differences and similarities among the figures. In a beginning activity, you may use several ordinary objects in the environment to introduce each shape. For example, use a facial-tissue box to represent a cube; a ball, a sphere; a can, a cylinder; and an ice-cream cone, a cone. Let the children discuss the uses of each item and tell why they think the items are shaped as they are. Introduce mathematical models of each solid, and place each model by its real life counterpart. Introduce the terms cube, sphere, cylinder, and cone. Pose these questions for the children to resolve: If a cube is bounced, what happens? What attributes of the sphere make it good for bouncing and rolling? Which shape slides when you push it? Why are many boxes in the shape of a cube? Why are many cans in the shape of a cylinder instead of a sphere or cube? Would a sphere or cube be a good shape for an ice-cream cone? Why or why not?

When the children can identify three-dimensional figures, guide them to relate plane figures to solid ones. (Actually, many activities with solid figures, plane figures, and lines and parts of lines run simultaneously.) For example, the sides of a cube are square shaped, and the top of a cone is circular shaped.

Other solids young children may explore are prisms and pyramids, examples of special solids called *polyhedra.* (You need not introduce this term with young children.) The flat surfaces of the polyhedra are *faces,* the faces meet along *edges,* and the edges meet at *corners.* (The proper mathematical term for corners is vertex, but children should be allowed to use "corners.") Help them to associate the plane figures they have studied with the faces of each polyhedron and to understand these solid figures have flat surfaces. You may informally introduce the term "plane" in connection with the flat surfaces of polyhedra, i.e., each face of a polyhedron is part of a plane.

Most children enjoy making polyhedra from flat patterns. For example, they can cut out the patterns and fold them to form, respectively, a cube and triangular

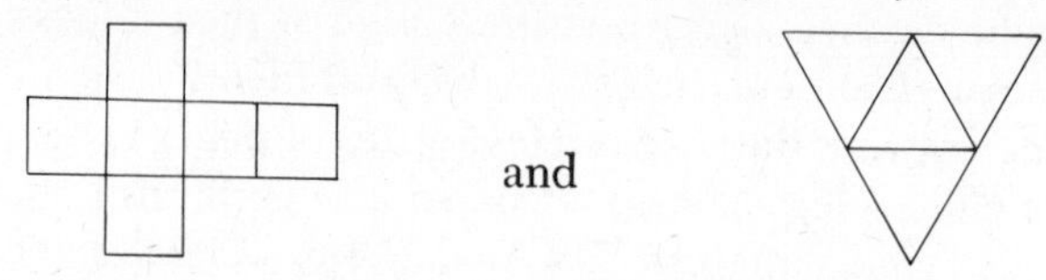

pyramid. Making these models before Christmas enables the children to decorate them for use as tree ornaments and room decorations.

Points, Line Segments, Lines, and Rays

Piaget's research indicates that treatment of geometric figures as sets of points and space as the set of all points should be delayed until children are nine or ten years old. Concepts of lines and subsets of lines, however, can be dealt with earlier in an intuitive, informal way. Prior experience with solid and plane figures can help a young child understand lines and parts of lines as geometric configurations in their own right.

"Point" is an undefined term in mathematics just as "set" and "element of a set" are. As young children work with topological notions of open and closed curves and interior and exterior of closed curves and Euclidean topics of plane and solid figures, they may intuitively use the word "point." To help a child understand a point is a location in space, describe certain locations in the classroom for the children to find. For example, tell them to look above the door at the top corner of the classroom, the point where two edges of the teacher's desk meet, the point on the floor one meter from one wall and two meters from an adjacent wall, or the point represented by the peg in the lower left-hand corner of their geoboards. After the children understand point is a location, have them represent points by drawing dots on paper, and introduce the notation of naming points with capital letters.

Introduce line segments by having children draw two dots on their paper to represent two distinct points and connect the two dots in several different ways. Responses may include:

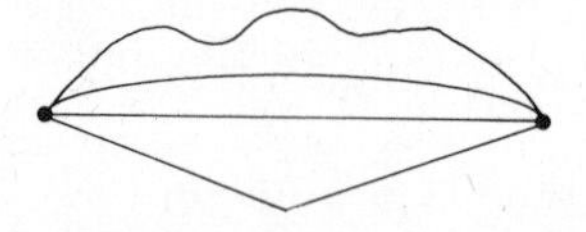

Some children will probably draw a line segment as one of the open curves. Focus on this special curve and tell them this curve is called a *line segment*. Ask them to name the two endpoints A and B, and introduce the notation $\overline{AB}$ as naming the line segment joining A and B. Point out that the line segment has two endpoints and can be named as $\overline{AB}$ or as $\overline{BA}$.

Next have the children draw line segments of varying lengths using a ruler. This task reinforces the fact that line segments are straight and helps a child understand line segments have length and can be measured. Children should also find the approximate length of several different curves joining two endpoints to discover the length of the line segment joining the two points is the shortest distance between them. They should also find examples of line segments in the real world, e.g., the edges of books, borders of window panes, marks for parking in a parking lot, and computer price markings on many grocery items.

Radii and diameters of circles, material generally introduced in third grade, are special line segments. To introduce the idea of a radius of a circle as well as the center of a circle, you might wish to take a bicycle to class and relate the spokes of a wheel to the radii of a circle and the hub to the center of a circle. As the children examine the wheel, ask questions such as "What is the shape of each wheel?" "How many spokes are on one wheel?" "How many radii do you think a circle can have?" "Why do you think so?" "Where is the center of this wheel?" "How many points do you think are at the center of a circle?" Radii and diameters of circles can also be seen when folding paper circular cut-outs to generate lines of symmetry. Each line segment determined by folding the paper in half is a diameter of the circle associated with the shape and a radius is a line segment from the center to the edge of the paper. Children should intuitively understand that a circle has infinitely many radii and diameters. When they understand these concepts, begin drawing representations of them.

A child's first encounter with a line is usually the number line used to represent the whole numbers 0, 1, 2, and so on. Help children develop an intuitive notion of line by having one child begin at one point and walk on a walk-on number line in one direction while another child begins at the same point and walks in the other direction. Emphasize a line is straight and never stops in either direction. Now rotate the number line on the floor about the point at which the students had begun their walk. Ask two other children to begin walking at the same point in opposite directions. Establish with the students this is another line through the same point. Continue rotating the number line about the point to help them generalize there are many (infinitely many) lines through a given point. This experience should help a child understand that one point does not determine a unique line. Guide the children to understand that two points are needed to name a line. Mark two spots on the floor with chalk. Ask the children to move the number line so that it passes through both points. They should see that there is only one such line. Illustrate this situation on the chalkboard by representing two distinct points, say A and B, and using a straightedge to draw the line through A and B. A name for the line is $\overleftrightarrow{AB}$ Children should also learn that any two points along the line can be used to name it. Represent point C on the line so that it is different from A and B.

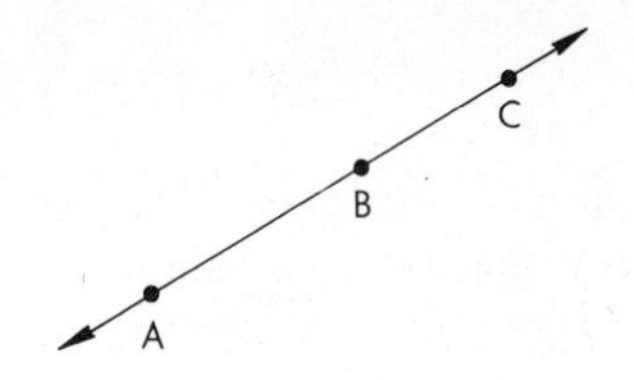

Select students to draw $\overleftrightarrow{BC}$ and $\overleftrightarrow{AC}$. They should discover that these lines are the same as $\overleftrightarrow{AB}$. Other names for the same line are $\overleftrightarrow{BA}$, $\overleftrightarrow{CA}$, and $\overleftrightarrow{CB}$.

Rays can also be illustrated by having a student walk along part of the number line. Designate a point on the line. Ask a student to start at the point and walk on the line in one direction from the point. Emphasize that rays have a starting point and continue in one direction only. Draw a representative ray on the chalkboard and introduce the notation AB.

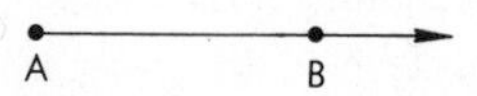

Children should find in their environment examples suggesting rays such as light rays emanating from a flashlight and a jet stream from a rocket.

Congruence

Intuitively, congruent figures are those which have the same size and shape. In beginning activities, children can create sets of congruent figures by drawing designs on a sheet of paper, placing blank sheets under the sheet of designs, and simultaneously cutting out the figures from the sheets. They should also sort attribute blocks into sets of congruent blocks. The geoboard and pegboard are useful in teaching congruence. Have children make a design on the geoboard or pegboard for a friend to copy. When they learn the concept of line segment, have them make sets of congruent line segments on their geoboards or dot paper. Measurement of length, area, and volume is based on congruent units. Encourage children to find congruent figures in their environment. Some examples are floor tiles, sheets of paper, and postage stamps.

Symmetry

Before children begin their study of symmetry, they should intuitively know what a line is and how to represent one on paper. Symmetry is an appropriate topic in second and third grades.

Introductory activities on symmetry may involve the students in folding and cutting paper. Ask them to copy what you do as you fold a sheet of paper and draw a figure along the folded edge. Ask them to cut their pictures out of the folded paper to find the shape.

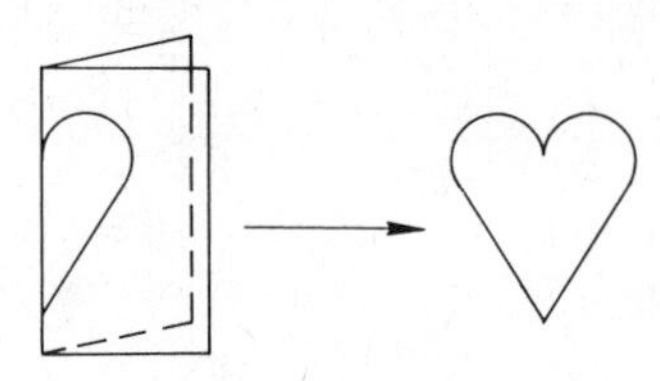

Let children describe what happened. Some may suggest that one half of the picture is drawn along the folded edge and the whole picture is shown when the design is cut out; others may suggest the two parts on each side of the fold are congruent. Tell them the fold in the paper represents the *line of symmetry* for the design. Draw another figure such as a semicircle along the folded edge and ask them to predict what shape will result when the design is cut out. Let students create their own designs.

Children can also explore line symmetry with mirror cards (source in Appendix), on the geoboard, and in art activities. This illustrates the use of mirrors on cards to complete designs:

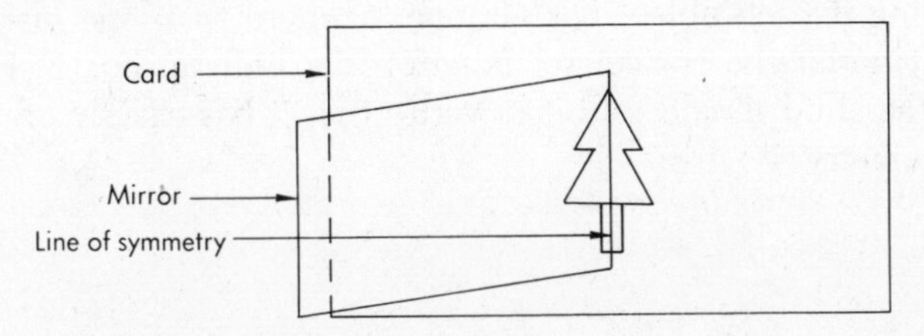

Children can explore symmetry on the geoboard by making symmetrical designs or by making half of a figure for a classmate to complete so that the design has line symmetry. The idea of symmetry can be reinforced in art activities. Children enjoy making splotches of paint on paper and then folding the paper to copy the pattern. The fold in the paper determines the line of symmetry for the design.

Children should also work with figures which are not symmetrical and those which have more than one line of symmetry. Present them with pictures of objects such as

to identify as having line symmetry or not. On figures that exhibit line symmetry, have them sketch what they think the line of symmetry is (as done in the first example) and then check their predictions by cutting out the figure and folding the paper. They should show various responses for folds of the rectangle, square, equilateral triangle, and circle. Use this opportunity to bring to their attention that some figures have more than one line of symmetry. For example, a square can be folded in four different ways to produce a line of symmetry.

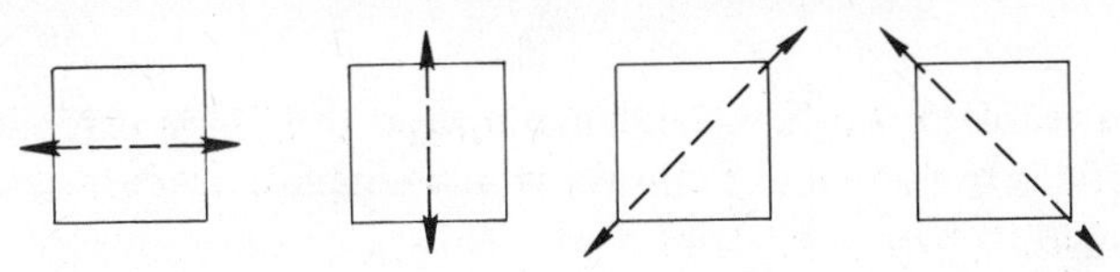

Children should suggest all possible lines of symmetry for the rectangle and the equilateral triangle and intuitively explore the fact that a circle has infinitely many lines of symmetry. Bright students should, additionally, find planes of symmetry in three-dimensional objects.

As an evaluative or a practice activity, make labels for three sets on the bulletin board: Not Symmetrical, One Line of Symmetry, More Than One Line of Symmetry. Let children take turns withdrawing figures, letters, and numerals from a box, determining if each object has no, one, or more than one line of symmetry, and pinning it under the correct set label on the bulletin board. If the figure is symmetrical, the child should fold it to verify that it has exactly one or more than one line of symmetry.

Polygons

After students have worked with shapes informally, they should learn to classify types of triangles and quadrilaterals and to identify other polygons. A polygon is a closed curve made from line segments each of which intersects two other line segments at their endpoints. Begin instruction by asking students to make a closed curve consisting of three line segments on their geoboards or dot paper. They should identify the figures as triangles. Next, instruct them to make a closed curve with four sides. Answers may include squares, rectangles, rhombi, or figures in which no side is congruent to another. Continue the activity by having them make closed curves with five, six, . . . , ten sides. Ask the students how all of these closed curves are alike. (They all consist of line segments.) Then tell them these closed curves are examples of *polygons*. Ask, "Can a polygon have two sides?" "Why not?" "What is the smallest number of sides a polygon can have?" "What do you think is the largest number of sides a polygon can have?"

Children should learn to classify triangles according to lengths of their sides. An equilateral triangle has all sides congruent, an isosceles triangle has at least two sides congruent, and a scalene triangle has no sides congruent. Bright students should identify a four-sided polygon as a quadrilateral, a five-sided polygon as a pentagon, a six-sided polygon as a hexagon, a seven-sided one as a heptagon, an eight-sided polygon as an octagon, a polygon with nine sides as a nonagon, and a ten-sided polygon as a decagon.

Further exploration should include environmental objects. Polygons can be related to shapes of road signs. Many tiles are hexagonal shaped. Some modern office buildings and the Pentagon in Washington, D.C., are pentagonal. Make an assignment for students to find which polygon is used in honeycombs.

Angles

After children learn to represent and name rays, they may extend their work to angles. An angle is the union of two rays which emanate from the same point. The angle represented by

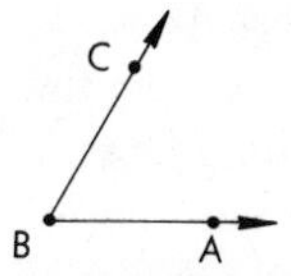

can be named ∠ABC or ∠CBA. Before students learn to measure angles with a protractor, they can informally classify angles at "square corners" as right angles, angles smaller than right angles as acute, and angles larger than right angles as obtuse angles.

Coordinate Geometry

When children have learned that arrays are arrangements with a certain number of rows and columns and can read the number of pegs or squares down and across, they can be taught coordinate geometry in which points are located by moving a certain number of units across and up from a given point. Children in second grade are usually ready to begin studying coordinate geometry. As a readiness activity, have students find locations on a geoboard. Ask them to place one finger over the peg in the lower left-hand corner. Then ask them to move to the right a certain number of pegs and up a certain number of pegs and place a finger on the other hand on the peg in the new location. Have children locate other pegs on the geoboard when you use "across and up" terminology; then select students to point out the locations for the other students to describe. You may also draw a large grid on the chalkboard, shade the square in the lower left-hand corner as the "beginning" square, and have various students draw a picture in each location you specify, e.g., make an "A" in the square four squares to the right and five squares up from the shaded square. Children should understand locations in the plane are determined by two movements—horizontal and vertical.

You may introduce students to coordinate geometry using dot paper. Tell them to begin at the bottom left-hand corner, draw a number line through the dots at the bottom of the sheet, and label the dots 0, 1, 2, and so on. Then instruct them to draw and label a vertical number line beginning again at the lower left-hand dot. Tell them these two number lines must be read to find points on the page. Emphasize that they should always begin at the "0" point on each number line. Ask them to begin at 0, move five units to the right and two units up, and make an "X" on the point.

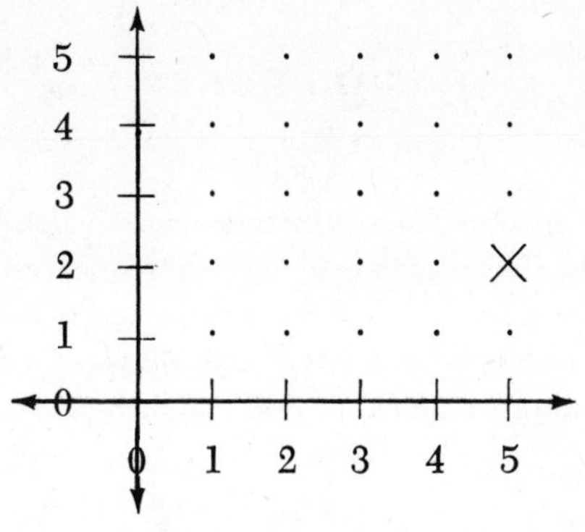

Describe other locations for students to mark, and check their work. Then introduce the notation for coordinates. Tell them the location of the "X" is (5, 2) because the point is five units to the right of and two units up from the beginning point. Describe more locations using the new notation, and let students find and label the points with coordinate notation. When they can locate points correctly, let them suggest a coordinate name for the beginning point. Represent it as (0, 0) and tell them the point is called the *origin*. Later negative numbers are used as possible coordinates.

Closing

Many topics in geometry can be taught without adhering to a rigid sequence of development. Geometry deals with the physical world, and young children enjoy exploring geometric ideas and profit from the study of this area of mathematics. Also, art which deals with form can be integrated with mathematics through geometry. At the early childhood level, the treatment of geometric concepts should be informal and intuitive. The work of Piaget indicates that topological concepts such as open and closed curves, inside and outside of closed figures, order, and nearness of an object should be taught to preoperational children before Euclidean geometry is introduced. Young children should study plane and solid figures, discuss their likenesses and differences, find representations of them in environmental objects, and learn their names. Lines and subsets of lines are also a part of the study of geometry in early childhood education, but children should not be required to learn formal definitions of these configurations as sets of points. Other topics taught during the early childhood years are symmetry, congruence, angles, and coordinate geometry. The number of ideas and extent to which they can be developed depend on the mathematical maturity of the students and the time required to deal with topics of arithmetic and algebra at the early childhood level.

Assignment

1. Examine a set of Mirror Cards and the accompanying manual. Work several examples from each set of cards. List several geometric concepts which can be taught using these materials, and describe how you would use them with young children.
2. Design an activity which involves all of the topological concepts described in this chapter.
3. Create lessons using a laboratory setting to develop each of the concepts of congruence and symmetry.

4. Outline a sequence of five or six lessons in which ideas of solid and plane Euclidean shapes are introduced and developed.
5. Create a game which reinforces concepts of coordinate geometry taught in class.
6. Read and summarize ideas in these articles in *The Arithmetic Teacher:*
   a) "Better Perception of Geometric Figures through Folding and Cutting" by Milagros D. Ibe, November 1970, pp. 583–86
   b) "Geometric Activities for Early Childhood Education" by George Immerzeel, October 1973, pp. 438–43
   c) "Geoboard Geometry for Preschool Children" by W. Liedtke and T. E. Kieren, February 1970, pp. 123–26
   d) "Experiences with Blocks in Kindergarten" by W. Liedtke, May 1975, pp. 406–12.

Laboratory Session

Preparation: Study chapter 12 of this textbook.

Objectives: Each student should:

1. generate open and closed curves,
2. find points inside, on, and outside closed curves,
3. represent polygons,
4. find lines of symmetry, and
5. represent congruent figures.

Materials: dot paper

Procedure: Each student should use the dot paper to illustrate each of the following: (If a situation is impossible to illustrate on dot paper, indicate such.)

1. a closed curve,
2. an open curve,
3. a closed curve which has four interior dots,
4. a closed curve which has four boundary dots,
5. a closed curve which has four exterior dots,
6. a right triangle,
7. a scalene triangle,
8. an isosceles triangle,
9. an equilateral triangle,
10. a quadrilateral with no pair of sides congruent,
11. a pentagon,
12. a hexagon,
13. a heptagon,
14. an octagon,
15. a nonagon,
16. a decagon,
17. two congruent line segments,

18. two congruent triangles,
19. a triangle which has no line of symmetry. What kind of triangle is this?
20. a triangle with exactly one line of symmetry. What is the name of this triangle?
21. a quadrilateral which has no line of symmetry,
22. a quadrilateral which has exactly two lines of symmetry,
23. a pentagon which has five lines of symmetry.

Bibliography

Copeland, Richard W. *How Children Learn Mathematics*, 2nd ed. New York: Macmillan Publishing Co., Inc., 1974.

Payne, Joseph N., ed. *Mathematics Learning in Early Childhood.* Thirty-seventh Yearbook. Reston, Virginia: National Council of Teachers of Mathematics, 1975.

13

Measurement

In this chapter, measurement of length, area, capacity, mass, temperature, money, and time are considered. The last part of the chapter deals with graphing. All of these are appropriate topics for young children. Work with measurement and graphs is practical and inherently motivating to children since it involves learning about themselves and the world about them and fosters social development since many measurement and graphing activities lend themselves to group work, factors all of which can contribute to a very enjoyable teaching and learning experience.

Measurement is the process of assigning a number to a set or object; the measure is the number actually assigned. Many preliminary activities, however, are appropriate for young children before they actually use an instrument to measure an object. When teaching the process of measurement, these steps should be considered:

1. Provide readiness for measurement by having children make informal comparisons of objects. Such comparisons include direct comparisons as well as those made by sight. Terminology such as "larger than," "smaller than," "the same size as," "heavier than," and so on should be introduced. Many children will intuitively understand this language by the time they are four or five years old, but due to their experiential deficiency in comparing objects, some children, visually handicapped ones in particular, must be taught these concepts.
2. Have students measure in nonstandard units. In this step, they actually assign a number, or measure, to a set or object. Children can measure lengths

by placing pencils or strips of paper end to end, area by covering a surface in sheets of paper or trading stamps, capacity in dried beans or rice, weight in books or pencils, and time by swings of a pendulum or by counting.

3. Discuss the need for a standard unit of measurement. As children measure objects in nonstandard units, help them understand the difficulty in telling someone about what is measured if the other person is not familiar with the unit of measurement used. Students should understand that measuring in standard units is essential for communication purposes. Metric units are used throughout the world.
4. Have children measure objects or sets in standard units. Length, capacity, mass, and temperature should be measured in metric and English units; time, in hours and minutes; and money value, in pennies, nickels, dimes, quarters, half dollars, and dollars.

Length

Readiness activities for linear measurement include some of the considerations of chapter 5—e.g., sorting and ordering objects by length and comparing lengths. These topics are expanded here. Such activities are generally appropriate for three-, four-, and five-year-olds and even some two-year-old children. Incidental teaching of the concepts of "longer than," "shorter than," and "the same length as" greatly reduces the need for organized activities addressing these concepts. For a given group of children, the readiness suggestions on length which follow may help some children learn beginning concepts of length while they serve as reinforcement of these concepts for other children.

Children should begin their study of length by comparing lengths of some movable objects. Have them sit around a table and present them with a set of sticks such as those found in Tinkertoys. Select four different lengths with two or three sticks of each length. Ask a child to choose one stick from the pile, and then ask the other children to suggest a way to find all the other sticks in the pile which have the same length as the chosen stick. If no child suggests a side-by-side matching procedure, do so yourself. Now, tell the child to put all the sticks the same length as his stick in one pile and those which do not match in another (discarded) pile. Emphasize the language "the same length as" and "not the same length as." When this is completed, select another child to choose a stick from the discarded pile and to repeat the procedure just described.

Young children should also order objects by length and learn to use the terms "longer than" and "shorter than" correctly. To help toward this end, gather them around the feltboard on which there are two rectangular strips. Ask the children which is longer and which is shorter. Reinforce the terms "longer than" and "shorter than" as you point to the appropriate strip. Now add a third strip (having length different from the other two) to the set and call on a child to order the three strips from shortest to longest. Ask him to touch the shortest one and show why it is shortest. (He should place it alongside the other two strips for direct comparison.) Similarly, ask the child to identify the longest and middle strips and to demonstrate the necessary comparisons. Now is an opportune time to teach the transitive property of order as described in chapter 5. Next, increase the number of strips to four and then five, arranging them in random order and position on the feltboard. Select children to order them from shortest to longest or longest to shortest and to point to all those strips longer than or shorter than one you specify. Ordering a set of Cuisenaire rods is another example of this sort. Also, related terms such as "taller than," "higher than," "bigger than," and "smaller than" should be introduced.

Once children understand the concepts of length and height, they can practice to achieve skill by ordering themselves, dolls, chairs, and tables according to height; pencils and crayons according to length; and items of food such as a carrot, bean pod, and celery stalk according to length. For evaluation, you might give each child a paper bag of six or seven straws, some the same length and some of different lengths. Instruct the child to reach in his bag, select a straw and place it in front of him, empty the bag, and place in one pile all the straws he thinks are the same length as the one he chose first, in another pile the straws he thinks are longer than the given one, and in a third pile those he thinks are shorter than the selected straw. When the children have made the sorting according to what they believe is correct, encourage them to check their work by side-by-side comparisons. Notice that in this activity young children are observing, comparing, classifying, and making and testing hypotheses—skills important in mathematics, science, and other subject areas as well as everyday life.

When the children can compare the lengths of movable objects, have them compare stationary objects by comparing them to a movable "tool." This helps them understand the function of a measuring instrument. To help toward this end, draw six or seven items on a poster, some having the same length and some not.

Provide paper strips having the same length as the items in the picture. Point to an object on the poster, and ask the children how to find out if there is another item in the picture the same length as the one to which you are pointing. The desired response is that the stationary objects must be compared to a movable object, such as a pencil, hand, Cuisenaire rod, or one of the paper strips, to make the necessary comparisons. Let the children mark all items in the picture having the same length with similar markings. In a follow-up activity, have them order the lengths of rectangles or line segments on ditto handouts. Encourage them to compare by sight before using a movable "tool." Be watchful of the frustration level of a learning disabled child on this perceptual task.

Reinforcement of the use of a movable object to make comparisons can be found in everyday activities in the room or center. For example, when a child builds a tower of blocks on the floor, ask him to build on the floor another tower the same height as the first one. Then ask him to build on a box or chair a tower the same height as the other two. Encourage him to compare the heights with some movable object such as his arm or a piece of string. Also, you might ask the children if the same tower could be built under a certain table in the room. Have them solve the problem without moving the tower or the table.

Postpone advancing further in the study of linear measurement with a child until he can conserve length. Diagnosis of this task was the subject of one of the Piagetian interviews in chapter 4. Arranging the items on the poster in various positions (some horizontal, some vertical, some otherwise) and having a child to build towers which match in length on the floor and chair as in the activities just described should promote conservation of length.

Once a child can conserve length and has compared and ordered lengths of objects, he is ready to measure the length of an object in nonstandard units such as paper clips and chalkboard erasers. Make sure that in each case the children lay the units end to end. A sample activity at this level would be to tape a picture of a dog to the floor and pose these questions for the children to answer: How many paper clips are needed to match the length of the dog in the picture? What is the length of the dog in erasers? Which is longer—a paper clip or a chalkboard eraser? Did it take more or fewer paper clips than erasers to measure the length of the dog? Which unit do you think is better for measuring the length of the dog; the length of a toothpick? Why? Such questioning helps children decide what an appropriate unit of measurement is for the object being measured. Encourage them to give answers using language such as "a little more than one eraser" or "between 8 and 9 paper clips long." Since work with fractional numbers parallels experiences in measurement, an answer such as "about $3\frac{1}{2}$ erasers long" is acceptable also.

It is instructive to have some groups measure lengths of objects in paper clips, for example, and other groups to measure lengths in sets of ten paper clips hooked together. (Let them create their own names for these units.) The idea of ten smaller units being the same measure as a longer unit prepares them for the metric system.

Introducing the Metric System of Length

When students are ready to measure lengths of objects in standard units, introduce metric units. This commonly occurs in first grade. The meter is the basic unit of length. The metric system, based on grouping by tens, should be related to the way our numeration system utilizes place value and base ten. Give some review on the fact 10 ones = 1 ten, 10 tens = 1 hundred, and 100 ones = 1 hundred. Centimeters, decimeters, and meters are related in the same way.

Cuisenaire rods can be used to teach the concepts of centimeter and decimeter. The white rod is one centimeter long and the orange rod is ten centimeters, or one decimeter, long. (The symbols cm and dm are representations of centimeter and decimeter, respectively.) To help students conceptualize these units of length, have them make a decimeter ruler by cutting out a strip of tagboard the same length as the orange Cuisenaire rod, placing white Cuisenaire rods along the strip, and marking the length of each white rod.

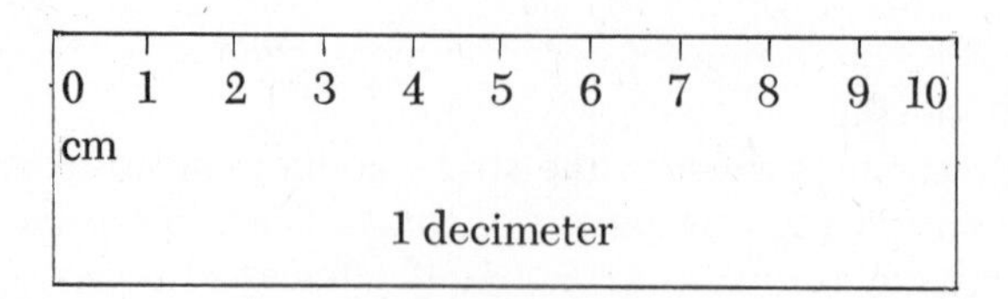

Students should label the centimeter units 0 through 10 and write "1 decimeter" on the ruler. They should then measure objects in the classroom in decimeters and centimeters and find objects which are about the same length as these units.

When children understand the relationship of centimeters and decimeters, introduce them to the standard unit, meter (symbolized as m). To help them understand measurement with a ruler and learn the relationship of centimeter, decimeter, and meter, have them make their own meter tape measures. They can do this task in much the same way they draw a number line and subdivide it. Ask them to mark off on narrow adding machine tape or seam binding ten line segments, each the same length as the decimeter ruler they have made, and ten centimeter units in each decimeter unit as they have done on their decimeter rulers. They could instead use the orange and white Cuisenaire rods to mark the decimeter and centimeter units on the meter stick. Students should label the 100 centimeters and 10 decimeters and write "1 meter" on their tape measures. In making the meter tape measure, they should learn that 100 centimeters measure the same distance as 1 meter as do 10 decimeters. Have them compare their tape measures to a commercial meter stick.

Provide many opportunities for students to measure objects to the nearest centimeter, decimeter, and meter. As children begin to measure length, insist that they place the "0" on the decimeter ruler or meter stick at the left edge of the object or line segment they are measuring.

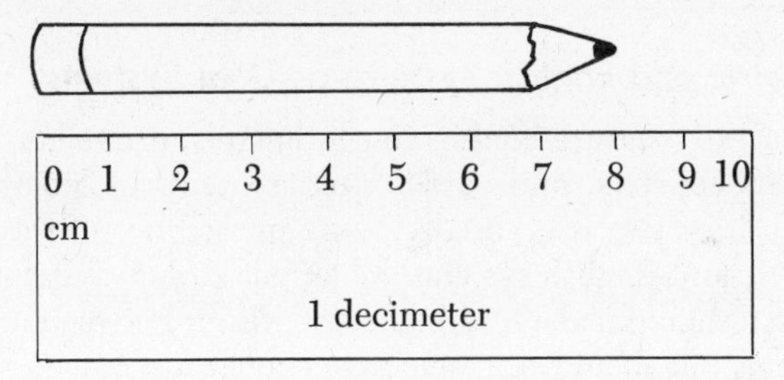

Children should, however, soon discover that the length of an object can be found by lining up one edge of the object with a numeral other than 0 on the ruler. In this case, a subtraction must be performed. The length of this line segment is 6 − 2, or 4 centimeters.

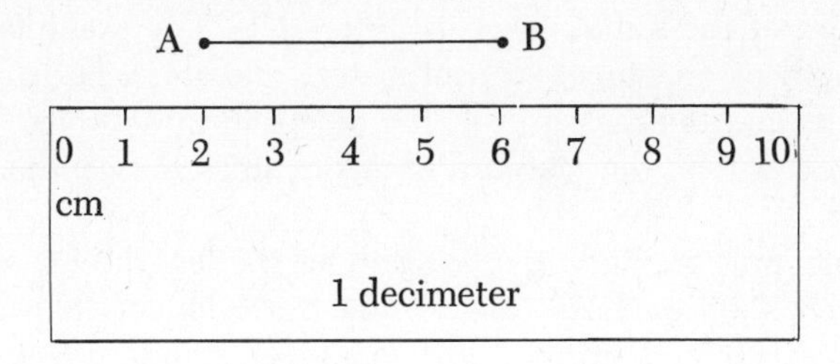

Just as you did with nonstandard measuring units, plan many activities to help children learn to use the best standard unit in different situations. For example, the width of the room is more efficiently measured in meters than decimeters or centimeters, but a decimeter or centimeter unit is well suited to measuring the length of a book or crayon respectively.

The metric system can also be related to our money system. The meaning of some of the common prefixes in the metric system is suggested by terms for money.

10 cents = 1 dime	$\longleftrightarrow$	10 centimeters = 1 decimeter
10 dimes = 1 dollar	$\longleftrightarrow$	10 decimeters = 1 meter
100 cents = 1 dollar	$\longleftrightarrow$	100 centimeters = 1 meter
10 mills* = 1 cent	$\longleftrightarrow$	10 millimeters = 1 centimeter
100 mills = 1 dime	$\longleftrightarrow$	100 millimeters = 1 decimeter
1000 mills = 1 dollar	$\longleftrightarrow$	1000 millimeters = 1 meter

The units millimeter ($\frac{1}{1000}$ meter), dekameter (10 meters), hectometer (100 meters), and kilometer (1000 meters) are seldom used at the early childhood level.

Children enjoy using the trundle wheel to measure longer distances such as the length of the hall or of the playground. The trundle wheel turns one meter from click to click as it is rolled along the object being measured.

The geoboard is useful in teaching students to measure lengths of line segments. The distance between two consecutive nails in a vertical or horizontal position is one unit. On metric geoboards, this distance is a certain number of centimeters. Children should represent line segments on the geoboard and determine their lengths by counting the units.

Teaching the Metric and English Systems as Dual Systems

After students learn to measure lengths in centimeters, decimeters, and meters, introduce them to inches, feet, and yards, standard units of length in the English system. The English system is gradually being phased out of the curriculum, but presently the metric and English systems are taught as dual systems.

Young children should make informal comparisons between units of the metric and English systems but should not memorize conversion factors relating the units of the two systems. They can compare a yardstick and a meter stick to determine that a meter stick is "a little longer" than a yardstick and can compare an inch and a centimeter to see that an inch is "a little more" than twice as long as a centimeter. Plan activities so that students discover that the larger the unit used, the smaller the measure of the object in those units, and the smaller the unit used, the larger the measure of the same object in those units. For example, have children measure the length of an object in centimeters and in inches and compare the length of the units and the measures of the object in those units. The measure in centimeters is greater than the measure in inches since the centimeter is a shorter unit than the inch.

Have students practice finding objects about the length of a centimeter, decimeter, meter, inch, foot, or yard, objects which are longer than a specified unit of length, and objects which are shorter than a given standard length. Let children use the measuring instruments to check their guesses.

---

* A mill is $\frac{1}{10}$ of a cent, but no coin having that value is made.

Important materials in teaching length are a meter stick, a ruler marked in centimeters, a yardstick, a foot ruler, and adding machine tape or seam binding for constructing a meter stick or tape measure.

Perimeter and Circumference

Perimeter, the distance around a plane figure, should be introduced after a child can measure length and add whole numbers. Students often have trouble with perimeter simply because an insufficient amount of time is spent finding perimeters of figures. Second- and third-grade teachers should have students actively involved in measuring perimeters of objects and should generally not be concerned with developing formulas for perimeter. Mathematically mature students may, however, make generalizations for finding perimeters of certain figures.

You might introduce the concept by having each student make a rectangle or square you specify on a geoboard and move around the shape with his finger. Tell the children the distance around the figure is its *perimeter*. Let them suggest a procedure for finding the perimeter of the rectangle or square. They should count the number of linear units around the figure.

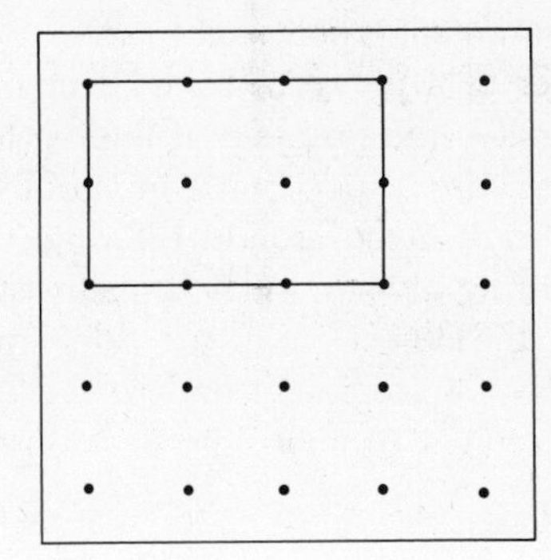

$$\begin{aligned} P &= 2 + 3 + 2 + 3 \\ &= 10 \text{ units} \end{aligned}$$

After they have worked with rectangles and squares, have them find the perimeter of a triangle.

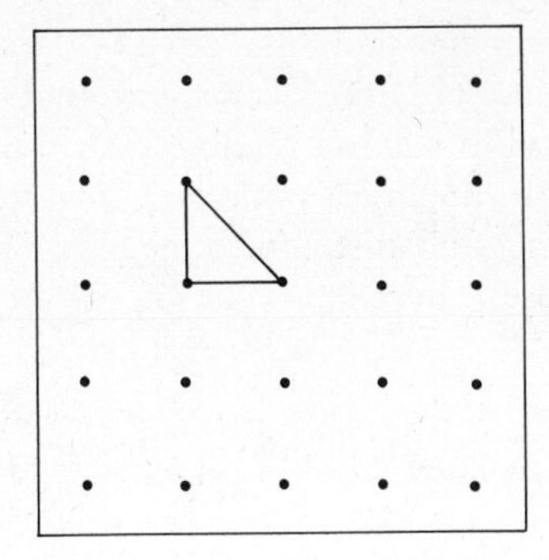

Two sides of this triangle have length 1 but the third side is "a little longer" than 1. The perimeter, then, is "a little more" than 3 units. Such terminology should be used at the early childhood level. Perimeters of more complicated polygons should be found also.

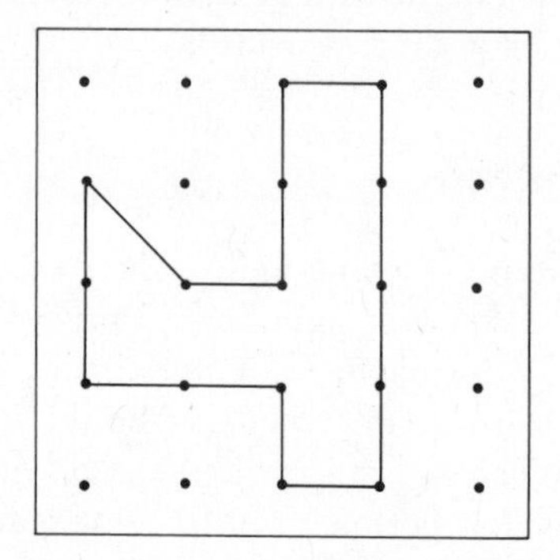

The perimeter here is "a little more" than 15 units.

After finding perimeters of figures on the geoboard and dot paper, students should measure perimeters of objects in the classroom. They enjoy measuring the distance around their desks, Valentine's cards, figures they draw themselves, and objects on the playground. If several students measure the same object, have them compare their measures so they will learn measurement is not exact.

Once students can measure the perimeter of a polygon, have them measure the circumference of a circle. They may approach the problem of finding the distance around a circular object such as a bicycle tire or pop bottle in various ways. They may use a tape measure to measure the distance around the object, wrap string around the object and then place the string along a ruler or meter stick, or mark a spot on the outer edge and then roll the object down the ruler until the mark touches the ruler again. These alternative strategies in finding the circumference illustrate that many aspects of measurement promote the development of problem-solving skills.

Area

Children often confuse perimeter and area. Area is the amount of surface a plane figure covers whereas perimeter is the distance around the figure. Area is usually introduced formally in third grade, but readiness for the concept begins in kindergarten. Prerequisite learning includes understanding the topological concepts of "inside," "outside," and "on" and distinguishing between a closed curve and the region determined by a closed curve.

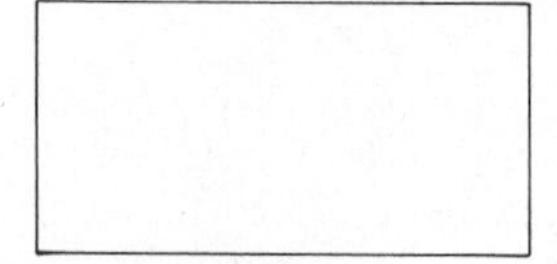

A rectangle is a closed curve.

A rectangular region has area.

Young children should also order regions by size before they actually start finding the number of units which cover an object. For example, you could distribute cut-outs of paws of the baby, mama, papa bear in "The Three Bears" story and have each child order the paws from smallest to largest or largest to smallest.

Before advancing further in work with area, children should be able to conserve area. The Piagetian diagnostic interviews on area in chapter 4 can be modified to serve as instructional activities on area conservation.

In beginning experiences in actually finding area, the children should cover surfaces with nonstandard units. You might have them use the materials just mentioned and count the number of baby-bear paws needed to cover the papa-bear paw.

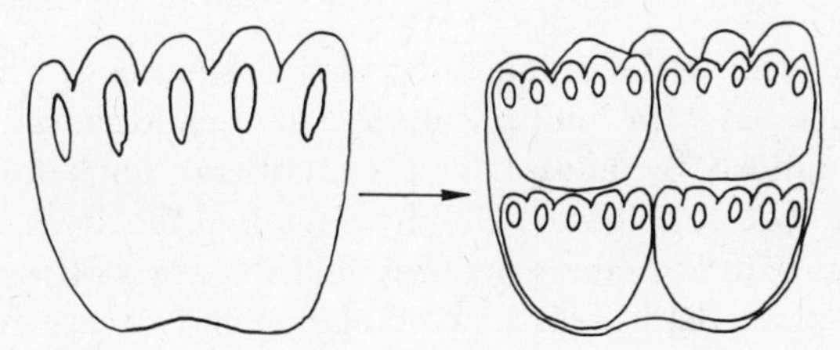

In another activity, three groups might measure the surface area of a table top—one group using sheets of newspaper for the basic unit, another group using sheets of notebook paper for the unit, and a third group using playing cards for the unit. Such experience should help them learn that area is associated with the covering of a surface and that the same surface can be assigned different measures. Work of this nature along with your questioning should also convince them of the need for a standard unit for measuring area.

Tangrams also help students learn to associate area with a covering of a surface and to break down a larger region into smaller parts. Have them, for example,

use the tangram pieces △, △, and □

to cover the rectangular region ▭ and the four pieces □

△, △, ▱, and to make ⏢

The geoboard, dot paper, and grid paper are useful aids in helping children find area in square units. Begin instruction by having students use rubber bands to cover a rectangular region with unit squares. Tell them each of these square regions has one square unit of area. They should see that the area of the rectangular region can be determined by counting the square units in the region.

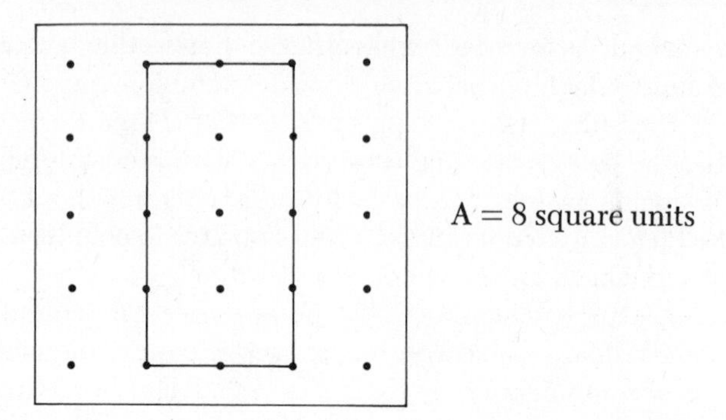

Some students may realize the arangement of unit squares is a rectangular array which can be associated with multiplication. These children understand that the area can be found quickly by multiplying the number of units down (height) with the number across (width, or base). Children can find the areas of squares, which are special rectangles, in the same way they find the areas of rectangular regions.

When students can find areas of rectangular and square regions, introduce them to the area of a triangular region by guiding them to see the triangular region as one half of a rectangular region. For example, when a child makes the following right triangle on the geoboard and "covers" it with a rectangle, he should see the

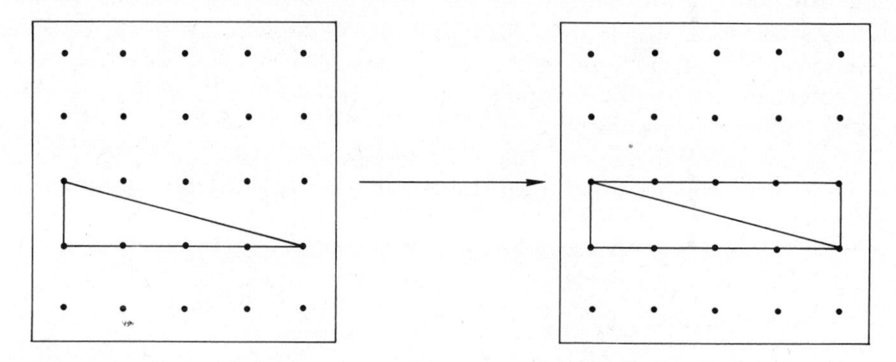

triangular region is 2 square units since the rectangular region consists of 4 square units. Children should also determine areas of triangles which are not right triangles.

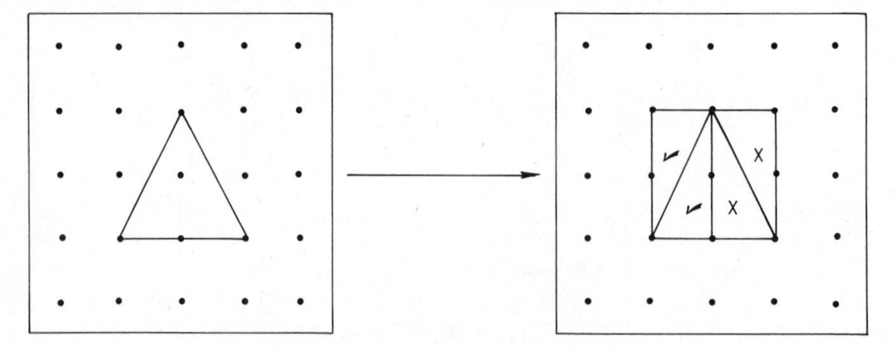

The rectangle marked with the ✓'s has an area of two square units. The same is true for the rectangle marked with the X's. The original triangular region, then, has area two square units.

After a student participates in structured activities using the geoboard, dot paper, or grid paper, he should find the area of surfaces of ordinary objects. Transparent grid paper works well for this task. The student places the grid paper over the object to be measured to determine the area. If the region to be measured is irregularly shaped, he may count the whole squares in the region, approximate the number of squares on the periphery of the region, and add the two numbers to find the area.

Capacity

Capacity is the amount of space inside a container. Very young children gain understanding about capacity in play activities such as putting smaller boxes or cans inside larger ones and pouring sand or beans from one container to another. To teach children to compare capacities, let them first compare the capacities of two cans having the same size and shape by pouring sand from one to the other. Then have them compare the capacities of containers which have different shapes. Pose the question: How can we tell if these cans hold the same amount of sand? It is hoped that someone will suggest that each of the cans can be filled and the contents poured into two containers which are the same size and shape. Let the children guess the relative sizes of their containers and check their guesses by actually pouring the contents into containers of like size and shape to allow for direct comparison. Many concepts of capacity can also be learned through cooking activities.

Before children measure the capacity of a container, they should conserve capacity. A diagnostic interview for this task was given in chapter 4.

The liter is the standard unit for measuring capacity in the metric system. The liter (symbolized as l) and milliliter (ml) should be the first metric units of capacity introduced. Students should learn that 1000 ml = 1 l. Help students conceptualize these units of capacity by having them make a liter box and a milliliter box. A box of milliliter capacity can be made by cutting out the following pattern of square centimeters and folding the sides up and taping them.

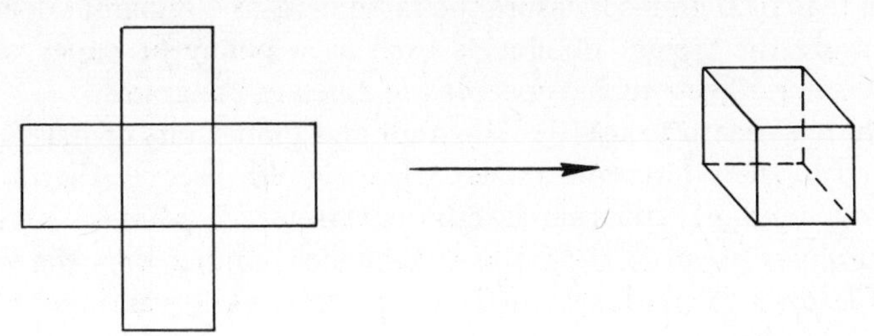

A liter box can be made by cutting out five decimeter squares from cardboard and gluing or taping them together. The liter box, then, is one cubic decimeter. Have students mark $\frac{1}{4}$ liter, $\frac{1}{2}$ liter, $\frac{3}{4}$ liter, and 1 liter on the box and use it to measure the capacity of other containers to the nearest $\frac{1}{4}$ liter.

Students should also learn how cups, pints, and quarts are related. They should also compare a quart and a liter to discover a liter is "a little more" than a quart. They can discover 1 teaspoon is approximately 5 milliliters by filling the milliliter box with sand or rice and pouring the contents into a teaspoon.

Materials for teaching capacity include a liter container, quart container, measuring cups, measuring spoons, milliliter box, and rice, sand, or beans to fill the containers.

Mass

To provide readiness for mass, have children make judgments as to which objects are heavier than others.* You might place pairs of real objects before the child and have him place them on a chart to complete the sentence "_______ is heavier than _______." His finished work may look like this:

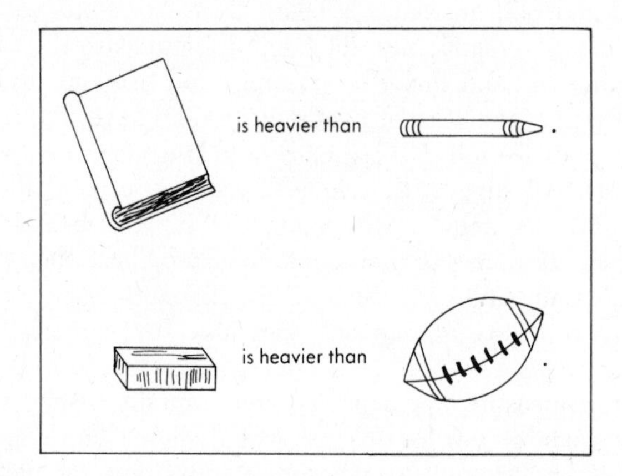

Beginning instruction with nonstandard units may include having students balance objects on a balance scale using nonstandard units of mass such as toothpicks and bottle caps.

When students are ready to measure weight in standard units, introduce the gram (g) and kilogram (kg), major units of mass in the metric system. Children should learn that 1000 grams measure the same mass as 1 kilogram does, grams are used to measure the weight of objects such as a penny or paper clip, and the weights of heavier objects such as people are given in kilograms.

After students learn to use the kilogram and gram units of mass, they should be introduced to the pound and ounce. They should discover that a kilogram is more than a pound (1 kilogram is approximately 2.2 pounds) and hence the weight of an object given in kilograms is a smaller number than the weight given in pounds. Children should know their own weight in kilograms and pounds.

---

* Although the term "mass" is technically correct, the familiar terms "weight" and "weigh" are still commonly used when referring to the mass of an object.

Important equipment for conducting a unit on weight includes balance scales, bathroom scales calibrated in kilogram and pound units, gram weights, commercial centimeter cubes each weighing one gram, an ounce weight, food and drink containers which have gram and ounce weights printed on them, and objects to weigh such as a nickel, paper clip, clothespin, and book.

Temperature

The scale on a thermometer can be related to a vertical number line with the units called degrees, and children can use this knowledge to read temperatures. Thermometers which have a Celsius scale, used in the metric system, and a Fahrenheit scale should be available. Observing the temperature inside and outside the classroom can be a daily occurrence in the primary grades and serves to apply subtraction skills: "How much colder is it outside than inside in Celsius; in Fahrenheit?" "How much warmer is it today than yesterday?" Having students record the temperature each day for a month and then use the data to compile a bar graph is appropriate for grades one through three. Students should generalize that the Celsius reading is always a smaller number than the Fahrenheit reading for a given temperature since the Celsius degree is larger than the Fahrenheit degree. Each student should also determine his approximate body temperature in Celsius and in Fahrenheit by holding the red bulb of a small thermometer between his fingers for several minutes. Small student thermometers having both scales are available commercially.

Work with temperature is also a part of cooking activities. Have children consider which foods need to be cooked at high temperatures and which at low temperatures and the amount of time needed to cook foods at certain temperatures. Temperature can also be a part of a science unit on weather and provides a basis for introducing negative integers in third grade.

Money

Counting by 1's, 5's, and 10's is useful in the study of money. To help young children develop the concept of monetary value, you might have them suggest what they could get in exchange for nonstandard units of value such as bubblegum, football cards, and toy cars. Of the standard units of money, children should be taught the smallest units first. You might begin by presenting to the children a penny, nickel, and dime and having them compare the relative sizes of the coins, to note whether their edges are rough or smooth, and identify the names of the pieces of money. After the physical characteristics are discussed and the names of the coins are given, help them learn equivalent values among them. Count five pennies and tell them this amount has the same value as one nickel. Continue counting to ten pennies and explain that the dime is equivalent in value to ten pennies. Also, help them associate two nickels with a dime. When children learn the names and values of these three coins, give them pennies to count and to make replacements with dimes and nickels.

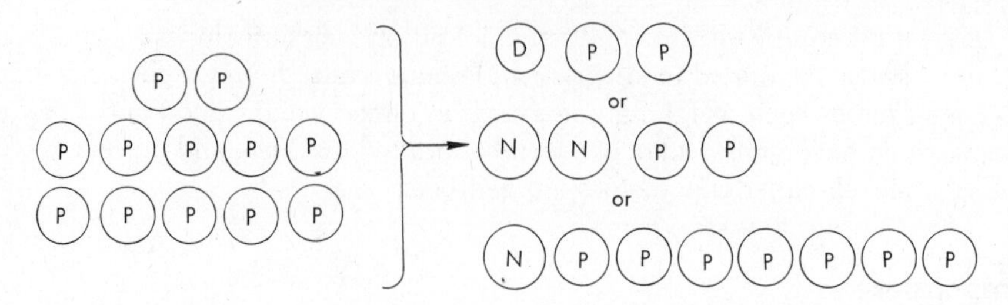

Let them describe their work after they have made the exchanges. Quarters, half-dollars, and dollars can be introduced in a similar way.

As pieces of money are introduced, let the children discuss what objects they can buy with them. For example, you might distribute various combinations of coins to the children and then present posters having pictures of items in everyday life and their accompanying price. Let each child tell whether he has enough money to buy the item.

Also, you might set up a "store" situation in which the children purchase items on which you have placed price tags. One or two children can operate the "cash registers" while the others act as customers. Real money can be used when small groups are involved, but play money is usually best when the entire class participates. In this activity, students extend their understanding of money as well as practice and apply addition and subtraction of whole numbers.

Children also need problem-solving work in finding equivalent values of coins. Provide play money and give the students problems such as these to solve:

1. Find all the ways there are of making change for a dime. Finish this chart.

	D	N	P
	1	0	0
10¢ =	0	2	0

2. Find all the money equivalents of 15 cents. Fill in this chart.

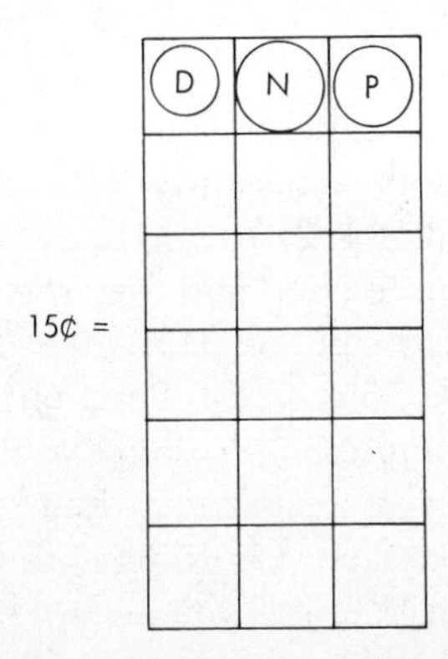

3. Can you find all the ways of making change for a quarter? Make your own chart. (Clue: There are 12 ways.) Work with a partner.

Time

The Piagetian task on conservation of time given in chapter 4 helps a teacher assess whether a student can coordinate rate and time. If children cannot perform this task, they need experiences dealing with time, rate, and distance. For example, have students measure a certain distance on the playground in meters using a trundle wheel and then let them line up and run that distance. Emphasize the fact that the distance is the same but the time needed to run the distance is different because some children run at a faster rate than others. A similar relationship holds in the simultaneous motion of the two hands on a clockface. It takes the minute hand one hour to make one complete revolution whereas the hour hand requires twelve hours to sweep the clockface one time.

Before introducing the clockface, review the students on counting through 12 and counting by fives through 60. Then present them with a clockface on which the numerals 1 through 12 and 5, 10, 15, . . . , 60 are listed. Also, have both hour and minute hands pointing toward 12. (It often helps to let the students construct two number lines labeled 0 through 12 and 0 through 60 and associate the number-line numerals with those on the number circle.)

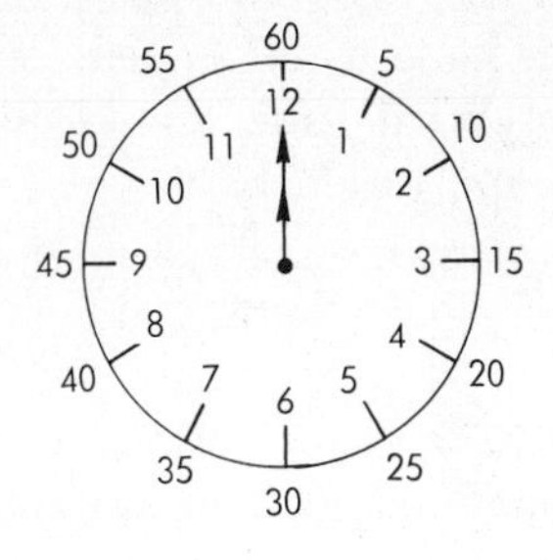

Tell the children that clocks do not show the outside numerals so they must learn to associate 5 with 1, 10 with 2, and so on. Point out that 12 is the starting point on a clockface. Then introduce them to the two hands. Explain that the shorter hand is the hour hand, the numerals 1 through 12 are used to read the hours, and one hour of time has passed when the hour hand moves from one numeral to the next. Begin at 12 and move the hour hand to 1, 2, 3, . . . , 12 as the students count orally. Tell them the longer hand is the minute hand and they should think of it as moving around the number circle labeled through 60. Move the minute hand around the clockface as the students count 5, 10, 15, . . . , 60. Point out, too, the position of the minute hand for one minute, two minutes, and so on. Now tell them that the hour hand moves from one numeral to the next in one hour while the minute hand moves all the way around the clockface. Coordinate the movement of the two hands by moving the hour hand from 12 to 1 as you make a complete revolution with the minute hand. Record on the chalkboard 1 hour = 60 minutes. Have students move the clock hands to show their motion in one hour. Ask questions such as "Which is the hour hand?" "Which hand moves faster?" "When the hour hand moves from 2 to 3, what does the minute hand do?" "When the hour hand moves from 7 to 8, how much time has passed?" "When the hour hand moves from 7 to 10, how many circles has the minute hand made?" "How many hours are read on the clock before the numerals begin to repeat?" "When the minute hand moves from the starting position to 30, how much of an hour has passed?" "What does the hour hand do during that time?" (Let students demonstrate the position of the hands.) "Which is more—one-half hour or 40 minutes?" "How can you tell?" "When the minute hand moves from the starting position to 15, how much of an hour has passed?"

When students understand the movement of the hands and the relationship of hours and minutes, help them tell time to the hour. Set the minute hand on 12 and the hour hand on 1, for example. Tell them the clockface shows 1 o'clock, or 1:00, which means 1 hour and no minutes. Have students illustrate other times and write and read the notation for those times. Let them make their own clocks out of paper plates or pizza board, represent various hours of the day, and discuss what they do at those times.

For practice in telling time to the hour, the children might enjoy a game (rhyme) such as this:

What Time Do We Have?

When the short hand is on 3 (place short hand on 3)
and the long hand is on 12 (place long hand on 12)
Think about it! What time do we have? (Children
say in unison, "3 o'clock!")

Repeat this rhyme with other hours, letting individual children manipulate the hands of the clock.

When they can tell time to the hour, introduce time to the minute. Place the minute hand on 12 and the hour hand on 2, for example. Move the minute hand to 6 and simultaneously the hour hand to halfway between 2 and 3. Introduce the symbolism 2:30, explaining that it means 30 minutes after 2 o'clock and is read as two-thirty. You may add that sometimes thirty minutes after an hour is referred to as half-past the hour because one-half hour is the same as thirty minutes. Continue with examples such as 4:15, 3:45, 8:10, and 9:42. Following work such as this, provide practice exercises. While students are studying time, both a regular clock and a digital clock should be available in the classroom.

Telling time on the clock is not the only aspect of time important in the early childhood curriculum. Here are some other considerations:

1. The calendar should be kept in a prominent place and frequent use of it should be made. Each student can place a star or happy face on his birthday, and the class can make a graph of the number of birthdays in each month.
2. Learning the seasons is important. Let the class discuss activities appropriate for each of the four seasons. Integrate time and temperature in an activity in which students match thermometer readings with seasons of the year.
3. Children should relate time and age. Ask them to bring from home several pictures of themselves from the time they were babies to the present and order the pictures from the time they were youngest to the time they were oldest.
4. Integrate time and music. Have children clap fast rhythms and slow rhythms.
5. Have students become conscious of a certain duration of time. For example, note the time before and after showing a filmstrip or playing a record.
6. Have children examine various types of timepieces such as a sand timer and sundial and learn how they are used.

Some books on time for preschool children include:

1. Behn, Harry. *All Kinds of Time.* New York: Harcourt, Brace and World, Inc., 1950.
2. Bendick, Jeannie. *The First Book of Time.* New York: Franklin Watts, Inc., 1963.
3. Brown, Margaret Wise. *A Child's Good Night Book.* New York: William R. Scott, Inc., 1943.
4. Brown, Margaret Wise. *Night and Day.* New York: Harper and Brothers, 1942.
5. Goudey, Alice E. *The Day We Saw the Sun Come Up.* New York: Charles Scribner's Sons, 1961.
6. Spier, Peter. *The Fox Went Out on a Chilly Night.* New York: Doubleday and Company, 1961.
7. Treasalt, Alvin. *Wake Up Farm.* New York: Lothrop, Lee and Shepard Co., 1955.

Graphing

Graphs are usually introduced, not as an isolated topic, but along with other topics in mathematics and other subjects such as science, social studies, and health. Graphs involve the gathering of data and show relationships in such a way that the results can easily be read and interpreted. Bar graphs and pictographs are particularly appropriate for young children. For bar graphs, either horizontal or vertical bars can be used. In a pictograph, pictures are used to represent unit values.

These are some ideas on teaching graphs:

1. As children are learning about the seasons of the year, let them make a graph of their favorite seasons. Write the names of the seasons on an overhead-projector transparency, and have each child draw a "happy face" on the transparency by his preferred season. Introduce the term "graph," and let children name it and discuss what is apparent from looking at it.

   Our Favorite Seasons—Mrs. Spencer's First Grade Class

   Fall

   Winter

   Spring

   Summer

   Other ideas include letting children graph their favorite pets, television shows, food, and colors.
2. Science and mathematics can be related through graphs. In a study of seeds, give each child a tangerine, for example, to eat and ask him to find out how many seeds it has. Instruct each child to then draw on a card (all cards should be the same size) a set equivalent to the set of seeds in his tangerine and to pin the card on the bulletin board under the column headed by that number. Constructing such a rudimentary pictograph would be appropriate when children are studying cardinal numbers.
3. Graphs can also be used in an integrated mathematics and social studies activity. For example, in a unit on families, have each child draw a picture of all the people in his family and then, as in the preceding activity, pin it on the bulletin board under the correct numeral.
4. Bar graphs can be introduced in a unit on transportation, for example. Tape pictures of different modes of transportation on a table in the room and give each child a cubical block. He places the block over the picture corresponding to the method of transportation he uses in getting to school. The children's crude bar graph may have this form:

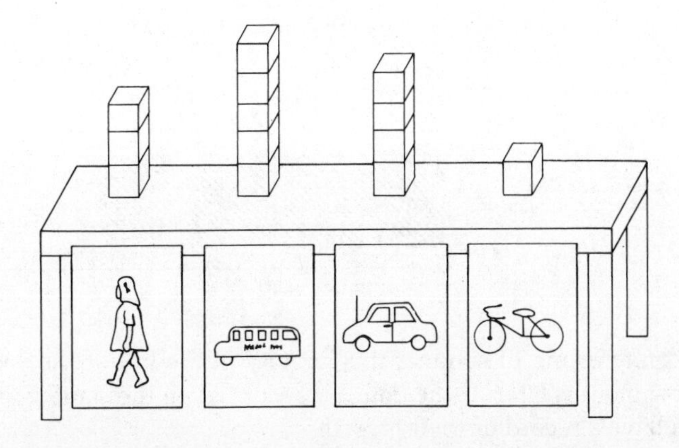

Such an activity is suitable for many three-, four-, and five-year-olds. Ask questions requiring them to "read" and interpret the results.

5. To give children experience in interpreting graphs, have them work problems such as these:
   a) Karen made a graph to show the number of pennies, nickels, and dimes she had saved. She counted her coins and shaded a square for each coin she had. This is the graph Karen made:

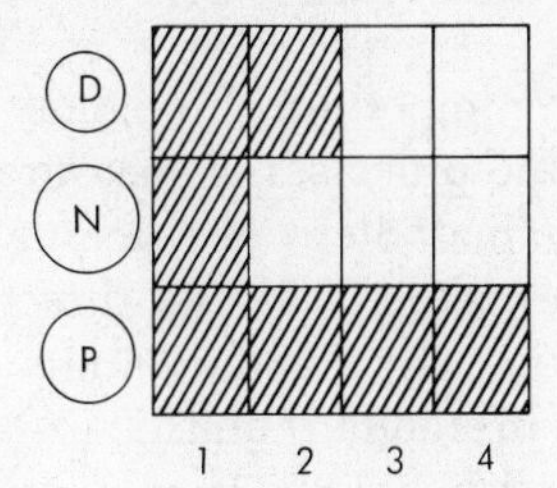

   Karen had ____ dimes, ____ nickels, and ____ pennies.

   b) Randy had saved 2 dimes, 3 nickels, and 6 pennies. Make a graph to show how many dimes, nickels, and pennies Randy had.

6. Graphs are useful in work with length and height. Let each child lie down on a strip of adding-machine paper while a partner cuts out a strip which matches the height of the child. Ask each student to measure the strips in centimeters, write his name and height on the strip, and tape it to the wall. The bottom of each strip should touch the floor. Such an arrangement suggests the use of a base line when preparing bar graphs. Now, present this problem to the students: How can we keep a record of the heights of the children as shown by the strips on the wall? It is hoped some child will suggest that the strips can be drawn on paper "but just made smaller." Such a response allows you, as teacher, to introduce the idea of a scale to represent longer distances. You might draw the bar vertically or present the information in this form using a number line:

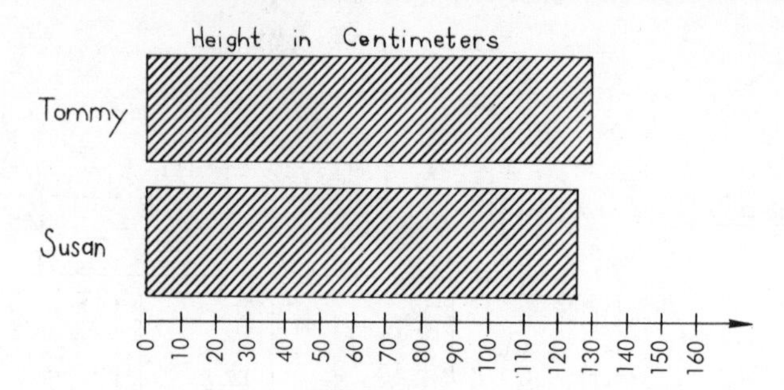

It is interesting to conduct this activity in the fall of the year and repeat it a few months later. If the data are recorded on the same graph, the children can have a record of their growth.

Scale drawings are important in the study of maps and globes in social studies.

Closing

Skills in measurement are important throughout one's life. Young children should be actively involved in the processes of measurement, and paper-and-pencil exercises should be kept minimal. Steps in teaching measurement concepts and skills include having children make visual and direct comparisons, measure objects in nonstandard units, discuss why standard units of measurement are needed, and actually measure objects in standard units.

The metric and English systems should presently be taught as dual systems, with the metric system being emphasized. The metric system should be related to our decimal numeration system and money system. Many aspects of measurement can be reinforced and extended in other subject areas, especially science, health, music, and art.

Graphing is also an important topic for young children and can be coordinated with measurement and other topics in mathematics and other subject areas such as science, social studies, and health.

Assignment

1. a) Read the April, 1973, issue of *The Arithmetic Teacher* and make a note of historical ideas of the metric system you can use when teaching this system to young children.

b) Some students may benefit from independent work on the history of measurement. Make a list of books and articles containing information on the history of the metric and English systems of measurement.

2. Begin a file of ideas for teaching length, perimeter and circumference, area, capacity, weight, temperature, and time. Include at least five ideas on each topic, and use various sources.
3. Write a laboratory activity on some aspect of measurement.
4. Write a sequence for teaching "telling time" on the clockface and for teaching length as presented in this chapter.
5. Examine current elementary mathematics textbooks for the approach used in teaching graphs and in teaching "telling time" and other topics in measurement. Make a list of the ideas you obtain.
6. Write an activity on graphing. Try to design it so that it can be used to integrate mathematics and some other subject area such as science, health, or social studies.
7. Prepare a card file of teaching ideas you obtain from these articles in *The Arithmetic Teacher:*
   a) "Let's Do It! The Neglected Decimeter" by Mary Montgomery Lindquist and Marcia E. Dana, October 1977, pp. 10–17,
   b) "Getting a Good Start in Teaching Metric Measurement Meaningfully" by Joan Doherty, May 1976, pp. 374–78,
   c) "Graphing as a Communication Skill" by James V. Bruni and Helene Silverman, May 1975, pp. 354–66,
   d) "Paper Folding and Cutting a Set of Tangram Pieces" by Steven S. Dickoff, April 1971, pp. 250–52.

Laboratory Session

Preparation:

1. Study chapter 13 of this textbook.
2. Make a milliliter box from the square-centimeter pattern given in this chapter and a liter box from a square-decimeter pattern.

Objectives: Measure and estimate lengths, weights, capacities, and temperatures and relate metric units of length, capacity, and mass.

Procedure: Students should work in pairs as they work the problems at each learning station. Then students should individually complete a set of practice problems.

LENGTH

Materials: centimeter rulers, meter tape measures, meter stick, yardstick, foot rulers, trundle wheel, string, box of paper clips, pencil.

1. a) Use the rulers, tape measure, meter stick, and string to find the length of each of the following:

Object Measured	Measure in cm	Measure in inches
paper clip		
pencil		
your height		
distance around the top of the wastepaper basket		

b) For which objects above would the meter and yard not be convenient units?
c) What problems might children encounter in finding lengths?

2. a) Study a paper clip. What part of it measures approximately one centimeter?
   b) Hook some together to make a chain about one decimeter long.
   c) How many paper clips would be needed to make a chain measuring one meter?
3. Without a ruler, try to draw a line segment
   a) 1 centimeter long,
   b) 5 centimeters long,
   c) 1 decimeter long.
   Use the rulers to check your judgment. Redraw the line segments if needed.
4. a) Place the yardstick and meter stick alongside each other. Which is longer?
   b) Compare a centimeter and an inch. Which is longer?
   c) Find approximately where one meter is above the floor in relation to your own body.
   d) Find at least two other objects in the classroom that are approximately one meter long.
5. a) Place a trundle wheel at the end of a meter stick. Roll the wheel down the meter stick until you hear a click. What distance do you measure from click to click on the trundle wheel?
   b) Use the trundle wheel to measure the width of the classroom to the nearest meter.
   c) For what measurement activities will the trundle wheel be especially useful to children?

MASS (WEIGHT)

Materials: balance scales; ounce weights; 100 to 200 commercial plastic centimeter cubes, each weighing one gram; bathroom scales calibrated in kilograms and pounds; jack ball; nickel; rice or sand; small book; clothespin; juice glass; dictionary.

1. Each of the centimeter cubes weighs one gram. Use the balance scales to check to see if 28 cubes balance a one-ounce weight. Which is more—an ounce or a gram?
2. Use the balance, centimeter cubes, and the ounce weights to find the weight of each object.
   a) jack ball Measure in grams ______; Measure in ounces ______
   b) juice glass Measure in grams ______; Measure in ounces ______
   c) small book Measure in grams ______; Measure in ounces ______
3. a) Weigh yourself using the bathroom scales provided. What do you weigh in kilograms? How many grams is that? What do you weigh in pounds? As a unit, which

is larger—a kilogram or a pound? To measure the same object, which measure (number) would be larger—the kilogram measure or the pound measure?

b) Find at least two objects in the classroom which weigh approximately one kilogram.

4. Go to the liquid-capacity center and hold the liter container filled with water in one hand. The weight of a liter of water is one kilogram.
5. Estimate the weight of each object and then weigh the object.
   a) nickel Estimate in grams _____; Measure in grams _____
   b) clothespin Estimate in grams _____; Measure in grams _____
   c) dictionary Estimate in kilograms _____; Measure in kilograms_____

CAPACITY

Materials: liter container with milliliter markings, quart container, pint jar, small drinking glass (about 8 ounces), aspirin jar, small jar lid, 1 teaspoon, frozen orange juice can, 1-cup measure with ounce markings, 1 milliliter box made from the pattern given in this chapter, rice or sand for pouring from one container to another.

1. Fill the ml box with sand or rice and pour the contents into the teaspoon measure. Continue until you fill the teaspoon. One teaspoon measures approximately the same capacity as _____ ml, and one tablespoon of capacity is approximately the same measure as _____ ml.
2. Use the ml markings on the liter container and the ounce markings on the cup measure to find the capacity of each object.
   a) small jar lid Measure in ml _____; Measure in ounces _____
   b) frozen orange juice can Measure in ml _____; Measure in ounces _____
   c) pint jar Measure in ml _____; Measure in ounces _____
3. Fill a quart container with water and pour it into the liter container. Which measures more capacity—a quart or a liter?
4. Estimate the capacity of each object, and then check your estimate by measuring the capacity.
   a) aspirin jar Estimate in ml _____; Measure in ml _____
   b) drinking glass Estimate in ml _____; Measure in ml _____
5. Answer the following.
   a) 1 ml = _____ l
   b) 1 l = _____ ml
   c) If you are serving lemonade in paper cups that have capacity of 250 ml each, how many people could have drinks from 10 liters of lemonade?

WEIGHT AND CAPACITY

Materials: 1 cubic decimeter (liter) box made from the pattern suggested in this chapter, at least 100 plastic centimeter cubes

1. a) Form a row of centimenter cubes across the edge of the bottom of the liter box you have made. How many centimeter cubes are needed to form the row?
   b) Form a column of centimeter cubes on the bottom of the box. How many cubes are there in the column? How many centimeter cubes cover the bottom of the liter box? (Remember the array interpretation of multiplication.)
   c) Stack centimeter cubes up the side of the liter box. How many are needed to reach the top of the box?

d) How many centimeter cubes does the entire box hold? What is the capacity of the liter box in cubic centimeters; in cubic decimeters?

2. a) Each of the plastic centimeter cubes weighs 1 gram (the same as 1 cubic centimeter of water weighs). If the liter box were filled with centimeter cubes or water, what would the weight of these cubes or this amount of water be in grams; in kilograms?

   b) What is the capacity in milliliters of one of the centimeter cubes? How many grams does 1 milliliter of water weigh?

TEMPERATURE

Materials: 4 to 8 student thermometers calibrated in Celsius and Fahrenheit units, 1 wall thermometer with Celsius and Fahrenheit scales.

1. Hold the red bulb of a small thermometer in your fingers for several minutes. What is your approximate body temperature in Celsius degrees; in Fahrenheit degrees?
2. Use the wall thermometer to answer these questions:
   a) What is the approximate room temperature in Celsius; in Fahrenheit?
   b) Which degree is larger—Celsius or Fahrenheit?
   c) At any given time, which reading would be given by a larger number—Celsius or Fahrenheit?
3. Answer the following:
   a) What is the boiling point of water on the Celsius scale? _____
   b) _____ °C = 212°F.
   c) What is the freezing point of water on the Celsius scale? _____
   d) _____ °C = 32°F.

PRACTICE PROBLEMS

When you complete work at the five learning stations, answer the following:

1. Complete these metric equivalents:
   a) 1 m = _____ mm b) 5 mm = _____ m c) 2 hm = _____ m
   d) 75 mm = _____ cm e) 75 mm = _____ dm f) 1 kg = _____ g
   g) 62 g = _____ kg h) 10 l = _____ ml i) 25 ml = _____ l
   j) 1 cubic decimeter = _____ cubic centimeters.
   k) 1 cubic centimeter of water weighs _____ g.
   l) 1 cubic decimeter of water weighs _____ g, or _____ kg.
2. Write $>$, $=$, or $<$ in the ○.
   a) 1 l ○ 1 quart
   b) 2 teaspoons ○ 8 ml
   c) 50°F ○ 20°C
   d) 50 g ○ 1 ounce
   e) 1 kg ○ 1 pound
   f) 5 cm ○ 3 inches

How well did you do? The answers are

1. a) 1000 b) 0.005 c) 200 d) 7.5 e) 0.75 f) 1000 g) 0.062 h) 10,000 i) 0.025 j) 1000 k) 1 l) 1000, 1
2. a) $>$ b) $>$ c) $<$ d) $>$ e) $>$ f) $<$

   If you missed more than one problem, review the metric prefixes and the discoveries you made in the five learning stations.

Bibliography

Brougher, Janet Jean. "Discovery Activities With Area and Perimeter." *The Arithmetic Teacher* 20 (May 1973): 382–85.

May, Lola June. *Teaching Mathematics in the Elementary School.* New York: The Free Press, 1974.

Reisman, Fredricka K. "Children's Errors in Telling Time and A Recommended Sequence." *The Arithmetic Teacher* 18 (March 1971): 152–55.

14

Number Theory

Some topics in number theory are appropriate for most "regular" first, second, and third grades but are not generally recommended for special education situations. Even and odd counting numbers are usually introduced in first grade and extended in second and third grades, and prime numbers are often a part of the third-grade curriculum. When studying number theory, a child must perform many arithmetic computations as he explores number patterns and relationships. The opportunity a student has to practice basic facts of operations on whole numbers and apply computational skills is a worthy by-product of investigations in number theory.

Even and Odd Whole Numbers

When students can count, identify the cardinal number of a set, join and separate sets, and add and subtract whole numbers, they are ready to classify even and odd counting numbers. Begin by giving each student ten plastic counters and five strips of yarn for making loops. Tell the students they are to try to separate the sets you specify into subsets of two and determine which sets can be separated into subsets of two and which cannot. Label two sets, "Can be separated into subsets of 2" and "Cannot be separated into subsets of 2," on the chalkboard so you can record the children's findings as they generate the subsets. Ask the students to select four counters and place loops around subsets of two. Since subsets of two can be formed, write the numeral "4" in the set labeled "Can be separated into subsets of 2." Next ask students to select five counters and try to form subsets of two. Since one is left over, write "5" in the set, "Cannot be separated into subsets of 2." Continue the classification through 10. Then ask students to classify 3, 2, and 1. A child's work and your accompanying board work are suggested here:

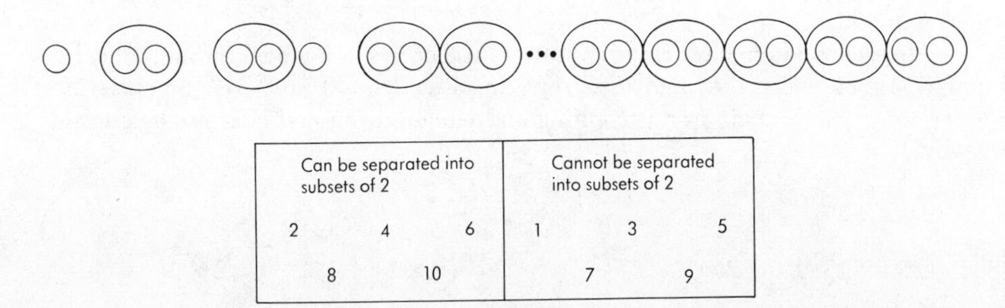

Can be separated into subsets of 2	Cannot be separated into subsets of 2
2 4 6 8 10	1 3 5 7 9

When the classification into the two sets is made, introduce the terms "even" and "odd." Even numbers are represented by sets which can be separated into subsets of two, and odd numbers are associated with sets which cannot be separated into subsets of two. After the new terminology is presented, ask, "What is the smallest even counting number?" "How much more is the second even counting number than the first?" "What seems to be the pattern in naming even numbers?" "How many counters are left over each time in the sets representing odd numbers?" "What is the smallest odd counting number?" "How much larger is the second odd number than 1?" "What is the pattern for finding the other odd numbers we have listed?" "Which do you think 11 would be—even or odd?" "How can you tell?" (To answer the last question, some may use the pattern of adding 2 to an odd number to get another one and thus claim 11 is odd because $9 + 2 = 11$. Others may suggest that a counter is left over when a set of eleven counters is separated into the subsets of two.) Encourage students to extend the even-odd classification to at least 20. To help students classify the whole number 0 as even, you can appeal to the pattern of adding 2 to or subtracting 2 from an even number to get another even number. The whole number 2 is even and 0 results when 2 is subtracted from 2; thus 0 falls in the even-number pattern.

A beginning activity can also be conducted with Cuisenaire rods. Give each student a set of Cuisenaire rods and ask him to try to build trains of only red rods to match rods representing 1 through 10. (The white rod represents 1 here, so the red rod is associated with 2.) Again, two sets result: rods which can be matched with a 2-rod train (these represent even numbers) and those which cannot (these represent odd numbers). Each rod associated with an odd number can be matched with a train of red rods and one white rod.

Once students understand the meaning of even and odd numbers and place value, design a symbolic activity to help them discover that any whole number is even if the ones digit is even and odd if the ones digit is odd. In this way they will be able to tell at a glance whether a number is even or odd. Ask them to list the even whole numbers through 20 in one column and the odd whole numbers through 19 in another column. Ask the students if they see a relationship in the way the numbers in each set are written. Guide them, if necessary, to see that 10, 12, 14, 16, 18, and 20 all have as ones digits 0, 2, 4, 6, or 8 which are themselves even. Similarly, 11, 13, 15, 17, and 19 have ones digits which represent odd numbers. Challenge children to use the pattern to classify examples such as 32 and 287 as even or odd.

In second grade after children learn to use arrays as a model for multiplication of whole numbers, they can use arrays to visualize odd and even numbers in a new way. Guide them in representing and naming even and odd numbers in this way:

$1 = (2 \times 0) + 1$ $\quad 2 = 2 \times 1$ $\quad 3 = (2 \times 1) + 1$ $\quad 4 = 2 \times 2$

$5 = (2 \times 2) + 1$ $\quad 6 = 2 \times 3$ $\quad 7 = (2 \times 3) + 1$

$8 = 2 \times 4$ $\quad 9 = (2 \times 4) + 1$ $\quad 10 = 2 \times 5$

Question them on the pattern: "The even numbers 2, 4, 6, 8, and 10 can all be represented by an array having two rows. What, then, is a factor of all these even numbers? Do you think all even numbers will have 2 as a factor? Does 30 have 2 as a factor? Would you classify 30 as even? Does 100 have 2 as a factor? Is 100 even? Look at the open sentence $0 = 2 \times \square$. What is the missing factor? Since 0 has 2 as a factor, is 0 even? Look at all the arangements representing odd numbers. How are they alike? How can all of them be written symbolically? What is a way to tell if a number is odd without drawing an array? Is 19 odd? (Yes, since $19 = (2 \times 9) + 1$.)" When students extend their skills in multiplication and division in third grade, advanced learners should use the pattern developed above to examine the formal definition of an even whole number (any number which can be written in the form $2 \times n$, n a whole number) and an odd whole number (any whole number which can be written as $(2 \times k) + 1$, k a whole number).

As a practice exercise for children to recognize and identify even and odd numbers, play thinking games such as "I'm thinking of the even number which comes just before 80. What is it?" Supply written exercises such as 15, ____, ____, 21 and ____, ____, ____, 40, 42, ____ for students to complete by counting by even or odd numbers. For further practice, have students circle their birthday dates on the calendar and identify how many even numbers and odd numbers are circled in each month. Involve students in applying even and odd numbers in ordinary occurrences in the classroom. For example, say, "There are twenty-five students in class today. Can all of you work in pairs? How many will be left over? Is 25 odd or is it even?" As an assignment, ask students to notice numerals on houses and buildings on opposite sides of a street and report how they are numbered.

Exploration of even-number and odd-number patterns in addition and subtraction should be a part of children's work when they complete the composite table of basic addition facts described in chapter 7. Guide students to examine the table for sums of combinations of even numbers represented by 0, 2, 4, 6, and 8. They should discover all of the sums are even. Next ask them to find the sums of even and odd numbers and the sums of two odd numbers. As they make their generalizations, construct an even-odd addition table on the chalkboard.

+	E	O
E	E	O
O	O	E

Subtraction facts can be read from the table in the same way they were read on the composite addition table, e.g., to find E - O on the table, locate E under O and read the missing addend O at the left. When students practice addition and subtraction basic facts, you may sometimes specify that they classify the two given numbers as even or odd, indicate whether the sum or difference is even or odd, and find the sum or difference.

In second or third grade, children should also study the composite table of basic multiplication facts for patterns in multiplying even and odd numbers. Ask them to record their generalizations in an even-odd multiplication table.

×	E	O
E	E	E
O	E	O

When a student learns the meaning of division, he may practice basic division facts, relate division to multiplication, and extend his knowledge of even and odd numbers all at the same time. Design open sentences such as these:

$$2\overline{)0}\ \text{with quotient}\ \square \quad \text{and} \quad 0 = \square \times 2$$

$$2\overline{)2}\ \text{with quotient}\ \square \quad \text{and} \quad 2 = \square \times 2$$

$$2\overline{)4}\ \text{with quotient}\ \square \quad \text{and} \quad 4 = \square \times 2$$

$$\vdots$$

$$2\overline{)18}\ \text{with quotient}\ \square \quad \text{and} \quad 18 = \square \times 2$$

Extend the list as a child's skill in division develops. These exercises should lead a student to discover a whole number is even if it is divisible by 2, i.e., the remainder is 0 when divided by 2. Guide students in renaming odd numbers in the form $(2 \times n) + 1$, n a whole number, by having them complete open sentences such as

$$1 = (2 \times \square) + 1,$$
$$3 = (2 \times 1) + \square,$$
$$7 = (2 \times \square) + 1,$$
$$\vdots$$
$$19 = (2 \times \square) + 1.$$

Mathematically mature students should be challenged to complete a table such as the following:

n	$2 \times n$	$(2 \times n) + 1$
0	0	1
3	6	
4		9
		15
	16	
9		

This work should help them understand any whole number can be multiplied by 2 to produce an even whole number, and an odd whole number results when 1 is added to an even number.

Prime and Composite Numbers

A prime whole number is a number which has exactly two factors, itself and one. A composite number has more than two factors. The unit one, neither prime nor composite, has exactly one factor, itself. These concepts, usually introduced in third grade after a student has had about a year's experience with multiplication, are important in later work with factoring composite numbers, generating lowest common multiples and greatest common factors, and finding the lowest common denominator of fractional numbers.

Cuisenaire rods and arrays offer introductory models for prime and composite counting numbers. Give each student a set of Cuisenaire rods and a piece of construction paper marked so that it is divided into three parts. Tell the children that the white rod represents 1 in this activity and that they will be separating the set of rods into three subsets determined by the number of equivalent-rod trains which can be matched with each rod. Ask the students to first place the red rod before them and build an equivalent-rod train to exactly match it. Children's responses should be

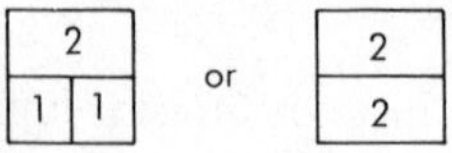

2		
1	1	

or

2
2

Thus there are two possible trains which match the 2 rod. Ask the students to place the 2 rod in one subdivision of the construction paper. Next instruct them to match the 3 rod with an equivalent-rod train. There are two correct responses again.

3		
1	1	1

3
3

Since the 3 rod, like the 2 rod, can be matched with exactly two trains, have students place it in the set with the 2 rod. Repeat the procedure with the 4 rod. Student responses should include

4			
1	1	1	1

,

4	
2	2

and

4
4

Since more than two equivalent-rod trains can be made to match the 4 rod, it belongs in a set different from the 2 and 3 rods. Students should place the 4 rod in a different subdivision of the construction paper. Continue the activity by having children classify 5, 6, . . . , 10. Then ask them to determine the number of trains they can build to match the 1 rod. Only the rod itself matches the white rod, so it belongs in the third subdivision on the construction paper. Summarize the student's work on the chalkboard and introduce the terms *prime* and *composite*.

Counting Numbers

Unit	Primes	Composites
1	2	4
	3	6
	5	8
	7	9
		10

Children should extend the classification to at least 20.

Pictorial experience with rectangular arrays is also useful in helping children conceptualize prime and composite numbers. Ask them to draw an array which has exactly two squares in it. Some may depict a 2×1 array and others may draw a 1×2 array, but these are the only two arrays which are correct. (Remember a tilted rectangle such as is not used for modeling multiplication.) Appealing to the student's prior experience is relating arrays to multiplication, record $2 = 2 \times 1$ and $2 = 1 \times 2$ on the chalkboard. Now ask the children to draw all possible rectangular arrays with three squares. Exactly two possibilities occur again. Record the work symbolically on the chalkboard: $3 = 3 \times 1$ and $3 = 1 \times 3$. Emphasize that 2 and 3 have exactly two factors. Continue with other numbers such as 6, 15, and 19. Students should understand that 6 and 15 can be represented by more than two rectangular arrays (have more than two factors) and are composite whereas 2, 3, and 19 are prime since they can be represented by exactly two arrays (have exactly two factors). Have them draw an array with exactly one square. One array results, and $1 = 1 \times 1$. Since 1 has only one factor, it is neither prime nor composite.

Once students understand what prime and composite numbers are, design activities and games to help them recognize and identify them and extend their knowledge. Some suggestions follow:

1. On a calendar have students circle the dates of holidays which occur on days given by prime numbers. Let groups of students chose a holiday, research it, and report their findings to the class.
2. In an art lesson, let students cut out numerals representing prime or composite numbers and make a design with them.
3. For a language-arts assignment, have children find the word "prime" in the dictionary and determine the various meanings it has. Separate the class into groups and have each group select a meaning and illustrate it.
4. Instruct students to write the prime (or composite) numbers between 1 and 25 in Roman numerals.
5. Have students write the prime (or composite) numbers between 1 and 25 in expanded notation.
6. Play "thinking" games such as "I'm thinking of the prime number between 20 and 25. What is it?" or "I'm thinking of the composite number which comes immediately before 12. What is it?"

Closing

Topics in number theory provide an opportunity for students to practice number facts and to extend their understanding of number patterns and relationships. "Normal" and mathematically mature children may benefit most from the study of topics such as even and odd numbers and prime and composite numbers, but some slow learners may profit as well, especially in the area of computational skills.

Assignment

1. Illustrate and explain why 1 and 0 are not prime and not composite.
2. Explain why 0 is an even number.
3. Consult books of activities and elementary mathematics textbooks, and make a list of games for reinforcing concepts of even and odd whole numbers and prime and composite whole numbers.
4. Do you think slow learners in mathematics would benefit from studying some of the topics discussed in this chapter? Explain.

Laboratory Session

Preparation: Study chapter 14 of this textbook.

Objectives: Each student should

1. illustrate odd and even counting numbers with plastic counters, a pegboard, grid paper, and Cuisenaire rods,
2. determine if sums and differences of odd and even counting numbers are odd or even, and
3. illustrate prime and composite counting numbers with a pegboard, grid paper, and Cuisenaire rods.

Materials: Cuisenaire rods, grid paper, pegboards, plastic counters, several strips of yarn

Procedure: Students should work in groups of two or three to solve these problems:

1. Use plastic counters to represent each number named below and attempt to separate the set into subsets of two to determine if the number is odd or even.
   a) 11 b) 18 c) 21
2. Repeat problem 1 using arrays with two rows on a pegboard.
3. Use arrays on grid paper to illustrate which numbers named in problem 1 are even and which are odd.
4. Use Cuisenaire rods to illustrate which sums and differences are even and which are odd.
   a) $7 + 6$ b) $6 + 14$ c) $15 + 9$ d) $12 - 9$ e) $13 - 5$ f) $18 - 12$
5. a) Use a pegboard to show all possible arrays which can be made with 12 pegs.
   b) List all factors of 12 using the result of 5(a).
   c) Graph all the factors of 12 on a number line.
   d) Is 12 prime or is it composite? Why?
6. With Cuisenaire rods, demonstrate that 13 is prime and 14 is composite.
7. Use arrays on grid paper to classify each number as prime, composite, or neither prime nor composite.
   a) 15 b) 11 c) 1

Bibliography

Jungst, Dale G. *Elementary Mathematics Methods: Laboratory Manual.* Boston: Allyn and Bacon, Inc., 1975.

15

Problem Solving

Problem solving represents the highest level of learning as pointed out in chapter 1, and the application step in the teaching sequence discussed in chapter 3 is a problem-solving one. When students relate number properties, place value, basic facts, and simple sums, differences, products, and quotients to computing with the algorithms described in chapters 9 and 10, they are engaging in problem solving. Problem-solving situations discussed in the present chapter include word problems and discovery situations in which the learner selects relevant mathematical ideas and applies them to the immediate problem. If students are to be competent problem solvers, they must have prerequisite concepts and skills such as understanding number; using the notation for number and number operations; knowing the basic facts of addition, subtraction, multiplication, and division; being able to compute with whole-number algorithms; understanding basic concepts of fractional numbers and geometry; and using standard measuring instruments. Mental computations and estimation procedures are also important in problem solving. Gaining proficiency in these topics promotes a learner's confidence in using mathematics to solve problems faced in everyday life.

Problem solving should be taught through experiences in mathematics, science, social studies, health, and other school subjects and activities. The same problems are not appropriate for all students. Gifted children may do extensive work in number theory and certain aspects of geometry while ideas in measurement and arithmetic may be more suitable for slow learners.

Story Problems

Although simple word problems can be used to get the attention of students when a new idea is introduced, the story problems considered at this time are solved in a systematic way after the child has become competent in subordinate concepts and skills. When constructing story problems, some points to keep in mind are:

1. Create stories that are interesting to and suitable for the students.
2. Use numbers in line with the pupil's level of concept and skill development.
3. Since some students cannot read at the level required in many verbal problems, read the problems to them. Picture stories in mathematics textbooks are helpful in that they represent mathematical situations but require virtually no reading skill.
4. In the early stages of solving word problems, let students enact the story with objects or pictures. Studies show that children's performance is better with the use of pictorial or manipulative aids than only with computation at the symbolic level.[1]
5. Provide problems that represent all of the interpretations of arithmetic operations. If the mathematics textbook deals with some interpretations inadequately, supplement with problem cards or worksheets.
6. Promote a child's understanding of the process involved by occasionally including examples in which letters instead of numbers are used. Process skills can also be enhanced if students perform computations with large numbers on a hand-held calculator instead of working through every operation with pencil and paper.
7. Provide some problems which have superfluous information and some which have too little information. Such problems encourage thoughtful attention when relating given data to the problem question.
8. Depict males and females in a variety of roles, and represent ethnic and racial groups fairly.

While most elementary mathematics textbooks supply many examples of word problems, many of these problems promote computational rather than problem-solving skills. Thus the burden of teaching children to develop strategies for solving word problems falls on the teacher. The following steps can be used when guiding students to solve word problems. As a child gains experience in solving verbal problems and recognizing types of situations, he can begin to omit some of the steps.

1. Identify the problem question.
2. Identify what is given in the problem.
3. Determine if the sets involved are to be joined (representing addition or multiplication), separated (representing subtraction or division), or compared (associated with subtraction).
4. Simulate the story by manipulating objects such as counters or by making drawings. The importance of active learning in early childhood can hardly be overemphasized.

5. Write the open sentence represented by step 4. This step is not new to the learner since translating an action on sets into mathematical language is a part of beginning work with the operations of arithmetic.
6. Solve the open sentence generated in step 5. Students should have already practiced the basic facts so this step, too, is review.
7. Use the answer in step 6 to answer the problem question.

Most students have little difficulty in solving problems involving only one operation. Those in which more than one operation must be used demand careful attention, however. The examples which follow illustrate the four operations of arithmetic (applied to whole numbers) and their various interpretations, various treatments of remainders in division, and situations in which more than one operation must be applied. The solutions given refer to the preceding seven steps or abbreviated versions of them, depending on whether the problem represents beginning or late work with the operation.

Addition

Addition problems are associated with a joining action.

Example: Al had seven coins in his coin collection. His parents gave him four more coins for his birthday. How many coins did Al have then?

Solution:

1. The problem question is "How many coins did Al have then?"
2. "He had 7 coins. He received 4 more." is given.
3. A set of 7 and a set of 4 are joined.
4.

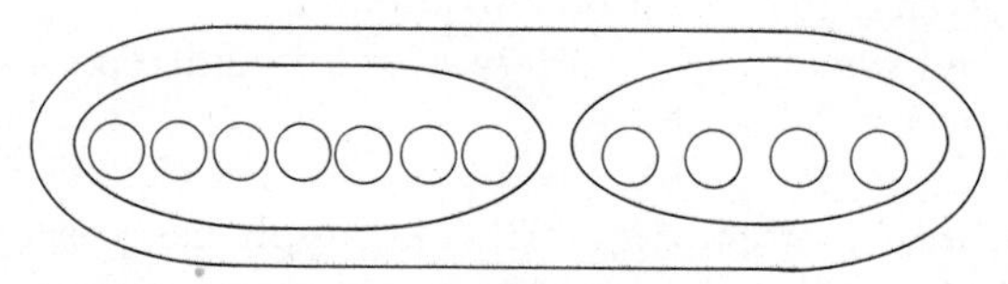

5. $7 + 4 = \square$
6. $7 + 4 = \boxed{11}$
7. Al has 11 coins in his collection.

Example: Susan enjoys running races with her friends. Last year she won a races and this year she has won b races. How many races has she won altogether?

Solution:

1. $a + b = n$
2. Susan has won $a + b$ races altogether.

Subtraction

Most verbal problems in subtraction can be interpreted as subtractive (take away), additive, or comparative. Students need not, however, label a subtraction problem as one of these types. Some research indicates that take-away subtraction is the easiest of the three types for kindergarten and first-grade children, but other studies show that students' performance on additive and comparison problem types improved with instruction.[2]

Examples illustrating the three interpretations of subtraction follow.

Example (subtractive): If you had 15 marbles and gave away 6 of them, how many marbles would you have left?

Solution:

1. The problem question is "How many marbles are left?"
2. "You had 15 marbles and gave away 6 of them" is the given information.
3. Six marbles have been removed from a set of 15.
4. The transformation is depicted this way:

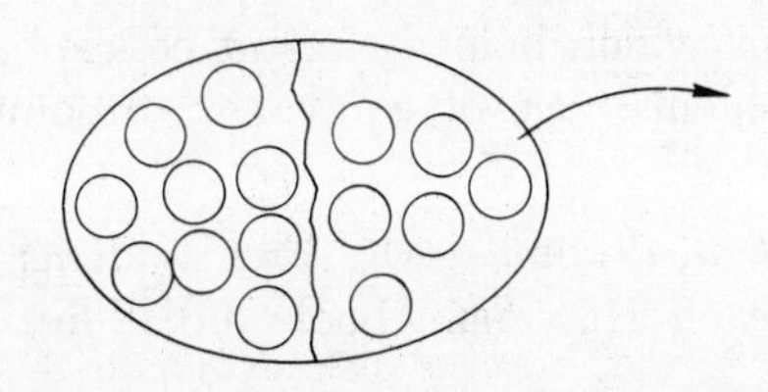

5. $15 - 6 = \square$
6. $15 - 6 = \boxed{9}$
7. Nine marbles are left.

Example (additive): If Kim has 57 cents in her piggy bank, how many more cents will she need to buy a coloring book which costs a dollar?

Solution:

1. A set of 100 is separated into two subsets, one having 57 elements and the other having n elements.
2. $100 - 57 = n$
3. $\begin{array}{r} 100 \\ -\ 57 \\ \hline 43 \end{array}$
4. Kim needs 43 more cents.

In problems of the comparison type, two different sets are compared and a one-to-one correspondence is made between one set and a proper subset of the other.

Example (comparison): If you completed 6 math task cards yesterday and 14 cards today, how many more did you work today than yesterday?

Help a student analyze the problem by questioning him: "Should the set of 6 and set of 14 be joined?" "Why not?" "Are the sets to be compared?" "Which set has fewer members?" "Can you show a one-to-one correspondence between the 6-member set and a subset of the 14-element set?"

Solution:

1. The problem question is "How many more problem cards did I work today than yesterday?"

2. "I completed 6 cards yesterday and 14 today." is the given information.
3. The sets of 6 and 14 are compared and 6 of the 14 elements are removed.
4.

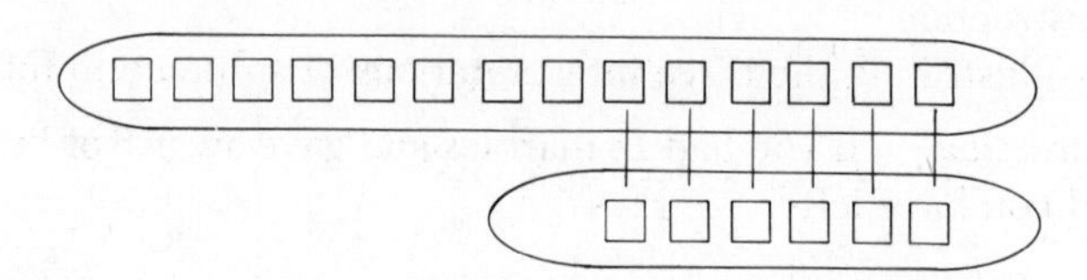

5. $14 - 6 = \square$
6. $14 - 6 = \boxed{8}$
7. I completed 8 more cards today than yesterday.

Multiplication

The same interpretations which help a student conceptualize multiplication are used in application problems: sets of equivalent, disjoint sets; arrays; and Cartesian product.

Example (set of equivalent, disjoint sets): Five students went to the library and checked out four books each. How many books did all five students check out?

Solution:

1. The number of books all five students checked out must be determined.
2. Five students checked out four books each.
3. Five sets of four must be joined.
4.

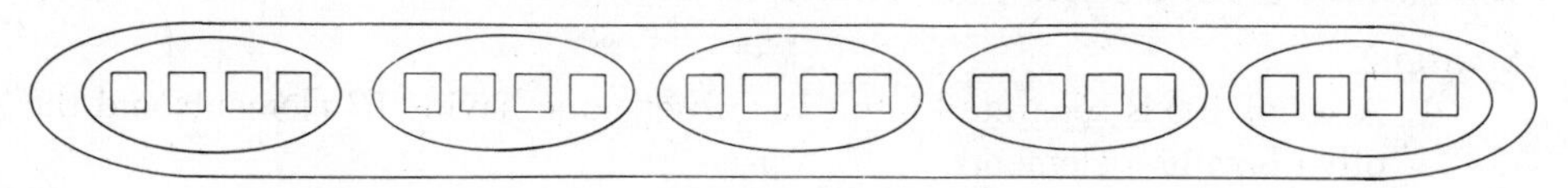

5. $5 \times 4 = \square$
6. $5 \times 4 = \boxed{20}$
7. Twenty books were checked out.

Example (array): The box has 4 rows and 7 columns of oranges. How many oranges are in the box?

Solution:

1. The number of oranges must be found.
2. The box contains 4 rows and 7 columns of oranges.
3. The problem involves an array.
4.

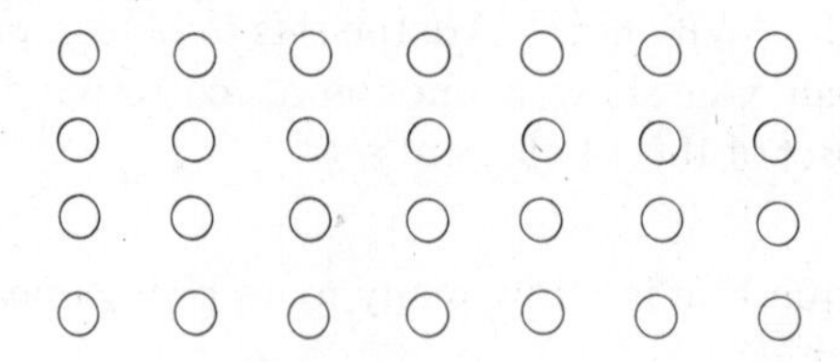

5. $4 \times 7 = \square$
6. $4 \times 7 = \boxed{28}$
7. The box has 28 oranges.

Example (Cartesian product): An ice-cream shop has 38 flavors of ice cream which are served in either a sugar cone or a plain cone. Five kinds of sandwiches are also served at the shop. How many choices do you have in getting your ice cream if you buy one scoop?

Solution:

1. Each flavor of ice cream must be paired with each kind of cone. (Notice the information about the sandwiches is not relevant to the problem question.)
2. $2 \times 38 = n$
3. $\begin{array}{r} 38 \\ \times 2 \\ \hline 76 \end{array}$
4. I have 76 choices when buying one scoop of ice cream.

Division

Word problems in division can apply the measurement (number of subsets) or partitive (number per subset) interpretation of quotient.

Example (measurement interpretation): At the bake sale Bob had 15 cents to buy cookies which cost 3 cents each. How many cookies could he buy?

Solution:

1. The problem question is "How many cookies could Bob buy?"
2. Bob had 15 cents and cookies cost 3 cents each.
3. A set of 15 must be separated into subsets of 3.
4.

5. $15 \div 3 = \square$
6. $15 \div 3 = \boxed{5}$
7. Bob can buy 5 cookies.

Example (measurement): If 30 students can ride in each bus, find the number of buses needed to carry 180 students on a school trip.

Solution:

1. A set of 180 is separated into subsets of 30.
2. $180 \div 30 = n$
3. $30 \overline{)180}$ with quotient 6
4. Six buses are needed for the school trip.

Example (partitive division): If 384 students assemble in 3 rooms, how many students will be in each room if the same number of students is in each room?

Solution:

1. $384 \div 3 = n$
2.
```
    128
 3)384
   300    100
    84
    60     20
    24
    24      8
          128
```
3. Each room has 128 students.

Division Problems with Remainders

The interpretation given to a remainder in a division problem depends on the problem situation involved. Children need experience in problems in which the remainder is treated in diverse ways.
Some examples follow:

The partitive example, "There are thirty students in our class. If four equivalent groups are formed, how many students will be in each group?", illustrates a case in which the remainder is not a part of the quotient. The answer is "Seven students will be in each group." Two children will not be in a group.

The quotient in the measurement example, "How many quarters are in $1.70?", is six. Since the two dimes are not enough to be equivalent to another quarter, the answer is six quarters.

The remainder becomes a part of the quotient in the example, "Karen's mother baked eighteen cupcakes for Karen and her three friends. How many cupcakes will each child get if the children have the same amount?" When each child receives four cupcakes, two are left to divide four ways. The answer, then, is $4\frac{1}{2}$.

In the measurement example, "A boy in a grocery store is putting 40 items on a shelf. How many shelves are needed if 15 items can be placed on each shelf?", the division results in a quotient of 2 and a remainder of 10. The answer, however, is 3 since two shelves are not enough to hold the 40 items.

Problems Involving More Than One Operation

When more than one operation must be used in the same problem, students may write different equations, each with only one operation, or one equation involving all the operations. Also, some children may solve the open sentences using some of the properties discussed in chapters 7 and 8. When different methods of solution arise, have students describe their strategies to the class.

The following are some sample problems requiring more than one arithmetic operation for solution:

1. Tommy's mother paid 69 cents a pound for 6 pounds of ground beef. If she gave the checker 5 dollars, how much change did she get back?

   Guide the students' thinking: "What is to be found in this problem?" "How much money did Tommy's mother have?" "Five dollars are equivalent to how many cents?" (Changing dollars to cents allows students to perform the operations on whole numbers.) "How much money did Tommy's mother spend?" "Are the 500 cents and the 6 sets of 69 cents to be joined?" "Why not?" The students should determine that 6 sets of 69 are joined and this money spent is taken away from 500 cents.

   Solution:

   a) $6 \times 69 =$ money spent
   $500 - (6 \times 69) =$ change received

   b) $\begin{array}{r} 69 \\ \times\ 6 \\ \hline 414 \end{array} \qquad \begin{array}{r} 500 \\ -\ 414 \\ \hline 86 \end{array}$

   c) Tommy's mother got back 86 cents.

2. A ball of string contains 50 decimeters of string. After 15 decimeters are used, how many strips 5 decimeters long can be cut from the remaining string?

   Help the students to relate the numbers in the problem. Ask questions such as "Are the sets of 50 and 15 separated or joined? How can you tell that 15 must be removed from a set of 50 before the strips of 5 are cut?"

   Solution:

   a) $(50 - 15) \div 5 = n$

   b) $50 - 15 = 35$
   $35 \div 5 = 7$

   c) 7 strips of string can be cut.

   Some students may solve step a) using the distributive property of division over addition:

   $$\begin{aligned}(50 - 15) \div 5 &= (50 \div 5) - (15 \div 5) \\ &= 10 - 3 \\ &= 7\end{aligned}$$

3. Kristen had 73 cents in her piggy bank but spent 28 cents of it. Her uncle gave her 46 more cents. How much money did Kristen have then?

   Guide the children to subtract first and then add and to enclose the operation to be performed first in parentheses.

   Solution:

   a) $(73 - 28) + 46 = n$ or
   $73 - 28 = a,\ a + 46 = b$

   b) $\begin{array}{r} 73 \\ -\ 28 \\ \hline 45 \end{array} \qquad \begin{array}{r} 45 \\ +\ 46 \\ \hline 91 \end{array}$

   c) Kristen had 91 cents.

Discovery Situations

Discovery situations which involve the student in formulating hypotheses, developing strategies, and testing hypotheses are well suited to the laboratory approach discussed in chapter 3. These problems should be taken from all areas of mathematics as well as from other disciplines. Topics can be chosen from logic, number theory, geometry, measurement, and number patterns. At this level of learning, activities should be pupil-centered with the teacher serving as a facilitator of learning by observing, listening, asking questions, and offering suggestions when necessary. Gifted learners should have ample opportunity to participate in independent work on problem-solving activities. Teachers may vary the class organization in discovery-oriented class sessions. Children may

1. work in groups of four or five on the same problem. The materials may be different from group to group, however. Students follow instructions as the teacher gives them orally, or the teacher may prepare a worksheet for the students.
2. work in pairs on the same problem. Such an arrangement is desirable when each student needs a partner to conduct the activity or when materials need to be shared. The teacher may present the problem orally and circulate among the students giving assistance when needed.
3. work in small groups, each group solving a different problem.[3] The teacher prepares task cards for the various groups, briefly describes what is involved in each group's work, and then lets the children move from one table to another as they complete the tasks. (The sample laboratory activity in chapter 3 utilizes this class organization.) Another possibility is to assign a different problem to each group and have the groups report on their work at the end of the period. Other students may suggest alternative strategies for solving the problems at that time.

Some samples of discovery problems for young children follow:

1. **Logic**

Objective: Using attribute blocks, play a two-difference game by completing a grid.

Prerequisite learning: knowledge of attributes of the blocks in a set of attribute blocks, experience with grids

Materials: a set of attribute blocks for each pair of children

Problem: Work with a partner to complete this grid:

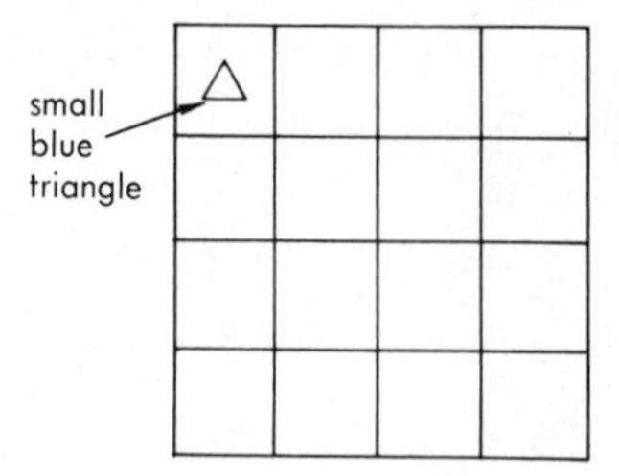

The first player should place the given block in the top left square. The second player puts a block which has two attributes different from that block either to the right of or below the given block. Players fill in the grid so that every block has two attributes different from each block directly above, below, and beside it.

Variations: Play a one-difference or three-difference game, or stack the blocks to form a three-dimensional grid. In the latter case, each block must be a certain number of attributes different from blocks under and on either side of it.

2. Length

Objectives: Measure the length of objects without a ruler.

Prerequisite learning: understanding of quantitative terms such as "longer than" and "shorter than"; measurement in nonstandard units.

Materials: cubical blocks, sheets of paper with ordinary objects drawn on them:

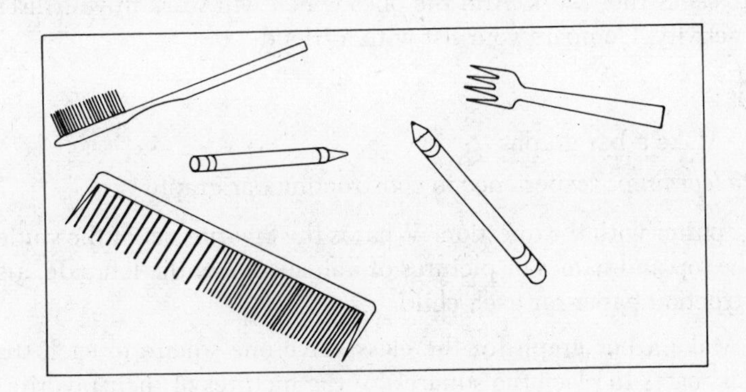

Problem: Use the cubes to answer these questions:

a) Which object is the longest? ____________

b) Name the shortest object. ____________

c) Are any objects the same length? ____________
If the answer is "yes" which ones are the same length? ____________

Variation: Draw line segments on paper and have students use centimeter cubes to determine the longest and shortest line segments and which ones, if any, are the same length. Then have students try to draw a line segment a certain number of centimeters long and check their work with the cubes.

3. Three-dimensional shapes

Objectives: Determine which plane patterns of six squares will fold into a cube. Make predictions and check them.

Prerequisite learning: identification of a cube

Materials: grid paper for each child

Problem:

a) Cut out these patterns on your grid paper.

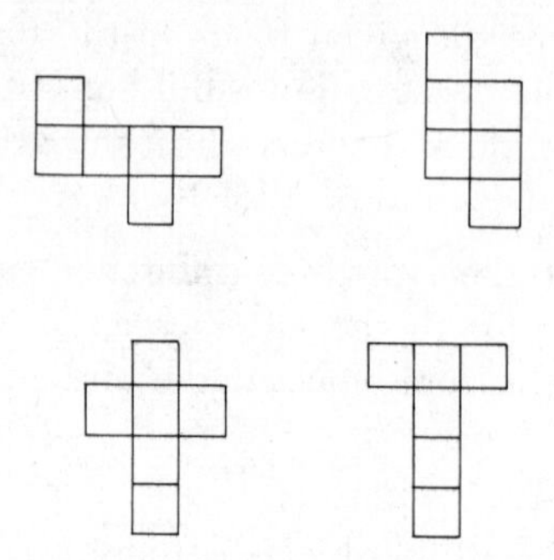

b) Try to fold each pattern into a cube. Which ones will work? Copy them on your paper.
c) Draw some six-square patterns you think will fold into a cube. Cut them out and check to see if they work. Add the ones which will work to your list in part b) of this activity. Compare your list with a friend.

4. Graphing

Objective: Make a bar graph.

Prerequisite learning: experience in constructing bar graphs.

Materials: poster with the question "What is the favorite pet of the students in this class?" at the top and names or pictures of animals down the left side, a square cut out of construction paper for each child.

Problem: Make a bar graph for the class. Give one square to each student, and ask your classmates to place the squares by the pictures of their favorite pets. Add the names of favorite pets that are not listed on the chart.

Use your graph to answer these questions.
1) What is the favorite pet of members of this class? ____________ How many students named this pet as their favorite? ____________
2) Which pet was named least often? ____________
How many students liked this pet best? ____________
3) How do you think a bar graph can be made without placing squares on a piece of paper?
4) Decide on another question to ask the students, collect the data, and make the bar graph.

5. Weight and shape

Objective: Relate the weight of objects to change in shape.

Prerequisite learning: use a balance scale. Weigh objects in grams.

Materials: balance scale, gram weights or gram centimeter cubes, ball of clay

Problem:
a) Use the balance scale to weigh the ball of clay. It weighs about _____ grams.
b) Press the ball flat into a pancake shape. The pancake weighs about _____ grams.

c) Reshape the clay into a snake-like object. It weighs about ____ grams.
d) What can you say about weight and shape?

6. **Volume and multiplication**

Objective: Find the number of cubes required to fill a container by multiplication.

Prerequisite learning: basic multiplication facts, simple products, place value, multiplication of three factors, length in the metric system

Materials: for each group, a box in the shape of a decimeter cube and 28 interlocking plastic centimeter cubes. (You may use any cubes and an ordinary box but make sure the box is such that a whole number of cubes will fit in the box for all three dimensions.)

Problem: Use the 28 cubes to find out how many cubes will fit in the box. (Students should discover that they can place ten cubes along one edge of the bottom of the box and ten cubes along an edge perpendicular to the other one and multiply 10 by 10 to find that 100 cubes would cover the bottom of the box. Then when 10 cubes are stacked up one edge of the box, they should see that $10 \times (10 \times 10)$, or 1000, cubes would fill the box.)

7. **Tessellations**

Objective: Determine which of certain regular polygons will fit together exactly to cover, or tessellate, a flat surface (plane).

Prerequisite learning: definition of regular polygon

Materials: scissors and paper for tracing.

Problem: Trace the following shapes and cut out ten of each. Paste the ten cut-outs of each shape on paper as if you were putting down tile on a floor or counter-top. The "tiles" must not overlap or have any gaps between them. Prepare to report your discoveries to the class.

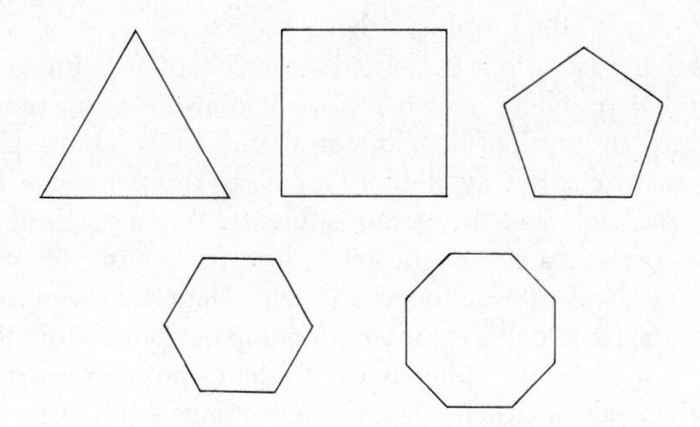

8. **Probability**

Objectives: Perform an experiment repeatedly, tally results, and make predictions.

Materials: for each pair of students, a spinner on a circular region with six congruent circular sectors—three in one color, two in another, and one in another

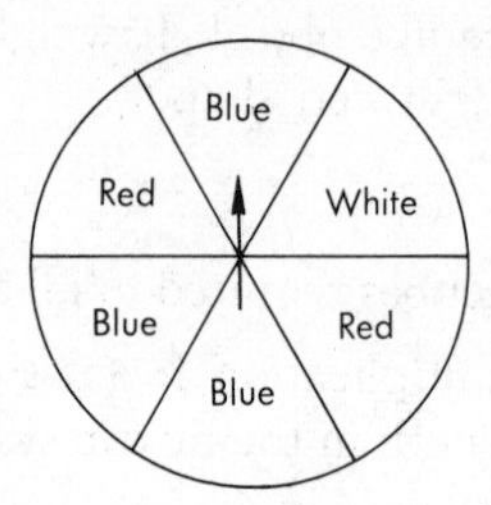

Problem: Write down the three colors. Spin the spinner 20 times while your friend tallies the number of times the spinner stops on each color. When you finish, tally the outcomes as your friend spins. Compare your results and answer these questions:

a) On which color did the spinner stop most often? _________ Do you think if you performed the experiment 20 more times the same result would happen? _______ Why do you think so?

____

b) If there were two of each color and you spun the spinner 20 times, what do you think the results would be? ______________________________

Why? ____

Closing

Helping students apply the concepts and skills they have developed to problem-solving situations can be one of the most enjoyable and challenging aspects of teaching elementary school mathematics. The early childhood educator is responsible for introducing children to the problem-solving process.

Problem-solving situations examined in this chapter include reasoning through and solving verbal problems which represent real-life happenings and discovery activities in which the student develops strategies for solving a problem for which an immediate answer is not available. Discovery problems can be used with the whole class or as enrichment for certain students. When teaching students to solve word problems, present a systematic procedure for solving them. This process includes identifying what is to be found and what data are given; determining if the sets involved are to be joined, separated, or compared; modeling the problem situation with objects or pictures (optional as a student's problem-solving skills mature); writing and solving the mathematical open sentence; and using the mathematical solution to answer the question in the problem.

In order for students to be successful in solving problems, they must understand the problem and be able to do the computation required in the problems. As problems are solved, encourage children to seek alternative ways of obtaining an answer, and let the students discuss their strategies in class.

Notes

1. Robert G. Underhill, "Teaching Word Problems to First Graders," *The Arithmetic Teacher* 25 (November 1977): 54.
2. Robert G. Underhill, *Teaching Elementary School Mathematics*, 2nd ed. (Columbus, Ohio: Charles E. Merrill Publishing Company, 1977), pp. 580–81.
3. Doyal Nelson and Joan Kirkpatrick, "Problem Solving," in Joseph N. Payne, ed. *Mathematics Learning in Early Childhood,* Thirty-seventh Yearbook (Reston, Virginia: National Council of Teachers of Mathematics, 1975), pp. 88–90.

Assignment

1. In the seventh discovery activity, Tessellations, determine which regular polygons tessellate the plane.
2. Try to work with a young child who is having difficulty in solving story problems. Use the seven steps given in this chapter to guide his thinking. What is your assessment of how the child understood the work?
3. Read and list major points in these articles in the November, 1977, issue of *The Arithmetic Teacher:*
   a) "Teaching Word Problems to First Graders" by Robert G. Underhill, pp. 54–56,
   b) "The Right Hemisphere's Role in Problem Solving" by Grayson H. Wheatley, pp. 36–39,
   c) "You Can Teach Problem Solving" by John F. LeBlanc, pp. 16–20,
   d) "Ideas about Problem Solving: A Look at Some Psychological Research" by Frank K. Lester, Jr., pp. 12–14,
   e) "On the Uniqueness of Problems in Mathematics" by Edith Robinson, pp. 22–26,
   f) "Research on Problem Solving: Implications for Elementary School Classrooms" by Marilyn N. Suydam and J. Fred Weaver, pp. 40–42.
4. Read and react to this article in *The Arithmetic Teacher:* "Problem-solving Activities Observed in British Primary Schools" by Rose Grossman, January, 1969, pp. 34–38.

Laboratory Session

Preparation: Study chapter 15 of this textbook.
Objective: Create word problems and discovery situations.
Materials: several K-3 mathematics textbook series

Procedure: Work with a partner to design a problem for each situation.

1. a) Consult at least one first-grade textbook and write an addition word problem and three subtraction word problems, one for each interpretation of subtraction. Use numbers which are in line with addition and subtraction skills at that level.
   b) Is there any interpretation of subtraction inadequately represented in the story problems in the textbooks you examined?
2. Modify problem 1(a) so that it reflects addition and subtraction concepts and skills taught in second grade.
3. How would you revise problem 2 so that it is appropriate for most third graders?
4. a) Examine one or two second-grade textbooks to ascertain what multiplication skills are commonly taught in second grade. Write multiplication story problems to illustrate the three interpretations of multiplication.
   b) Is any interpretation of multiplication glaringly lacking in the problems the books give?
5. How should problem 4(a) be revised to reflect multiplication skills taught in third grade?
6. a) Construct a measurement-division story problem and a partitive-division story problem appropriate for third grade.
   b) Create a word problem from which the division $\begin{array}{r} 9 \\ 5\overline{)47} \\ \underline{45} \\ 2 \end{array}$ might arise, but the answer would be 10.
   c) Create a word problem associated with the division $\begin{array}{r} 8\frac{1}{3} \\ 3\overline{)25} \\ \underline{24} \\ 1 \end{array}$.
7. Create a word problem which can be associated with the equation $500 - (234 + 155) = n$.
8. Review one of the textbook series to determine the extent to which word problems are treated. Does the teacher's manual offer suggestions for teaching them?
9. Use a textbook series and your own thinking to
   a) create a discovery problem card appropriate for kindergarten or first grade,
   b) create a discovery problem card appropriate for second or third grade.

Bibliography

Grossnickle, Foster E., and Reckzeh, John. *Discovering Meanings in Elementary School Mathematics,* 6th ed. New York: Holt, Rinehart and Winston, Inc., 1973.

Nelson, Doyal, and Kirkpatrick, Joan. "Problem Solving," in Joseph N. Payne, ed. *Mathematics Learning in Early Childhood,* Thirty-seventh Yearbook. Reston, Virginia: National Council of Teachers of Mathematics, 1975.

Underhill, Robert G. "Teaching Word Problems to First Graders." *The Arithmetic Teacher* 25 (November 1977): 54–56.

————. *Teaching Elementary School Mathematics,* 2nd ed. Columbus, Ohio: Charles E. Merrill Publishing Company, 1977.

16

Closing

A good program in mathematics for young children should deal with all aspects of mathematics—vocabulary, logical thinking, arithmetic, geometry (both metric and nonmetric), and patterns. To help you identify at a glance what performance level in mathematics many students acquire by the time they finish third grade, the following list is given. These behaviors can also be considered as informational objectives in mathematics learning for the early childhood level.

1. Count to a million.
2. Read and write numerals to a million.
3. Specify place, face, and total value given in numerals.
4. Use the final addition algorithm to add numbers represented by four-digit numerals.
5. Use the final subtraction algorithm (decomposition or equal-additions when renaming is required) to subtract numbers represented by four-digit numerals.
6. Multiply two numbers, one represented by a one-digit numeral and the other by a two-digit numeral, with the final multiplication algorithm.
7. Know the basic division facts.
8. Compute higher level division problems with the subtractive algorithm or distributive method or both.
9. Estimate sums, differences, products, and quotients.
10. Solve word problems which apply the student's skills in the four operations of arithmetic.

11. Read and write common fractional numbers, determine equivalences, and order fractional numbers.
12. Tell time to the minute.
13. Specify money values and demonstrate money equivalents.
14. Measure length, weight, area, liquid capacity, and temperature in standard metric and English units using common measuring instruments.
15. Recognize and identify common two- and three-dimensional shapes, describe characterizing properties of them, and find examples of the shapes in the physical world.
16. Identify symmetric and congruent figures by informal investigations.
17. Identify, name, and represent points, line segments, lines, rays, and angles.
18. Classify whole numbers as even or odd and identify prime whole numbers by informal investigations.

In this book, sequences of learning tasks which culminate in these objectives and alternative teaching strategies, techniques, and materials for introducing and developing these topics have been considered. It is hoped that the suggestions given will help you make curricular and instructional decisions which help children learn mathematics.

As a teacher you can use research findings of mathematics educators and psychologists to help you understand how children learn mathematics and to plan mathematics experiences which are appropriate for children at different levels of intellectual development and which take into account various backgrounds and visual, auditory, and physical learning styles of students. We have seen that young children should experience many hands-on activities and enjoy working with a variety of materials. You should encourage creativity and language and problem-solving skills by giving your students opportunity to formulate their own questions and to discuss and compare alternative strategies for solution of problems. Children need varying degrees of guidance in their explorations and inquiries. You should be sure to summarize and unify their activities and seize opportunities to reinforce mathematical concepts and skills in their play activities. Such experiences should help to provide children with a strong basis for future work in mathematics.

Although research studies can provide helpful information in teaching mathematics, they have not determined all of the answers to what mathematics should be taught and how it should be taught. In your daily dealings with children, you are encouraged to conduct your own research on what techniques work best for you, which topics in the mathematics curriculum seem to be stumbling blocks for many children and how you deal with these problems, and what methods of instruction are most effective with exceptional children and to report your findings in publications such as *The Arithmetic Teacher*, *Early Years*, and *Learning*. In the meantime, good luck to you in your work as an educator of young children!

Assignment

1. List at least five major topics in mathematics usually taught in preschool, kindergarten, and each primary grade.
2. Complete a term project dealing with mathematics instruction at the grade-level of your choice. Assume your class is mainstreamed; include provisions for the exceptional child's needs in your project. These are some examples of types of projects you might prepare.
   a) a learning/listening center. The center should contain related learning experiences sequenced to provide for at least three levels of difficulty and should be designed so a child or small group of children can work independently of the teacher.
   b) a two-week unit of study. Include an overall description of the unit plan, and specify daily objectives, materials, procedure (given in brief form), and evaluation. Be sure to reference each idea taken from sources other than yourself.
   c) interest center. Create an interest center organized around a central theme such as outer space, protection of the environment, or energy conservation.
   d) games and puzzles. Make several mathematics games and puzzles appropriate for the age level of your choice.

Appendix

MATHEMATICS MATERIALS LIST

Abacus
Creative Publications
3977 East Bayshore Road, P.O. Box 10328
Palo Alto, California 94303

Attribute Materials, Games, and Problems
Webster Division, McGraw Hill Book Company
Manchester Road
Manchester, Missouri 63011

Base-ten Blocks and Teacher's Guide
Creative Publications
3977 East Bayshore Road, P.O. Box 10328
Palo Alto, California 94303

Chip-trading Primary Set
Scott Resources, Inc.
Box 2121
Fort Collins, Colorado 80521

Counting Frames
Selective Educational Equipment, Inc.
3 Bridge Street
Newton, Massachusetts 02195

Cuisenaire Rods, Squares, and Cubes
Cuisenaire Co. of America, Inc.
12 Church Street
New Rochelle, New York 10805

Elementary Mathematics Concepts with Calculators
Texas Instruments
Education and Communications Center
P.O. Box 3640, M/S84
Dallas, Texas 75285

Fractions Are as Easy as Pie
Milton Bradley Company
Springfield, Massachusetts 01101

Geoboards, Activity Cards, and Teacher's Guide
Creative Publications
3977 East Bayshore Road, P.O. Box 10328
Palo Alto, California 94303

Kinesthetic Numeral Cards and Counting Discs
Instructo Corporation
Paoli, Pennsylvania 19301

Mathematical Balance, Activity Cards, and Teacher's Guide
(Scale balances when child makes the correct mathematical sentence.)
Invicta International Educators
Innovative Educational Services, Inc.
P.O. Box 29096
New Orleans, Louisiana 70189

Mathfacts Games (for drill in basic facts of addition, subtraction, multiplication, and division)
Milton Bradley Company
Springfield, Massachusetts 01101

Math Readiness
Vocabulary and Concepts, Addition and Subtraction
(cassette tapes)
Activity Records
Educational Activities, Inc.
Freeport, Long Island, New York 11520

Metric Materials (transparent centimeter grids, trundle wheel, clear liter cube, interlocking centimeter cubes, scales, meter stick, etc.)
La Pine Scientific Company
Department D4, 6009 S. Knox Avenue
Chicago, Illinois 60629

Mirror Cards
Webster Division, McGraw Hill Book Company
Manchester Road
Manchester, Missouri 63011

Montessori Shapes and Stencils
La Pine Scientific Company
Department D4, 6009 S. Knox Avenue
Chicago, Illinois 60629

Parquetry Blocks
Playskool—a Milton Bradley Company
Chicago, Illinois 60618

Pegboards
Developmental Learning Materials
3505 North Ashland Avenue
Chicago, Illinois 60657

Plastic Counters
Milton Bradley Company
Springfield, Massachusetts 01101

Play Tray Currency Card Set
Judy-Ed Aids
The Judy Company
Minneapolis, Minnesota 55401

Ruth Cheves Program for Telling Time
Teaching Resources
Boylston Street
Boston, Massachusetts 02116

Tangrams and Associated Materials
Creative Publications
3977 East Bayshore Road, P.O. Box 10328
Palo Alto, California 94303

Time Teller I (card file, includes pre-test, suggested progress chart, and post-test for telling time)
Teech-Um Company
Box 4232
Overland Park, Kansas 66204

Unifix Cubes and Related Materials
Didax Inc.
3 Dearborn Road, P.O. Box 2258
Peabody, Massachusetts 01960

Walk-on Number Line
Instructo Corporation
Paoli, Pennsylvania 19301

Index